Soil Biology

Series Editor: Ajit Varma

9

Volumes published in the series

Volume 1
A. Singh, O.P. Ward (Eds.)
Applied Bioremediation and Phytoremediation
2004

Volume 2
A. Singh, O.P. Ward (Eds.)
Biodegradation and Bioremediation
2004

Volume 3
F. Buscot, A. Varma (Eds.)
Microorganisms in Soils: Roles in Genesis and Functions
2005

Volume 4
S. Declerck, D.-G. Strullu, J.A. Fortin (Eds.)
In Vitro Culture of Mycorrhizas
2005

Volume 5
R. Margesin, F. Schinner (Eds.)
Manual for Soil Analysis –
Monitoring and Assessing Soil Bioremediation
2005

Volume 6
H. König, A. Varma (Eds.)
Intestinal Microorganisms of Termites
and Other Invertebrates
2006

Volume 7
K.G. Mukerji, C. Manoharachary, J. Singh (Eds.)
Microbial Activity in the Rhizosphere
2006

Volume 8
P. Nannipieri, K. Smalla (Eds.)
Nucleic Acids and Proteins in Soil
2006

Barbara J.E. Schulz
Christine J.C. Boyle
Thomas N. Sieber (Eds.)

Microbial Root Endophytes

With 29 Figures, 4 in Color

Springer

PD Dr. Barbara J. E. Schulz
Technical University of Braunschweig
Institute of Microbiology
Spielmannstraße 7
38106 Braunschweig
Germany
e-mail: b.schulz@tu-bs.de

Dr. Christine J. C. Boyle
Augustastraße 32
02826 Görlitz
Germany
e-mail: c.boyle@tu-bs.de

Dr. Thomas N. Sieber
Swiss Federal Institute of Technology
Department of Environmental Sciences
Institute of Integrative Biology
Forest Pathology and Dendrology
8092 Zürich
Switzerland
e-mail: thomas.sieber@env.ethz.ch

ISSN 1613-3382

ISBN 978-3-642-07013-6 e-ISBN 978-3-540-33526-9

Springer is a part of Springer Science + Business Media
springer.com

Editor: Dr. Dieter Czeschlik, Heidelberg, Germany
Desk Editor: Dr. Jutta Lindenborn, Heidelberg, Germany
Cover design: *design&production*, Heidelberg, Germany

Preface

Healthy plant roots are not only colonized by mycorrhizal fungi and rhizobial bacteria, but also by a myriad of other microorganisms, including endophytic bacteria and fungi. Comparatively little is known about these endophytic microorganisms, which do not cause apparent disease, but colonize root tissues inter- and/or intracellulary. Although there had been previous research studying both bacterial and fungal endophytes, it was in the mid-1980s that numerous investigators began studying these groups of microorganisms more intensively. Initially, most work on endophytes centered on the diversity of isolates and correlations with ecological factors. Recently it has become clear that some of these interactions with endophytic bacteria and fungi can be latently pathogenic and/or mutualistic. In mutualistic interactions, the endophyte may improve growth of the host, convey stress tolerance, induce systemic resistance, or supply the host with nutrients. On the other hand, most endophytes are also able to grow saprotrophically, e.g., from surface-sterilized tissues on media containing dead organic substrates. Thus, it has become obvious that endophytes have multiple life history strategies and that these can be extremely plastic, as will become clear to the readers of the subsequent 19 chapters.

This book is the first to deal with bacterial and fungal root endophytes, their diversity, life history strategies, interactions, applications in agriculture and forestry, and also with methods for isolation, cultivation, and both conventional and molecular methods for identification and detection. The first chapter deals with the question: What are endophytes? However, it also introduces the reader to the subjects treated in the subsequent chapters. We hope that readers will not only find this book informative, but will also be provoked to further study these fascinating interactions, and in particular to better understand the mechanisms regulating them. It will become apparent that we are still far from understanding the factors that determine whether a plant-microbial interaction remains asymptomatic, leads to disease, or is mutualistic.

We would like to thank our colleagues for their contributions and their work to make this book a successful unity, to Jutta Lindenborn of Springer for her friendly help and advice, and to Ajit Varma for the invitation to edit a book in this series.

Braunschweig,
Görlitz and Zürich,
June 2006

Barbara Schulz
Christine Boyle
and *Thomas Sieber*

Contents

Part II Endophytic Fungi

Contributors

Anand, Richa
Faculty of Land and Food Sciences, Faculty of Forestry, University of British Columbia, Vancouver, British Columbia, Canada V6T 1Z4; Current address: Department of Forest Mycology and Pathology, Swedish University of Agricultural Sciences (SLU), Box 7026, 750 07 Uppsala, Sweden

Bacon, Charles W.
Richard B. Russell Research Center, ARS, United States Department of Agriculture, Toxicology and Mycotoxin Research Unit, SAA, P.O. Box 5677, Athens, GA 30604, USA

Bärlocher, Felix
63B York Street, Department of Biology, Mount Allison University, Sackville, New Brunswick, E4L 1G7, Canada

Bayman, Paul
Departamento de Biologia, Universidad de Puerto Rico - Rio Piedras, PO Box 23360, San Juan, PR 00931, USA

Berg, Gabriele
Graz University of Technology, Department of Environmental Biotechnology, Petersgasse 12, 8010 Graz, Austria

Bloemberg, Guido V.
Leiden University, Institute of Biology, Wassenaarseweg 64, 2333AL Leiden, The Netherlands

Boyle, Christine
Augustastraße 32, 02826 Görlitz, Germany

Broughton, William J.
Université de Genève, Laboratoire de Biologie Moléculaire des Plantes Supérieures, Sciences III, 30 Quai Ernest-Ansermet, 1211 Genève 4, Switzerland

Brundrett, Mark C.
School of Plant Biology, Faculty of Natural and Agricultural Sciences, The University of Western Australia, Crawley, WA 6009, Australia

Cairney, John W.G.
Centre for Plant and Food Sciences, Parramatta Campus, University of Western Sydney, Locked Bag 1797, Penrith South DC, NSW 1797, Australia

Carvajal, Margarita M. Camacho
Leiden University, Institute of Biology, Wassenaarseweg 64, 2333AL Leiden, The Netherlands

Chanway, Chris
Faculty of Land and Food Systems, Faculty of Forestry, University of British Columbia, Vancouver, British Columbia, Canada V6T 1Z4

Currah, Randolph S.
Department of Biological Sciences, University of Alberta, Edmonton, AB, T6G 2E9, Canada

Deakin, William J.
Université de Genève, Laboratoire de Biologie Moléculaire des Plantes Supérieures, Sciences III, 30 Quai Ernest-Ansermet, 1211 Genève 4, Switzerland

Elsas, Jan D. van
Groningen University, Department of Microbial Ecology, Biological Center, Kerklaan 30, 9751 NN, Haren, The Netherlands

Girlanda, Mariangela
Dipartimento di Biologia Vegetale and IPP - Torino, Viale PA Mattioli 25, 10125 Torino, Italy

Grünig, Christoph R.
Swiss Federal Institute of Technology, Department of Environmental Sciences, Institute of Integrative Biology, Forest Pathology and Dendrology, 8092 Zürich, Switzerland

Hallmann, Johannes
Biologische Bundesanstalt für Land- und Forstwirtschaft, Institut für Nematologie und Wirbeltierkunde, Toppheideweg 88, 48161 Münster, Germany

Jansson, Hans-Börje
Departamento de Ciencias del Mar y Biología Aplicada, Universidad de Alicante, Apdo. 99, 03080 Alicante, Spain

Kloepper, Joseph W.
Department of Entomology and Plant Pathology, Auburn University, Auburn, AL 36849, USA

Lopez-Llorca, Luis V.
Departamento de Ciencias del Mar y Biología Aplicada, Universidad de Alicante, Apdo. 99, 03080 Alicante, Spain

Luppi, Anna Maria
Dipartimento di Biologia Vegetale and IPP, Viale PA Mattioli 25, 10125 Torino, Italy

Otero, J. Tupac
Universidad Nacional de Colombia-Palmira, Departamento de Ciencias Agrícolas, AA 237, Palmira, Valle del Cauca, Colombia

Overbeek, Leo S. van
Plant Research International B.V, Droevendaalsesteeg 1, 6708 PB, Wageningen, The Netherlands

Paul, Leslie
Faculty of Land and Food Sciences, Systems, University of British Columbia, Vancouver, British Columbia, Canada V6T 1Z4

Perotto, Silvia
Dipartimento di Biologia Vegetale and IPP - Torino, Viale PA Mattioli 25, 10125 Torino, Italy

Rice, Adrianne V.
Northern Forestry Centre, Canadian Forest Service, Natural Resources Canada, 5320-122 St., Edmonton, AB, T6H 3S5 Canada

Ryu, Choong-Min
Plant Biology Division, The Samuel Roberts Noble Foundation, 2510 Sam Noble Parkway, Ardmore, OK 73401, USA

Saad, Maged M.
Université de Genève, Laboratoire de Biologie Moléculaire des Plantes Supérieures, Sciences III, 30 Quai Ernest-Ansermet, 1211 Genève 4, Switzerland

Salinas, Jesús
Departamento de Ciencias del Mar y Biología Aplicada, Universidad de Alicante, Apdo. 99, 03080 Alicante, Spain

Schulz, Barbara
Institute of Microbiology, Technical University of Braunschweig, Spielmannstraße 7, 38106 Braunschweig, Germany

Sieber, Thomas N.
Swiss Federal Institute of Technology, Department of Environmental Sciences, Institute of Integrative Biology, Forest Pathology and Dendrology, 8092 Zürich, Switzerland

Vicente, José Gaspar Maciá
Departamento de Ciencias del Mar y Biología Aplicada, Universidad de Alicante, Apdo. 99, 03080 Alicante, Spain

Vuurde, Jim van
Plant Research International B.V, Droevendaalsesteeg 1, 6708 PB, Wageningen, The Netherlands

Yates, Ida E.
Richard B. Russell Research Center, ARS, United States Department of Agriculture, Toxicology and Mycotoxin Research Unit, Athens, GA 30604, USA

Abbreviations

ACC	1-aminocyclopropane-1-carboxylate
AHL	acyl homoserine lactones
AM	arbuscular mycorrhiza
AMF	arbuscular mycorrhizal fungi
ARA	acetylene reduction activity
AUDPC	area under the disease progress curve
BAC	bacterial artificial chromosome
BCA	biocontrol agents
BRD	root border cells
BrdU	bromide oxyuridine
Cfu	colony forming units
CLSM	confocal laser scanning microscopy
CMA	corn meal agar
CMV	Cucumber mosaic virus
CPS	capsular polysaccharides
CTAB	cetyl trimethyl ammonium bromide
DGGE	denaturing gradient gel electrophoresis
DIC	differential interference microscopy
DON	deoxynivalenol
DSE	dark septate endophytes
DSF	dark septate fungi
DSM	dark sterile mycelia
ECFP	enhanced cyan fluorescent protein
ECM	ectomycorrhizae
ECM	extracellular material
EGFP	Enhanced GFP
ELISA	enzyme-linked immunosorbent assay
EPS	extra-cellular polysaccharides
EYFP	Enhanced Yellow Fluorescent Protein
FISH	fluorescence in situ hybridization
GFP	Green fluorescent protein
GSP	general secretory pathway
GUS	β-glucuronidase

IAA	indole acetic acid
ICR	induced systemic resistance
ISSR	inter-simple sequence repeat
ITS	internal transcribed spacer
ITS-RFLP	internal transcribed spacer-restriction fragment length polymorphism
LPS	lipo-polysaccharides
MLH	multilocus haplotypes
MRA	*Mycelium radicis atrovirens*
NDFA	nitrogen derived from the atmosphere
PAH	polyaromatic hydrocarbon
PAR	photosynthetically active radiation
PBM	peribacteroid membrane
PCR	polymerase chain reaction
PCR-RFLP	polymerase chain reaction -restriction fragment length polymorphism
PDA	potato dextrose agar
PGPR	plant growth promoting rhizobacteria
PR	pathogenesis-related
PVP	polyvinylpyrrolidone
RAPD	random amplified polymorphic DNA
RDNA	16S rRNA gene
RFLP	restriction fragment length polymorphism
RFP	red fluorescent protein
RGR	relative growth rate
SAR	systemic acquired resistance
SEM	Scanning electron microscopy
SPB	sterile phosphate buffer
SPS	Surface polysaccharides
SSCP	single strand conformational polymorphism
TGGE	temperature gradient gel electrophoresis
TNV	tobacco necrosis virus
ToMoV	Tomato mottle virus
TRFLP	terminal restriction fragment length polymorphism
TSA	tryptic soya agar
VAM	vesicular arbuscular mycorrhiza
VOC	volatile organic compound
VWT	variable white taxon
WA	Western Australia

1 What are Endophytes?

Barbara Schulz, Christine Boyle

1.1 Introduction and Definitions

Taken literally, the word endophyte means "in the plant" (endon Gr. = within, phyton = plant). The usage of this term is as broad as its literal definition and spectrum of potential hosts and inhabitants, e.g. bacteria (Kobayashi and Palumbo 2000), fungi (Stone et al. 2000), plants (Marler et al. 1999) and insects in plants (Feller 1995), but also for algae within algae (Peters 1991). Any organ of the host can be colonised. Equally variable is the usage of the term "endophyte" for variable life history strategies of the symbiosis, ranging from facultatively saprobic to parasitic to exploitive to mutualistic. The term endophyte is, for example, used for pathogenic endophytic algae (Bouarab et al. 1999), parasitic endophytic plants (Marler et al. 1999), mutualistic endophytic bacteria (Chanway 1996; Adhikari et al. 2001; Bai et al. 2002) and fungi (Carroll 1988; Jumpponen 2001; Sieber 2002; Schulz and Boyle 2005), and pathogenic bacteria and fungi in latent developmental phases (Sinclair and Cerkauskas 1996), but also for microorganisms in commensalistic symbioses (Sturz and Nowak 2000).

Some authors also designate the interactions of mycorrhizal fungi with the roots of their hosts as being endophytic (reviewed by Sieber 2002). However, we concur with Brundrett (2004; see Chap. 16 by Brundrett), who distinguishes mycorrhizal from endophytic interactions; the former having synchronised plant-fungus development and nutrient transfer at specialised interfaces. Nevertheless, as we will see in this book, distinctions between mycorrhizal and non-mycorrhizal fungi are not always clear-cut [see Chaps. 9 (Bayman and Otero), 12 (Girlanda et al.), 13 (Rice and Currah), 14 (Cairney), and 15 (Schulz)]. Not only can mycorrhizal fungi become pathogenic, but, for example, dark septate endophytes (DSE) can assume mycorrhizal functions [Jumpponen and Trappe 1998; see Chaps. 7 (Sieber

Barbara Schulz: Technical University of Braunschweig, Institute of Microbiology, Spielmannstraße 7, 38106 Braunschweig, Germany, E-mail: b.schulz@tu-bs.de

Christine Boyle: Augustastraße 32, 02826 Görlitz, Germany

Soil Biology, Volume 9
Microbial Root Endophytes
B. Schulz, C. Boyle, T. N. Sieber (Eds.)

and Grünig), and 15 (Schulz)]. In addition, there are also cases in which fungal root endophytes seem to be saprobes, e.g. *Oidiodendron maius* (see Chap. 13 by Rice and Currah) and *Phialocephala fortinii* (Jumpponen and Trappe 1998; Jumpponen et al. 1998; see Chap. 15 by Schulz).

Although there are diverse uses for the word endophyte, "endophytes" are most commonly defined as those organisms whose "...infections are inconspicuous, the infected host tissues are at least transiently symptomless, and the microbial colonisation can be demonstrated to be internal..." (Stone et al. 2000). Although these authors used this definition to describe fungal endophytes, it is equally applicable to bacterial endophytes.

It is important to remember that the definition describes a *momentary status*. Thus it includes an assemblage of microorganisms with different life history strategies: those that grow saprophytically on dead or senescing tissues following an endophytic growth phase (Stone 1987; see Chap. 8 by Bacon and Yates), avirulent microorganisms as well as latent pathogens and virulent pathogens in the early stages of infection (Sinclair and Cerkauskas 1996; Kobayashi and Palumbo 2000). Unfortunately, taken literally, it can include all pathogens at some stage of their development. Since the plant host responds to at least some infections with mechanical defence reactions (Narisawa et al. 2004; see Chap. 15 by Schulz), there is merit to Petrini's additional characterisation of endophytic interactions as not "causing apparent harm" (Petrini 1991), which presumably refers to an absence of macroscopically visible symptoms. Aware of the determinative discrepancies, we will nevertheless use the term "endophyte" to describe those bacteria and fungi that can be detected at a particular moment within the tissues of apparently healthy plant hosts (Schulz and Boyle 2005).

1.2 Colonisation

In spite of the fact that bacteria are prokaryotes and fungi are eukaryotes, they share many attributes of their associations with plant hosts, e.g. both colonise root tissues inter- and intra-cellularly, and often systemically (Table 1.1). They do, however, differ somewhat in their modes of colonisation. Bacteria primarily colonise intercellularly (Hinton and Bacon 1995; Hallmann et al. 1997), though they have also been found intracellularly, e.g. *Azoarcus* spp. (Hurek et al. 1994). They are frequently found in the vascular tissues of host plants (Kobayashi and Palumbo 2000), which is advantageous for distribution, whereas asymptomatic colonisation by fungi may be inter- and intra-cellular throughout the root. Although DSE sometimes colonise the vascular cylinder in asymptomatic interactions (Barrow 2003), such colonisation is frequently associated with pathogenicity (Bacon and Hinton 1996; Schulz and Boyle 2005).

Table 1.1. Characteristics of the interactions of bacterial vs. fungal endophytes with plant roots (see also all other book chapters)

Criteria	Bacteria	Fungi
Host spectrum	Broad, depends on habitat, host, season	Broad, depends on habitat, host, season
Mode and site of infection	Passive through wounds and other tissue openings or active with enzymes or vectors, e.g. insects	Active: through stomata, cell wall or wounds
Nutritional source during first stage infection	Host exudates, dead cortex cells, plant debris	Storage material in the spores, dead cortex cells, plant debris, host exudates
Nutritional source during colonisation, i. e. during a "steady state" status	Components of the symplast and apoplast	Components of the symplast and apoplast
Growth in root	Inter- and/or intra-cellular, slow, low colonisation densities	Inter- and/or intracellular, often extensive
Growth from roots into the shoot	Yes	Sometimes
Systemic growth in roots	Possible	Possible
Tissue colonised	Primarily intercellular, also vascular tissue	Usually not within vascular tissue
Specialised structures for nutrient access	Nodules, glands	Sometimes
Physiological status	Only little data available	Balanced antagonisms, active interaction
Outcome of the interaction	Commensalism, mutualism or latent pathogenicity	Commensalism, mutualism or latent pathogenicity
Benefits for the microbial symbiont	A reliable supply of nutrients and protection from environmental stresses, passive transfer and spread between hosts via vectors, e.g. insects	A reliable supply of nutrients and protection from environmental stresses, advantages for reproduction and colonisation at host senescence
Potential benefits for the plant symbiont	Induced resistance, improved growth (N-fixation, phytohormones), synthesis of metabolites antagonistic to plant pathogens and parasites	Induced resistance, improved growth (phytohormones, improved access to minerals and nutrients), synthesis of metabolites antagonistic to predators and antagonists
Reproduction	Usually passive transfer and spread between hosts via vectors, e.g. insects, but also active, e.g. Pseudomonads	Active and passive following host senescence, sometimes with vectors

The assemblages of fungi that colonise plant roots are diverse (Vandenkoornhuyse et al. 2002). In contrast to endophytic growth in the aboveground plant organs, endophytic growth of fungi within the roots has frequently been found to be extensive (Stone et al. 2000; Schulz and Boyle 2005; see Chap. 11 by Lopez-Llorca et al.). Root colonisation can be both inter- and intra-cellular, the hyphae often forming intracellular coils, e.g. DSE (Jumpponen and Trappe 1998; Stone et al. 2000; Sieber 2002), the basidiomycete *Piriformospora indica* (Varma et al. 2000), or *Oidiodendron maius* (see Chap. 13 by Rice and Currah) and *Heteroconium chaetospira* (Usuki and Narisawa 2005), which can even form characteristic ericoid mycorrhizal infection units (see Chap. 14 by Cairney). DSE may also form ectendomycorrhiza (Lubuglio and Wilcox 1988) and ectomycorrhizal-like structures (Wilcox and Wang 1987; Fernando and Currah 1996; Kaldorf et al. 2004; see Chap. 15 by Schulz).

Many orchid roots are systemically and mycoheterotrophically colonised by fungi of the genus *Rhizoctonia* (Ma et al. 2003; see Chap. 16 by Brundrett) and *Leptodontidium* (Bidartondo et al. 2004). In some cases, e.g. *Fusarium verticillioides* (= *F. moniliforme*), colonisation by an avirulent strain was found to be systemic and intercellular, whereas pathogenic strains also colonised intracellularly (Bacon and Hinton 1996). Latent pathogens, e.g. *Cryptosporiopsis* sp. (Kehr 1992; Verkley 1999) may occasionally penetrate the vascular bundles (Schulz and Boyle 2005).

Bacteria usually invade the roots passively, e.g. at open sites on roots such as lateral root emergence or wounds (Kobayashi and Palumbo 2000), even achieving systemic colonisation from a single site of entry (Hallmann et al. 1997). Although colonisation densities of nonpathogenic endophytic bacteria are rarely as high as those of pathogenic bacteria, they are highest in the root tissue; perhaps because this is the primary site of infection (Kobayashi and Palumbo 2000; Hallmann et al. 1997; see Chap. 2 by Hallmann and Berg).

1.3 Assemblages and Adaptation

Both fungal and bacterial endophytes have been isolated from the roots of almost all hosts studied to date [Petrini 1991; Stone et al. 2000; Kobayashi and Palumbo 2000; Sieber 2002; see Chaps. 2 (Hallmann and Berg), 3 (Kloepper and Ryu), and 7 (Sieber and Grünig)]. The assemblages of endophytes that colonise a particular host vary both with habitat and host, some even being adapted to very specialised habitats, e.g. the aquatic fungi that colonise submerged roots (see Chap. 10 by Bärlocher). Recent molecular methods enable better analyses of the geographical distribution of given groups of microorganisms, for example that of the DSE [Jumpponen 1999, see Chaps. 7 (Sieber and Grünig), 12 (Girlanda et al.), and 15 (Cairney)].

Both diversity and colonisation density frequently increase during the course of the vegetation period (Smalla et al. 2001), since horizontal transmission predominates (Carroll 1988, 1995; Petrini 1991; Guske et al. 1996; Hallmann et al. 1997; Arnold and Herre 2003, see Chap. 2 by Hallmann and Berg). Particularly asexual sporulation increases in autumn at the end of the vegetation period.

Communities of endophytes inhabiting a particular host may be ubiquitous, or have what is frequently referred to as host specificity (e.g. Carroll 1988; Petrini 1996; Stone et al. 2000; Berg et al. 2002; Cohen 2004). We concur with Carroll (1999) and Zhou and Hyde (2001) that the term "specificity" should be reserved for organisms that will only grow in one host (Schulz and Boyle 2005). If this is not the case, this phenomenon could be termed host preference (Carroll 1999) or host-exclusivity (Zhou and Hyde 2001). Whether the interaction represents specificity, preference or exclusivity, an adaptation of host and endophyte to one another has occurred. The adaptation may not only be to a particular host, but to endophytic growth in one plant organ, e.g. in the roots in contrast to the shoots [Petrini 1991; Hallmann et al. 1997; Sieber 2002; Schulz and Boyle 2005; see Chaps. 2 (Hallmann and Berg), and 7 (Sieber and Grünig)].

It is often extremely difficult to know whether or not a particular fungus or bacterium that has been detected in healthy plant tissue has actually been growing within the host tissue or has been incidentally isolated, i. e. is normally found on other substrates. As reviewed by Schulz and Boyle (2005) and in Chap. 17 by Hallmann et al., there are four methods presently in use for detecting and identifying fungi and bacteria in plant tissue: (1) histological observation (see Chap. 6 by Anand et al.), most recently in combination with molecular methods (see Chap. 18 by Bloemberg and Carvajal), (2) surface sterilisation of the host tissue and isolation of the emerging fungi on appropriate growth media, (3) detection by specific chemistry, e.g. immunological methods (see Chap. 18 by Bloemberg and Carvajal), or (4) by direct amplification of fungal DNA from colonised plant tissues [Vandenkoornhuyse et al. 2002; see Chaps. 17 (Hallmann et al.), and 19 (van Overbeek et al.)], having first ascertained that there are no fungal residues on the plant surface (Arnold et al. 2006). Methods for quantification are reviewed by Sieber (2002), Schulz and Boyle (2005) and in Chap. 17 (Hallmann et al.).

1.4 Life History Strategies

Organisms detected at any one moment in asymptomatic plant tissue and arbitrarily named "endophytes" include microorganisms with different life history strategies. Endophytes represent, both as individuals and collectively, a continuum of mostly variable associations: mutualism, commensalism, latent pathogenicity, and exploitation. The phenotypes of the interactions are often plastic, depending on the genetic dispositions of the two partners, their developmental stage and nutritional status, but also on environmental factors (see Chap. 12 by Girlanda et al.). The role of genetic disposition was demonstrated by Freeman and Rodriguez (1993): a single mutation resulted in loss of a virulence factor, transforming a pathogenic fungus, *Colletotrichum magna*, into an endophyte. Similarly, avirulence genes and the machinery of pathogenicity may be lacking or suppressed in bacterial endophytes (Kobayashi and Palumbo 2000).

Just as fungi have been found to develop ectomycorrhiza in one host and what appear to be ericoid mycorrhiza in another host (Villarreal-Ruiz et al. 2004), a mycorrhizal fungus can grow endophytically in the roots of a non-host (see Chap. 12 by Girlanda et al.). The importance of a particular combination of host and microorganism as well as their reciprocal influences also becomes apparent when a fungal or bacterial pathogen is inoculated into a non-host and is no longer virulent, colonising as an asymptomatic endophyte (Carroll 1999; Kobayashi and Palumbo 2000; Schulz and Boyle 2005) The influence of the host plant in determining the mycorrhizal, endophytic or even pathogenic character of a DSE association is likely to be a prime factor. In plant communities, the multiple mutualistic potential of these fungi, establishing hyphal links or inoculum reservoirs, may favour inter-plant interactions (see Chap. 12 by Girlanda et al.).

Interactions are frequently complex, involving more than two partners. Endophytic bacteria and fungi may interact not only with the plant host, but also with other organisms, including mycorrhizal fungi (see Chap. 9 by Bayman and Otero) and metazoa. For example, nematophagous fungi, which are ubiquitous organisms in soils, not only can switch from a saprophytic to a parasitic stage to kill and digest living nematodes, but can also grow endophytically in plant roots (see Chap. 11 by Lopez-Llorca et al.).

Mutualistic interactions involving fungi and bacteria that endophytically colonise plant roots benefit the microbial partner with a reliable supply of nutrients as well as protection from environmental stresses. As reported in this book, benefits for the host plant may include improved growth [see Chaps. 6 (Anand et al.), 13 (Rice and Currah), 15 (Schulz), and 19 (van Overbeek et al.)], induced resistance [see Chaps. 3 (Kloepper and Ryu), 4 (Berg and Hallmann), 6 (Anand et al.), and 15 (Schulz)], biocontrol of plant parasitic nematodes (see Chap. 11 by Lopez-Llorca et al.) and of fungi in agriculture [see Chaps. 3 (Kloepper and Ryu), 4 (Berg and Hallmann), 15 (Schulz)] and forestry (see Chap. 6 by Anand et al.), as well as microbial synthesis of metabolites antagonistic to predators [Schulz et al. 2002; Schulz and Boyle 2005; see Chaps. 6 (Anand et al.), 8 (Bacon and Yates), 15 (Schulz), and 19 (van Overbeek et al.)] When synthesized in agricultural crops in situ, mycotoxins synthesised by endophytes, e.g. *Fusarium verticillioides* in maize, are potentially problematic for human consumption of these crops (see Chap. 8 by Bacon and Yates).

Factors responsible for improving plant growth are the microbial synthesis of phytohormones [Tudzynski 1997; Tudzynski and Sharon 2002; Kobayashi and Palumbo 2000; see Chaps. 6 (Anand et al.) and 15 (Schulz)], access to minerals and/or other nutrients from the soil (Caldwell et al. 2000; Barrow 2003; see Chap. 13 by Rice and Currah), bacterial fixation of atmospheric nitrogen, which has been demonstrated not only for the nodule-forming members of the Rhizobiaceae, but also for non-nodule-forming bacteria, e.g. *Acetobactor* and *Azoarcus* (Reinhold-Hurek and Hurek 1998; see Chap. 5 by Saad et al.). In the associations of nitrogen-fixing rhizobia with legumes, some of the same signalling molecules are involved as in the interactions of mycorrhizal fungi with their hosts, e.g. flavonoids and nod-factors (Lapopin and Franken 2000; Martin et al. 2001; Mirabella et al. 2002; Imaizumi-Anraku et al. 2005; see Chap. 5 by Saad et al.). And as has recently been shown, plastid membrane proteins involved in the first signalling interactions are crucial for the entry of both symbionts into the host roots (Imaizumi-Anraku et al. 2005).

1.5 Balanced Antagonism

According to Heath (1997), only a few fungi are actually capable of causing disease in any one plant, since they must first cross several barriers and overcome other plant defences. This must also be true for bacteria. Thus, one question has motivated many investigations: how does the endophyte manage to exist, and often to grow, within its host without causing visible disease symptoms? We have proposed a working hypothesis based on observations from the interactions studied thus far (Schulz et al. 1999; Schulz and Boyle 2005). Asymptomatic colonisation is a balance of antagonisms between host and endophyte (Fig. 1.1). Endophytes and pathogens both possess many of the same virulence factors: the endophytes studied thus far produced the exoenzymes necessary to infect and colonise the host (Sieber et al. 1991; Petrini et al. 1992; Ahlich-Schlegel 1997; Boyle et al. 2001; Lumyong et al. 2002), even though only some of these endophytes are presumably

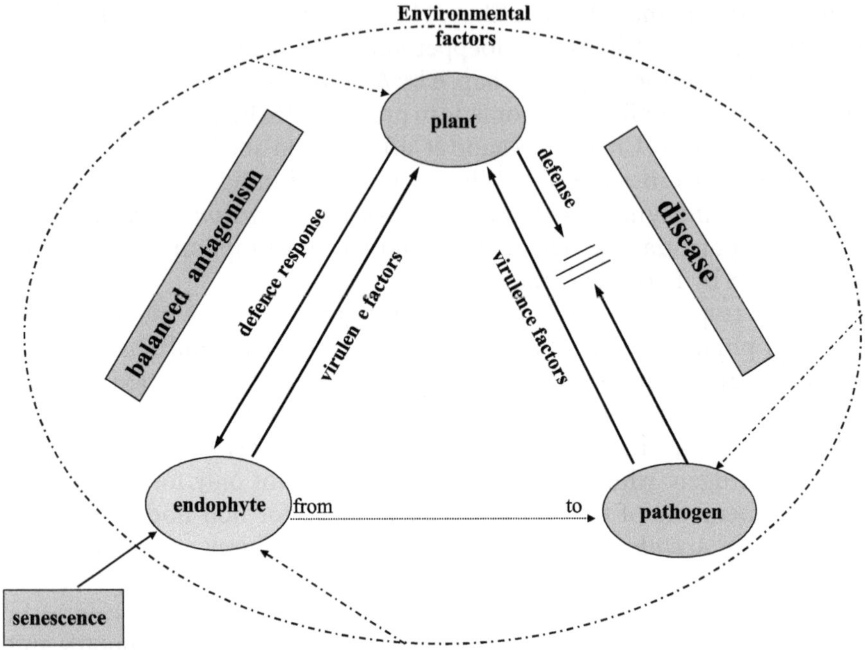

Fig. 1.1. Hypothesis: a balance of antagonisms between endophytic virulence and plant defence response results in asymptomatic colonisation (reproduced with permission from Schulz and Boyle 2005)

latent pathogens. The majority can produce phytotoxic metabolites (Schulz et al. 2002; Schulz and Boyle 2005). The host can respond with the same defence reactions as to a pathogen, i.e. with preformed and induced defence metabolites [Yates et al. 1997; Schulz et al. 1999; Mucciarelli et al. 2003; see Chaps. 3 (Kloepper and Ryu), and 11 (Lopez-Llorca et al.)], and general defence responses (Narisawa et al. 2004; Schulz and Boyle 2005). As long as fungal virulence and plant defence are balanced, the interaction remains asymptomatic. In all of these interactions we are referring to a *momentary status*, an often fragile balance of antagonisms.

If the host-pathogen interaction becomes imbalanced, either disease results or the fungus is killed. In some cases, the virulence of weak pathogens such as *Pezicula* spp. (Kehr 1992) is sufficient for disease development only when the host is stressed or senescent. Whether the interaction is balanced or imbalanced depends on the general status of the partners, the virulence of the fungus, and the defences of the host – both virulence and defence being variable and influenced by environmental factors, nutritional status and developmental stages of the partners. Although this hypothesis has been developed to explain the interactions of fungal endophytes with their hosts, further studies may well provide evidence that it is also applicable to endophytic bacteria.

Balanced antagonistic interactions are plastic in expression, depending on the *momentary status* of host and endophyte, but also on biotic and abiotic environmental factors and on the tolerance of each of the partners to these factors. In particular, many endophytes seem to be masters of phenotypic plasticity: infecting as a pathogen, colonising cryptically, and finally sporulating as a pathogen or saprophyte. This necessitates a balance with the potential for variability, which means that these endophytic interactions are creative, having the potential for evolutionary development – the symbioses can evolve both in the direction of more highly specialised mutualisms and in the direction of more highly specialised parasitisms and exploitation. Indeed, there is evidence that mycorrhizal fungi may have evolved from the endophytic activity of saprophytic fungi (see Chap. 16 by Brundrett), but also that plastids that have evolved from endosymbiotic bacteria facilitate further symbioses with other bacterial and fungal symbionts (Imaizumi-Anraku et al. 2005).

1.6 Conclusions

The usage of the term "endophyte" is as broad as its literal definition and spectrum of potential hosts and inhabitants. The most common usage of the term "endophyte" for organisms whose infections are internal and inconspicuous, and in which the infected host tissues are at least transiently symptomless, is equally applicable to bacterial prokaryotes and fungal eukaryotes.

Endophytes include an assemblage of microorganisms with different life history strategies: those that, following an endophytic growth phase, grow saprophytically on dead or senescing tissue, avirulent microorganisms, incidentals, but also latent pathogens and virulent pathogens at early stages of infection. These parasitic interactions may vary from mutualistic to commensalistic to latently pathogenic and exploitive. Phenotypes of the interactions are often plastic, depending on the genetic dispositions of the two partners, their developmental stage and nutritional status, but also on environmental factors.

We have proposed a working hypothesis based on observations from the interactions studied thus far to explain asymptomatic microbial colonisation as a balance of antagonisms between host and endophyte (Fig. 1.1; Schulz and Boyle 2005). This often fragile balance of antagonism is a *momentary status* and depends on the general status of the partners, the virulence of the fungus and defences of the host, environmental factors, nutritional status, as well as the developmental stages of the partners.

References

Adhikari TG, Joseph CM, Yang G, Philips DA, Nelson LM (2001) Evaluation of bacteria isolated from rice for plant growth promotion and biological control of seedling disease of rice. Can J Microbiol 47:916–924

Ahlich-Schlegel K (1997) Vorkommen und Charakterisierung von dunklen, septierten Hyphomyceten (DSH) in Gehölzwurzeln. Dissertation, Eidgenössische Technische Hochschule, Zürich, Switzerland

Arnold AE, Herre EA (2003) Canopy cover and leaf age affect colonisation by tropical fungal endophytes: ecological pattern and process in *Theobroma cacao* (Malvaceae) Mycologia 95:388–398

Arnold AE, Henk DA, Eells RL, Lutzoni F, Vilgalys R (2006) Diversity and phylogenetic affinities of foliar fungal endophytes in loblolly pine inferred by culturing and environmental PCR. Mycologia (in press)

Bacon CW, Hinton NS (1996) Symptomless endophytic colonisation of maize by *Fusarium moniliforme*. Can J Bot 74:1195–1202

Bai Y, Aoust F, Smith D, Driscoll B (2002) Isolation of plant-growth-promoting *Bacillus* strains from soybean root nodules. Can J Microbiol 48:230–238

Barrow JR (2003) Atypical morphology of dark septate fungal root endophytes of *Bouteloua* in arid southwestern USA rangelands. Mycorrhiza 13:239–247

Berg G, Roskot N, Steidle A, Eberl L, Zock A, Smalla K (2002) Plant-dependent genotypic and phenotypic diversity of antagonistic rhizobacteria isolated from different *Verticillium* host plants. Appl Environ Microbiol 68:3328–3338

Bidartondo ML, Burghardt B, Gebauer G, Bruns TD, Read DJ (2004) Changing partners in the dark: isotopic and molecular evidence of ectomycorrhizal liaisons between forest orchids and trees. Proc R Soc London B 271:1799–1806

Bouarab K, Potin P, Correa J, Kloareg B (1999) Sulfated oligosaccharides mediate the interaction between a marine red alga and its green algal pathogenic endophyte. Plant Cell 11:1635–1650

Boyle C, Götz M, Dammann-Tugend U, Schulz B (2001) Endophyte-host interactions III. Local vs. systemic colonisation. Symbiosis 31:259–281

Brundrett MC (2004) Diversity and classification of mycorrhizal associations. Biol Rev 79:473–495

Caldwell BA, Jumpponen A, Trappe JM (2000) Utilization of major detrital substrates by dark-septate, root endophytes. Mycologia 92:230–232

Carroll GC (1988) Fungal endophytes in stems and leaves: from latent pathogen to mutualistic symbiont. Ecology 69:2–9

Carroll GC (1995) Forest endophytes: pattern and process. Can J Bot 73:1316–1324

Carroll GC (1999) The foraging Ascomycete. XVI International Botanical Congress, St Louis, MN

Chanway CP (1996) Endophytes: they're not just fungi! Can J Bot 74:321–322

Cohen SD (2004) Endophytic-host selectivity of *Discula umbrinella* on *Quercus alba* and *Quercus rubra* characterized by infection, pathogenicity and mycelial compatibility. Eur J Plant Pathol 110:713–721

Feller IC (1995) Effects of nutrient enrichment on growth and herbivory of dwarf red mangrove (*Rhizophora mangle*). Ecol Monogr 65:477–505

Fernando AA, Currah RS (1996) A comparative study of the effects of the root endophytes *Leptodontidium* and *Phialocephala fortinii* (Fungi imperfecti) on the growth of some subalpine plants in culture. Can J Bot 74:1071–1078

Freeman S, Rodriguez RJ (1993) Genetic conversion of a fungal plant pathogen to a non-pathogenic, endophytic mutualist. Science 260:75–78

Guske S, Boyle C, Schulz B (1996) New aspects concerning biological control of *Cirsium arvense*. IOBC/wprs Bulletin 19:281–290

Hallmann J, Quadt-Hallmann A, Mahaffee WF, Kloepper JW (1997) Bacterial endophytes in agricultural crops. Can J Microbiol 43:895–914

Heath MC (1997) Evolution of plant resistance and susceptibility to fungal parasites. In: Carroll GC, Tudzynski P (eds) The mycota V Part B Plant relationships. Springer, Berlin Heidelberg New York, pp 257–276

Hinton DM, Bacon CW (1995) *Enterobacter cloacae* is an endophytic symbiont of corn. Mycopathologia 129:117–125

Hurek T, Reinhold-Hurek B, Van Montagu M, Kellenberger E (1994) Root colonization and systemic spreading of *Azoarcus* sp. strain BH72 in grasses. J Bacteriol 176:1913–1923

Imaizumi-Anraku H, Takeda N, Charpentier M, Perry J, Miwa H, Umehara Y, Kouchi H, Murakami Y, Mulder L, Vickers K, Pike J, Downie JA, Wang T, Sato S, Asamizu EE, Tabata S, Yoshikawa M, Murooka Y, Wu G, Kawaguchi M, Kawasaki S, Parniske M, Hayashi M (2005) Plastid proteins crucial for symbiotic fungal and bacterial entry into plant roots. Nature 433:527–531

Jumpponen A (1999) Spatial distribution of discrete RAPD phenotypes of a root endophytic fungus, *Phialocephala fortinii*, at a primary successional site on a glacier forefront. New Phytol 141:333–344

Jumpponen A (2001) Dark septate endophytes – are they mycorrhizal? Mycorrhiza 11:207–211

Jumpponen A, Trappe JM (1998) Performance of *Pinus contorta* inoculated with two strains of root endophytic fungus, *Phialocephala fortinii*: effects of synthesis system and glucose concentration. Can J Bot 76:1205–1213

Jumpponen A, Mattson KG, Trappe JM (1998) Mycorrhizal functioning of *Phialocephala fortinii* with *Pinus contorta* on glacier forefront soil: interactions with soil nitrogen and organic matter. Mycorrhiza 7:261–265

Kaldorf M, Renker C, Fladung M, Buscot F (2004) Characterization and spatial distribution of ectomycorrhizas colonizing aspen clones released in an experimental field. Mycorrhiza 14:295–306

Kehr RD (1992) Pezicula canker of *Quercus rubra* L, caused by *Pezicula cinnamomea* (DC) Sacc II Morphology and biology of the causal agent. Eur J For Pathol 22:29–40

Kobayashi DY, Palumbo JD (2000) Bacterial endophytes and their effects on plants and uses in agriculture. In: Bacon CW, White JF (eds) Microbial endophytes. Dekker, New York, pp 199–236

Lapopin L, Franken P (2000) Modification of plant gene expression. In: Kapulnik Y, Douds DD (eds) Arbuscular mycorrhizas: physiology and function. Kluwer, Dordrecht, pp 69–84

Lubuglio KF, Wilcox HE (1988) Growth and survival of ectomycorrhizal and ectoendomycorrhizal seedlings of *Pinus resinosa* on iron tailings. Can J Bot 66:55–60

Lumyong S, Lumyong P, McKenzie EHC, Hyde K (2002) Enzymatic activity of endophytic fungi of six native seedling species from Doi Suthep-Pui Nathional Park, Thailand. Can J Microbiol 48:1109–1112

Ma M, Tan TK, Wong SM (2003) Identification and molecular phylogeny of *Epulorhiza* isolates from tropical orchids. Mycol Res 107:1041–1049

Marler M, Pedersen D, Mitchell OT, Callaway RM (1999) A polymerase chain reaction method for detecting dwarf mistletoe infection in Douglas fir and western larch. Can J For Res 29:1317–1321

Martin F, Duplessis S, Ditengou F, Lagrange H, Voiblet C, Lapeyrie F (2001) Developmental cross talking in the ectomycorrhizal symbiosis: signals and communication genes. New Phytol 151:145–154

Mirabella R, Franssen H, Bisseling T (2002) LCO signalling in the interaction between rhizobia and legumes In: Scheel D, Wasternack C (eds) Plant signal transduction. Oxford University Press, Oxford, New York, pp 250–271

Mucciarelli M, Scannerini S, Bertea C, Maffei M (2003) In vitro and in vivo peppermint (*Mentha piperita*) growth promotion by nonmycorrhizal fungal colonisation. New Phytol 158:579–591

Narisawa K, Usuki F, Hashiba T (2004) Control of Verticillium Yellows in Chinese cabbage by the dark septate endophytic fungus LtVB3. Phytopathology 94:412–418

Peters AF (1991) Field and culture studies of *Streblonema - Macrocystis* new species Ectocarpales Phaeophyceae from Chile, a sexual endophyte of giant kelp. Phycologia 30:365–377

Petrini O (1991) Fungal endophytes of tree leaves. In: Andrews J, Hirano S (eds) Microbial ecology of leaves. Springer, New York Berlin Heidelberg, pp 179–197

Petrini O (1996) Ecological and physiological aspects of host specificity in endophytic fungi In: Redlin SC, Carris LM (eds) Endophytic fungi in grasses and woody plants. APS, St Paul, MN, pp 87–100

Petrini O, Sieber T, Toti L, Viret O (1992) Ecology, metabolite production, and substrate utilization in endophytic fungi. Nat Toxins 1:185–196

Reinhold-Hurek B, Hurek T (1998) Life in grasses: diazotrophic endophytes. Trends Microbiol 6:139–144

Schulz B, Boyle C (2005) The endophytic continuum. Mycol Res 109:661–687

Schulz B, Römmert A-K, Dammann U, Aust H-J, Strack D (1999) The endophyte-host interaction: a balanced antagonism. Mycol Res 103:1275–1283

Schulz B, Boyle C, Draeger S, Römmert A-K, Krohn K (2002) Endophytic fungi: a source of biologically active secondary metabolites. Mycol Res 106:996–1004

Sieber TN (2002) Fungal root endophytes In: Waisel Y, Eshel A, Kafkafi U (eds) The hidden half. Dekker, New York, pp 887–917

Sieber TN, Sieber-Canavesi F, Petrini O, Ekramoddoullah AK, Dorworth CE (1991) Characterization of Canadian and European *Melanconium* from some *Alnus* by morphological cultural and biochemical studies. Can J Bot 69:2170–2176

Sinclair JB, Cerkauskas RF (1996) Latent infection vs. endophytic colonisation by fungi. In: Redlin SC, Carris LM (eds) Endophytic fungi in grasses and woody plants. APS, St Paul, MN, pp 3–30

Smalla K, Wieland G, Buchner A, Zock A, Parzy J, Roskot N, Heuer H, Berg G (2001) Bulk and rhizosphere soil bacterial communities studied by denaturing gradient gel electrophoresis: plant dependent enrichment and seasonal shifts. Appl Environ Microbiol 67:4742–4751

Stone JK (1987) Initiation and development of latent infections by *Rhabdocline parkeri* on Douglas-fir. Can J Bot 65:2614–2621

Stone JK, Bacon CW, White JF (2000) An overview of endophytic microbes: endophytism defined. In: Bacon CW, White JF (eds) Microbial endophytes. Dekker, New York, pp 3–30

Sturz AV, Nowak J (2000) Endophytic communities of rhizobacteria and the strategies required to create yield enhancing associations with crops. Appl Soil Ecol 15:183–190

Tudzynski B (1997) Fungal phytohormones in pathogenic and mutualistic associations In: Carroll GC, Tudzynski P (eds) The mycota V. Springer, Berlin Heidelberg New York, pp 167–184

Tudzynski B, Sharon A (2002) Biosynthesis, biological role and application of fungal hormones. In: Osiewacz HD (ed) The mycota X Industrial applications, Springer, Berlin Heidelberg New York, pp 183–211

Usuki F, Narisawa K (2005) Formation of structures resembling ericoid mycorrhizas by the root endophytic fungus *Heteroconium chaetospira* within roots of *Rhododendron obtusum* var. *kaempferi*. Mycorrhiza 15:61–64

Vandenkoornhuyse P, Baldauf SL, Leyval C, Straczek J, Young JPW (2002) Extensive fungal diversity in plant roots. Science 295:2051

Varma A, Singh A, Sahay NS, Sharma J, Roy A, Kumari M, Raha D, Thakran S, Deka D, Bharti K, Hurek T, Blechert O, Rexer K-H, Kost G, Hahn A, Maier W, Walter M, Strack D, Kranner I (2000) *Piriformospora indica*: an axenically culturable mycorrhiza-like endosymbiotic fungus In: Hock B (ed) The mycota, vol IX. Fungal associations. Springer, Berlin Heidelberg New York, pp 125–150

Verkley G (1999) A monograph of the genus *Pezicula* and its anamorphs. Stud Mycol 44:5–171

Villarreal-Ruiz L, Anderson IC, Alexander IJ (2004) Interaction between an isolate from the *Hymenoscyphus ericae* aggregate and roots of *Pinus* and *Vaccinium* New Phytol 164:183–192

Wilcox HE, Wang CJK (1987) Ectomycorrhizal and ectendomycorrhizal associations of *Phialophora finlandia* with *Pinus resinosa*, *Picea rubens*, and *Betula alleghaniensis*. Can J For Res 17:976–990

Yates I, Bacon CW, Hinton DM (1997) Effects of endophytic infection by *Fusarium moniliforme* on corn growth and cellular morphology. Plant Dis 81:723–728

Zhou D, Hyde K (2001) Host-specificity, host-exclusivity and host-recurrence in saprobic fungi. Mycol Res 105:1449–1457

2 Spectrum and Population Dynamics of Bacterial Root Endophytes

Johannes Hallmann, Gabriele Berg

2.1 Introduction

Since the first reports regarding the existence of bacteria residing in plant roots (Trevet and Hollis 1948; Philipson and Blair 1957), numerous publications have described the spectrum and population dynamics of indigenous endophytic root endophytes for various plant species (Bell et al. 1995; Gardner et al. 1982; Hallmann et al. 1997a, 1999; Hallmann 2003; Mahaffee and Kloepper 1997a, 1997b; McInroy and Kloepper 1995; Misaghi and Donndelinger 1990; Sturz et al. 1997). But what do we really know about the spectrum and population dynamics of those bacteria residing in the endorhiza? What are the potential risks associated with these bacteria? Answers to these questions will not only improve our understanding of plant/endophytic bacterial interactions, but are a prerequisite for any future commercialisation. This chapter reviews our current knowledge of (1) population density, bacterial spectrum and bacterial diversity of endophytic root bacteria, (2) factors influencing the population dynamics of indigenous and introduced bacterial endophytes, (3) bacterial interactions with biotic and abiotic factors, and (4) potential risks associated with endophytic bacteria.

2.2 Population Density

Population densities of indigenous endophytic bacteria in roots are found to be about 10^5 cfu g^{-1} fresh root weight (Hallmann et al. 1997a). This is higher than in any other plant organ, as the population density usually decreases acropetally, with average densities of 10^4 cfu g^{-1} fresh weight in

Johannes Hallmann: Federal Biological Research Centre for Agriculture and Forestry, Institute for Nematology and Vertebrate Research, Toppheideweg 88, 48161 Münster, Germany, E-mail: j.hallman@bba.de

Gabriele Berg: Graz University of Biotechnology, Department of Environmental Technology, Petersgasse 12, 8010 Graz, Austria

Soil Biology, Volume 9
Microbial Root Endophytes
B. Schulz, C. Boyle, T. N. Sieber (Eds.)

the stem and 10^3 cfu g^{-1} fresh weight in leaves. Generative organs such as flowers, fruits and seeds are colonised by even lower numbers and, in many cases, are below the detection limit. However, depending on plant species, methodology and other factors, reported population densities can vary significantly. Common densities reported for indigenous endophytic bacteria in roots range from 10^4 to 10^6 cfu g^{-1} for cotton and sweetcorn (McInroy and Kloepper 1994), 10^3 to 10^6 cfu g^{-1} for sugar beet (Jacobs et al. 1985), 4.0×10^2 to 1.3×10^4 cfu g^{-1} for cotton (Hallmann et al. 1997a, Misaghi and Donndelinger 1990), 10^5 cfu g^{-1} for potato (Krechel et al. 2002) and 10^5 cfu g^{-1} for pine seedlings (Shishido et al. 1995). Some authors have even reported population densities up to 10^{10} cfu g^{-1} without negative effects on plant growth (Dimock et al. 1988; McInroy and Kloepper 1994). However, these high densities are considered exceptional; population densities above 10^7 cfu g^{-1} are known from bacterial plant pathogens to cause pathogenicity (Tsiantos and Stevens 1986; Grimault and Prior 1994). Within the first 3 weeks after seeding the population density generally increases to an optimum carrying capacity of about 10^5 cfu g^{-1} and then remains at this level for the rest of the growth period (Mahaffee and Kloepper 1997a, 1997b; McInroy and Kloepper 1995; Krechel et al. 2002). Even artificial inoculation of a highly concentrated bacterial suspension into the root tissue does not increase population density in the long run (Frommel et al. 1991). However, the above figures account only for culturable bacteria, which are only part of the total endophytic bacterial community. Their relative population size is still unknown as reliable data are lacking; but from other environments it is known that culturable bacteria represent between 0.001% and 15% of the total bacterial population.

2.3 Bacterial Spectrum

Studying the bacterial spectrum requires bacterial identification. Before 1990 this was laborious, using a combination of morphological and physiological characters often supported by ready-made tests such as API identification strips (Arrow Scientific, Lane Cove, Australia), or Biolog (Biolog, Hayward, CA) tests. With the fatty acid method described by Sasser (1990), bacterial identification and thus analysis of bacterial spectra became considerably more efficient, allowing time-course studies in which hundreds of bacteria could be identified (McInroy and Kloepper 1995; Mahaffee and Kloepper 1997a, 1997b). Nowadays, sequencing of 16S rDNA genes and alignment with public databases allows rapid and accurate species identification (Krechel et al. 2002; Reiter et al. 2002; Sessitsch et al. 2004). While this approach still relies on bacterial isolation (see Chap. 17 by Hallmann et al.), newly developed molecular methods use universal or specific primers to characterise either the endophytic community, certain bacterial groups or particular species, therefore enabling complete analysis of culturable as well as non-culturable bacteria (see Chap. 19 by van Overbeek et al.).

As mentioned for bacterial density, the spectrum of indigenous endophytic bacteria is affected by similar biotic and abiotic factors. However, it also seems to be influenced by niche specialisation, differences in colonisation pathway (seed-borne, rhizosphere, above-ground plant organs), or some form of mutual exclusion operating between different bacterial populations (Hallmann 2001; Sturz et al. 1997). Table 2.1 gives an overview of endophytic bacteria commonly isolated from the roots of some cultivated plants. Although far from complete, it shows the broad spectrum of bacterial endophytes that occur in plant roots. Table 2.2 provides a more comprehensive list of bacterial species isolated from roots, comprising 219 species representing 71 genera, with *Bacillus* and *Pseudomonas* being the most common genera.

Table 2.1. Spectrum of endophytic bacteria most commonly isolated from roots of various cultivated plant species

Plant	Bacterial taxa	Reference
Alfalfa	*Pseudomonas*, *Erwinia*-like bacteria	Gagné et al. 1987
Carrot	*Agrobacterium*, *Pseudomonas*, *Staphylococcus*	Surette et al. 2003
Clover	*Agrobacterium*, *Bacillus*, *Methylobacterium*, *Pseudomonas*, *Rhizobium*	Sturz et al. 1997
Cotton	*Bacillus*, *Burkholderia*, *Clavibacter*, *Erwinia*, *Phyllobacterium*, *Pseudomonas*	Chen et al. 1995; Hallmann et al. 1999; Misaghi and Donndelinger 1990
Cucumber	*Agrobacterium*, *Bacillus*, *Burkholderia*, *Chryseobacterium*, *Clavibacter*, *Curtobacterium*, *Enterobacter*, *Micrococcus*, *Paenibacillus*, *Phyllobacterium*, *Pseudomonas*, *Serratia*, *Stenotrophomonas*	Mahaffee and Kloepper 1997a, 1997b; McInroy and Kloepper 1995
Grapevine	*Enterobacter*, *Pseudomonas*, *Rahnella*, *Rhodococcus*, *Staphylococcus*	Bell et al. 1995
Maize	*Agrobacterium*, *Arthrobacter*, *Bacillus*, *Burkholderia*, *Corynebacterium*, *Curtobacterium*, *Enterobacter*, *Micrococcus*, *Phyllobacterium*, *Pseudomonas*, *Serratia*	Lalande et al. 1989; McInroy and Kloepper 1995
Potato	*Agrobacterium*, *Arthrobacter*, *Bacillus*, *Chryseobacterium*, *Enterobacter*, *Micrococcus*, *Pantoea*, *Pseudomonas*, *Stenotrophomonas*, *Streptomyces*	Krechel et al. 2002; Sturz 1995

Table 2.1. (continued)

Canola	*Acidovorax, Agrobacterium, Aureobacterium, Bacillus, Cytophaga, Chryseobacterium, Flavobacterium, Micrococcus, Rathayibacter, Pseudomonas*	Germida et al. 1998; Graner et al. 2003; Misko and Germida 2002
Red clover	*Agrobacterium, Bacillus, Methylobacterium, Pantoea, Pseudomonas, Rhizobium, Xanthomonas*	Sturz et al. 1997
Rice	*Serratia, Azoarcus*	Gyaneshwar et al. 2001; Hurek et al. 1994
Rough lemon	*Bacillus, Corynebacterium, Enterobacter, Pseudomonas, Serratia*	Gardner et al. 1982
Soybean	*Bacillus*	Bai et al. 2002
Sugar beet	*Bacillus, Corynebacterium, Erwinia, Lactobacillus, Peudomonas, Xanthomonas*	Jacobs et al. 1985
Sugar cane	*Acetobacter*	Cavalcante and Döbereiner 1988
Tomato	*Bacillus, Burkholderia, Chryseobacterium, Kluyvera, Micrococcus, Pseudomonas, Serratia*	Munif 2001
Wheat	*Bacillus, Flavobacterium, Microbispora, Micrococcus, Micromonospora, Nacardiodes, Rathayibacter, Streptomyces*	Coombs and Franco 2003; Germida et al. 1998
Diverse species	*Streptomyces*	Sardi et al. 1992

Table 2.2. Endophytic bacteria isolated from roots of various plants

Acetobacter diazotrophicus[13], *A. pasteurianus*[9]

Acidovorax delafieldii[7,9,10], *A. facilis*[10]

Acinetobacter baumannii[10], *A. calcoaceticus*[10], *A. wolffi*[5], *A. radioresistens*[10]

Aeromonas salmonicida[9]

Agrobacterium radiobacter[7,8,9,10,12], *A. rhizogenes*[12], *A. rubi*[7,9], *A. tumefaciens*[12,15]

Alcaligenes piechaudii[9,12], *A. xylosoxydans*[9,12]

Arthrobacter atrocyaneus[10,11], *A. aurescens*[11], *A. crystallopodietes*[12], *A. ilicis*[15], *A. mysorens*[10,12], *A. nicotianae*[12], *A. pascens*[10,11]

Aureobacterium barkeri[12], *A. esteroaromaticum*[10,11], *A. liquefaciens*[11], *A. saperdae*[12], *A. testaceum*[12]

Azospirillum amazonense[13], *A. brasiliense*[9,13], *A. lipoferum*[13]

Bacillus alvei[12], *B. amyloliquefaciens*[12], *B. azotoformans*[12,14], *B. brevis*[6,12,14,15], *B. cereus*[7,12], *B. circulans*[10,14], *B. citinusporus*[11], *B. coagulans*[12,15], *B. fastidiosus*[2], *B. gordonae*[6], *B. insolitus*[2,14], *B. laterosporus*[6,7,11,12], *B. lentus*[12], *B. licheniformis*[6], *B. longisporus*[6], *B. macerans*[12], *B. megaterium*[6,7,10,11,12,14,15], *B. mycoides*[11], *B. pumilus*[6,7,11,12], *B. sphaericus*[7,12], *B. subtilis*[7,12,14], *B. throphaeus*[6], *B. thuringiensis*[12]

Table 2.2. (continued)

Bordetella avium[14], *B. bronchiseptica*[8]
Brevibacterium acetylicum[11]
Brevundimonas diminuta[9], *B. vesicularis*[7,9]
Burkholderia cepacia[7,8,9,12], *B. gladioli*[8,10,12], *B. pickettii*[7,8,9,12], *B. solanacearum*[12]
Cellulomonas cellulans[12], *C. flavigena*[10], *C. turbata*[12,15]
Chryseobacterium balustinum[9,11], *C. indologenes*[7,11], *C. meningosepticum*[7,11]
Citrobacter freundii[5], *C. koseri*[12]
Clavibacter michiganensis[2,7,11,12,15]
Comamonas acidovorans[7,8,9,10,11], *C. terrigena*[2], *C. testosteroni*[9,12,14]
Curtobacterium allidum[15], *C. citreum*[14], *C. flaccumfaciens*[2,10,12,14,15], *C. luteum*[14,15], *C. pusillum*[2,12,15]
Cytophaga aquatilis[10], *C. johnsonae*[9,10,11]
Enterobacter aerogenes[5], *E. asburiae*[8,10,12], *E. cancerogenus*[12], *E. cloacae*[2,5,8,12], *E. intermedius*[11], *E. taylorae*[7,8,12], *E. sakazakii*[5]
Erwinia carotovora[9,12], *E. chrysanthemi*[9]
Escherichia coli[12,14], *E. vulneris*[12]
Flavimonas oryzihabitans[12]
Flavobacterium aquatile[11], *F. indologenes*[12], *F. meningosepticum*[12], *F. resinovorum*[9,10,11]
Gluconobacter oxidans[9]
Gordonia polyisoprenivorans[3]
Herbaspirillum spp.[13]
Hydrogenophaga flava[12], *H. pseudoflava*[10,12]
Kingella kingae[15]
Klebsiella ozaenae[2], *K. planticola*[12], *K. pneumoniae*[2,12], *K. terrigena*[12]
Kluyvera ascorbata[12], *K. cryocrescens*[8,9,10,12]
Kocuria kristinae[11]
Lactobacillus kefir[10]
Leuconostoc pseudomesenteroides[10]
Methylobacterium extorquens[14], *M. fuijsawaense*[12], *M. mesophilicum*[12], *M. radiotolerans*[12], *M. rhodinum*[14]
Microbacterium imperiale[12], *M. laevaniformans*[12]
Micrococcus agilis[12], *M. halobius*[9,10,11], *M. kristinae*[12], *M. luteus*[8,9,10,11,12], *M. lylae*[10,12], *M. roseus*[12], *M. varians*[10,12,15]
Micromonospora endolithica[3], *M. peucetica*[3]
Moraxella bovis[2], *M. phenylpyruvica*[10]
Mycobacterium aichiense[3], *M. bohemicum*[3], *M. cookii*[3], *M. flavescens*[3], *M. heidelbergense*[3], *M. palustre*[3]
Neisseria flavescens, *N. mucosa*[9]

Table 2.2. (continued)

Nocardia pseudobrasiliensis[3]
Nocardiodes albus[4]
Ochrabactrum anthropi[12]
Paenibacillus alvei[10], *P. pabuli*[12], *P. polymyxa*[12]
Pantoea agglomerans[2,5,11,12,14,15], *P. ananas*[12], *P. stewartii*[9]
Paracoccus denitrificans[7]
Pedicoccus pentosaceus[9]
Photobacterium angustum[9]
Phyllobacterium myrsinacearum[7,8,9,12,14], *P. rubiacearum*[7,8,9,10,12]
Plesiomonas shigelloides[10]
Providencia spp.[5]
Pseudomonas aeruginosa[5,9], *P. cepacia*[5], *P. chlororaphis*[6,7,10,11,12], *P. cichorii*[2,6,12,15], *P. coronofaciens*[12], *P. corrugata*[2,6,10,14,15], *P. flectens*[10], *P. fluorescens*[5,6,7,9,11,12], *P. fragi*[14], *P. fulva*[14], *P. marginalis*[2,6,9,12], *P. mendocina*[6,9,12], *P. pseudoalcaligenes*[6], *P. putida*[2,5,6,7,9,11,12], *P. rubrisubalbicans*[9,12], *P. saccharophila*[8,10,12], *P. savastanoi*[6,10,12], *P. stutzeri*[9], *P. syringae*[2,6,9,11,12], *P. tolaasii*[14,15], *P. vesicularis*[12,14], *P. viridiflava*[6,11]
Rahnella aquatilis[2]
Rhizobium etli[10], *R. japonicum*[12], *R. leguminosarum*[14], *R. loti*[14], *R. meliloti*[15]
Rhodococcus coprophilus[3], *R. luteus*[2]
Salmonella cholerasuis[7]
Serratia liquefaciens[5,12], *S. marcescens*[12], *S. plymuthica*[12]
Shigella spp.[5]
Shingobacterium heparinum[11]
Sphingomonas capsulata[9], *S. paucimobilis*[9,12,15], *S. thalpophilum*[15]
Staphylococcus capitis[12], *S. epidermidis*[12], *S. haemolyticus*[11]
Stenotrophomonas maltophilia[9,11,12]
Streptomces argenteolus[4], *S. bikiniensis*[4], *S. caviscabies*[4], *S. cyaneus*[11], *S. galilaeus*[4], *S. halstedii*[11], *S. maritimus*[4], *S. pseudovenezuelae*[4], *S. scabies*[4], *S. setonii*[4], *S. tendae*[4], *S. thermolineatus*[3], *S. violaceusniger*[11]
Variovorax campestris, V. paradoxus[7,9,12]
Vibrio cholerae[9]
Xanthobacter agilis[9,10]
Xanthomonas agilis[7], *X. campestris*[2,8,9,12,14,15], *X. oryzae*[15]
Yersinia frederiksenii[12], *Y. pseudotuberculosis*[10]

[a]References:
[1]Adhikari et al. 2001; [2]Bell et al. 1995; [3]Conn and Franco 2004; [4]Coombs and Franco 2003; [5]Gardner et al. 1982; [6]Germida and Siciliano 2001; [7,8,9]Hallmann et al. 1997b, 1998, 1999; [10]Hallmann 2003; [11]Krechel et al. 2002; [12]McInroy and Kloepper 1995; [13]Reis et al. 2000; [14,15]Sturz et al. 1997, 1999

2.4 Bacterial Diversity

Diversity indices allow the compression of bacterial spectrum and bacterial density into a single number to facilitate comparisons between plant species, habitats, etc., as well as elucidation of changes in community relationships (Mahaffee and Kloepper 1997a, 1997b). Of the many diversity indices used in ecology, only few can be applied to endophytic bacteria. Three of the more commonly used indices are genera richness, Hill's modified Shannon's index N1, and Hill's modified Simpson's index N2, with N2 more than N1 representing very abundant species (Ludwig and Reynolds 1988). Using these indices for the endorhiza of field-grown cucumber, Mahaffee and Kloepper (1997a, 1997b) observed that all three indices tended to increase over the growing season, reaching their highest values at the final sampling date 70 days after planting. However, density varied between two consecutive years, indicating the importance of climatic conditions for the community structure of bacterial root endophytes. For potatoes, Krechel et al. (2002) observed the highest bacterial diversity at flowering. During that period the plant undergoes massive physiological changes that probably increase nutrient availability and thus bacterial diversity. Other biotic and abiotic factors that influence bacterial diversity are discussed later in this chapter. There is little question, though, that a better understanding of the population dynamics of bacterial root endophytes will enhance our ability to take advantage of their beneficial potential to enhance plant growth and health.

2.5 Factors Influencing Colonisation

The enormous spectrum and high diversity of endophytic bacteria found in different plant species begs the question: what biotic and abiotic factors influence bacterial colonisation?

2.5.1 Methodology

Methodology is especially important in describing the spectrum of culturable bacteria, as different methods will give different results. Key factors affecting the bacterial spectrum recovered are (1) length of surface sterilisation, (2) concentration of the sterilising agent and (3) the method itself. Comparison of the trituration method with the pressure bomb technique (see Chap. 17 by Hallmann et al.) revealed significantly larger populations of endophytic bacteria in cotton roots using the first method (Hallmann et al. 1997b). However, total number of bacterial genera and species recovered was greater using the pressure bomb technique, mainly because a higher number of less commonly occurring species was recovered. These results suggest that the two techniques sample two different microhabitats, i.e. the pressure bomb technique more effectively recovering mainly vascular colonists while the trituration method recovers both vascular colonisers and bacteria residing in the root cortex. An even higher bacterial spectrum can be identified using molecular methods that detect culturable as well as non-culturable bacteria (see Chap. 17 by Hallmann et al.).

2.5.2 Geography

Geographical regions are believed to differ in their bacterial spectra. Munif (2001) compared the bacterial spectrum of tomato roots grown under temperate conditions in Bonn, Germany, with that of tomato roots grown under tropical conditions in Bogor, Indonesia. In Germany, 38 species comprising 21 genera were isolated, whereas in Indonesia, 50 species comprising 32 genera were isolated. Twenty-four bacterial species were exclusively isolated from tomato roots in Germany and 38 species exclusively from tomato roots in Indonesia. However, 14 species were isolated from both regions. Interestingly, the most abundant species in both geographical regions were *Pseudomonas putida* and *Bacillus megaterium*. These two bacterial species are also commonly reported in other regions of the world (Tables 2.1, 2.2). How is it possible that a bacterial species colonises such different regions and still dominates the endophytic spectrum? And are populations from different regions distinguishable? Using molecular fingerprint techniques, fluorescent *Pseudomonas* strains collected worldwide could be regionally grouped suggesting that adaptation to their specific environment has occurred (Cho and Tiedje 2000). But adaptations also seem to occur within a given environment. For example, *Pseudomonas fluorescens* strains isolated from the endorhiza of potato are distinguishable from *P. fluorescens* strains isolated from the rhizosphere of the same plant (Berg et al. 2005).

2.5.3 Plant Species

The effect of plant species on the bacterial spectrum being recovered has only been marginally studied. Plant specificity of the bacterial community was shown for the rhizosphere by Smalla et al. (2001) and Berg et al. (2002).

As endophytic bacteria represent a subset of the rhizosphere colonisers, one would expect that different plant species growing side by side would be colonised by a different spectrum of endophytic bacteria. This hypothesis was confirmed by McInroy and Kloepper (1995), who showed that the bacterial spectrum of sweet corn and cotton grown next to each other differed. Although the total number of genera was similar for both plant species, some genera, such as *Alcaligenes*, *Aureobacterium*, *Cellulomonas*, *Comamonas*, *Erwinia*, *Ochrobactrum*, and *Yersinia*, were recovered only from cotton roots, while other genera, such as *Arthrobacter*, *Citrobacter*, *Flavimonas*, *Microbacterium* and *Stenotrophomonas*, were recovered exclusively from sweetcorn roots. Plant-species-specific factors such as root architecture, surface structure, and composition of the root exudates as well as non-plant factors such as mycorrhization or wounding probably influence the bacterial spectrum prior to colonisation. Following root colonisation, size of the intercellular space, nutrient composition within the apoplastic fluid, and the plant's response to endophytic colonisation are probably the main factors determining bacterial selectivity and thus the bacterial spectrum found inside the roots.

2.5.4
Plant Genotype

Does the plant genotype affect the spectrum of endophytic bacteria? What about cultivars resistant to bacterial plant pathogens? For the latter, no differences were seen in the population densities of endophytic bacteria of two grapevine cultivars resistant and susceptible to *Agrobacterium tumefaciens*, the causal agent of crown gall (Bell et al. 1995). However, resistance varied in some colonised cultivars (Conn et al. 1997), suggesting that not only the host genotype but also the associated bacterial endophytes may contribute to plant resistance, as suggested by Bird et al. (1980). Differences in the endophytic spectrum were also reported to occur in potato tubers of four potato cultivars (Sturz et al. 1999). Although the two bacterial species *Curtobacterium flaccumfaciens* and *Pseudomonas cichorii* occurred in all four cultivars, cultivar-specific preferences were observed. For example, *C. flaccumfaciens* dominated the endophytic spectrum of potato 'Kennebec' but was rarely found in 'Butte', whereas *P. cichorii* was dominant in 'Butte' and less frequently isolated from the other three cultivars. In addition, some bacterial species were exclusively isolated from one cultivar, but not from others. Similar to plant species, plant genotypes also vary in their biochemical composition, which may thus affect bacterial spectra. Similarly, Graner et al. (2003) found noticeable differences in endophytic bacterial populations of oil-seed rape cultivars that differ in their susceptibility to the fungal wilt pathogen *Verticillium longisporum*. Genotype-specific differences were also shown for wheat cultivars: modern cultivars supported a more diverse endophytic community than ancient land races (Germida and Siciliano 2001). Furthermore, endophytic bacteria isolated from different canola cultivars had different carbon utilisation profiles (Misko and Germida 2002).

2.6 Interactions

Ecological theory states that when a disruptive force such as pathogen infestation affects a community, the diversity of that community initially increases and its membership becomes highly variable before reaching a new equilibrium (Mahaffee and Kloepper 1997a; Petratis et al. 1989). But does this also apply to endophytic bacteria of roots? Plant roots are continuously challenged by soil-borne pathogens, mutualistic symbionts and various soil conditions. As a result, plant defence mechanisms might be induced. How do these parameters affect the endophytic bacterial spectrum?

2.6.1 Plant Pathogens

While the beneficial effects of endophytic bacteria to control plant pathogens are well documented [see Chaps. 3 (Kloepper and Ryu) and 4 (Berg and Hallmann)], very little is known about how plant pathogens affect the spectrum and diversity of bacterial root endophytes. Regarding fungal pathogens, Mahaffee et al. (1997a, 1997b) reported that root infection by *Rhizoctonia solani* promoted colonisation of the two introduced bacterial endophytes *Enterobacter asburiae* JM22 and *Pseudomonas fluorescens* 89B-27. Interactions between plant pathogenic bacteria and endophytic bacteria have so far been described only for above ground plant organs. Inoculation of potatoes with *Erwinia carotovora* subsp. *atroseptica* caused an increase in endophytic bacterial diversity of infected plants compared with healthy plants (Reiter et al. 2002); similarly, infestation of citrus by *Xylella fastidiosa* was positively correlated with the occurrence of *Methylobacterium* spp. (Araújo et al. 2002). A third group of plant pathogens, plant parasitic nematodes, significantly increased the total number of bacterial root endophytes (Hallmann et al. 1998; Hallmann 2003), and also affected the bacterial spectrum (Hallmann 2003). In cotton roots infested with the root-knot nematode *Meloidogyne incognita*, 11 species were exclusively isolated from roots of infested plants, while 15 species occurred only in non-infested plants.

2.6.2 Plant Symbionts

Interactions between endophytic bacteria and mutualistic plant symbionts have been best studied for nodule bacteria. Sturz et al. (1997) reported that bacterial density and diversity was lower in root nodules of red clover than in tap roots, but root nodules yielded a higher number of species exclusively colonising this specific niche (Sturz et al. 1997). Coinoculation of *Rhizobium leguminosarum* BV *trifolii* with endophytic bacteria promoted root nodulation. In other cases, endophytic bacteria reduced or inhibited root colonisation by nitrogen-fixing bacteria (Bacilio-Jiménez et al. 2001), indicating that those interactions seem to be strain-specific. Little is yet known about the interaction between endophytic bacteria and other symbionts such as fungal endophytes and mycorrhizal fungi. Mycorrhizal fungi are at least known to influence rhizosphere bacteria (Azacón-Aguilar and Barea 1992) and similar effects might apply to endophytic bacteria.

2.6.3 Plant Defence Mechanisms

A question frequently asked is: why are endophytic bacteria not inhibited by a plant defence response? And if there is a plant defence response, how does this affect the bacterial spectrum? To answer these questions, potato plant resistance was experimentally induced by *Rhizobium etli* G12 (Hallmann 2003). *R. etli* G12 was applied to one-half of a split potato root system and the endophytic bacterial spectrum was analysed for the other ("induced") half of the split root system and compared with that of non-treated plants. Total bacterial density was significantly higher in treated (1.6×10^4 cfu g^{-1}) than in non-treated (3.2×10^3 cfu g^{-1}) roots. Seventeen bacterial species were isolated from treated roots compared with 14 species from non-treated roots. The results indicated that, under the conditions of bacteria-mediated induced plant defence responses, the density and spectrum of bacterial root endophytes is increased.

2.6.4 Agricultural Practices

Besides the plant itself and plant-associated microorganisms, agricultural practices can also influence the spectrum and population dynamics of bacterial root endophytes. Seghers et al. (2004) showed that different fertiliser treatments influenced the endophytic community of maize roots, whereas the tested herbicide treatments did not. High N-fertilisation inhibited the colonisation of sugarcane by *Acetobacter diazotrophicus* (Fuentes-Ramírez et al. 1999), while application of nitrogen-containing chitin as an organic amendment supported endophytic species in cotton roots that otherwise did not occur (Hallmann et al. 1999). For the latter, it was shown that the endophytic spectrum was completely different from that of the rhizosphere, indicating that the bacterial composition of the rhizosphere is not the only factor determining the endophytic spectrum. Differences in plant biochemistry due to the organic amendment, such as enhanced chitinase and peroxidase concentrations (Hallmann 2003), might have changed the plants' preference for certain bacterial endophytes.

2.7 Potential Human Pathogens Among Root Endophytes

By definition, endophytic bacteria reside within the plant without causing visible disease symptoms (see Chap. 1 by Schulz and Boyle). However, the distinction between harmless and harmful bacteria is not always clear. Potentially harmful bacteria might colonise the plant latently or reside as dormant stages, becoming harmful only at later growth stages or when a critical density, known as "quorum sensing", is reached (Eberl 1999).

Since the beneficial attributes of endophytic bacteria are covered elsewhere in this book [see Chaps. 3 (Kloepper and Ryu), 4 (Berg and Hallmann), and 6 (Anand et al.)], we will focus on potentially human pathogenic bacteria. In general, mechanisms of pathogenicity and antagonism are very similar, and sometimes only the expression of one metabolite or total bacterial numbers dictates whether the bacteria are harmful or harmless (Suckstorff and Berg 2003; Berg et al. 2006). For example, type III secretion systems, known for their role in bacterial pathogenicity, are present in many plant-associated *Pseudomonas* strains and may confer induced resistance, plant growth promotion or biological control (Preston et al. 2001). A few bacterial species isolated from inside plant roots are also known to be pathogenic to humans These include strains of *Bacillus cereus*, *Burkholderia cepacia*, *Serratia marcescens* and *Stenotrophomonas maltophilia*, which not only have excellent antagonistic properties against plant pathogens, but are also known to cause human diseases, especially of debilitated or immunosuppressed individuals (LiPuma 2003; Minkwitz and Berg 2001; Vandamme and Mahenthiralingam 2003; Wolf et al. 2002). *Staphylococcus* species, also known to be human pathogens, have been isolated from the potato endorhiza (Krechel et al. 2002; Reiter et al. 2002). In some cases, human-pathogenic isolates differ from the harmless isolates occurring in plants by the expression of certain toxins, in other cases no such distinction has yet been found. Reiter et al. (2002) showed, based on 16S rDNA sequences, high homology of endophytic bacterial isolates from potato leaves with those of human pathogens such as *Enterobacter amnigenus*, *Enterobacter cloacae*, *Stenotrophomonas maltophilia*, *Staphylococcus xylosus* and *Ochrobactrum anthropi*. *E. cloacae*, *S. maltophilia*, and *O. anthropi* have also been reported to occur in plant roots (McInroy and Kloepper 1995).

Nevertheless, the presence of human-pathogenic bacteria in plant roots can be a health concern and therefore should be carefully monitored. As a result of several *Salmonella* outbreaks attributed to alfalfa sprout contamination in Finland and the United States (van Beneden et al. 1999), alfalfa seeds were found to be contaminated and surface sterilisation did not eliminate the enteric pathogen. Gandhi et al. (2001) detected a *Salmonella* Stanley strain inside alfalfa sprouts to a depth of 18 μm without finding cells on the surface, thus confirming the endophytic properties of this pathogen. Strains isolated from patients infected with *Salmonella enterica* during an alfalfa-sprout-associated outbreak and labelled with the green fluorescent protein (GFP) successfully colonised alfalfa seedlings (Dong et al. 2003). An inoculation with very few cells was sufficient to colonise the plant interior; however, strains of *S. enterica* differed greatly in their ability to invade the plant interior and to colonise alfalfa roots. This demonstrates the need for further investigations to ensure food safety in the future.

2.8 Conclusions

In conclusion, of the plants thus far studied, the spectrum and diversity of endophytic bacteria in the roots varies greatly. What about the endophytic bacterial spectrum of plants growing under extreme climatic conditions, such as halophytes and xerophytes? Survival mechanisms developed by those bacteria may have some interesting industrial or pharmaceutical applications. Newly developed cultivation-independent methods have made clear that there is much more diversity among endophytic bacteria than at first expected. The major factors influencing bacterial diversity and colonisation have been discussed and their potential to manage endophytic communities towards increased benefits for plants and human health have been outlined. However, the potential risks of endophytic bacteria, especially of those strains known also to be potential human pathogens, need further exploration.

References

Adhikari TB, Joseph CM, Yang G, Phillips DA, Nelson LM (2001) Evaluation of bacteria isolated from rice for plant growth promotion and biological control of seedling disease of rice. Can J Microbiol 47:916–924

Araújo WL, Marcon J, Maccheroni W Jr, Van Elsas JD, Van Vuurde JW, Azevedo JL (2002) Diversity of endophytic bacterial populations and their interaction with *Xylella fastidiosa* in citrus plants. Appl Environ Microbiol 68:4906–4914

Azacón-Aguilar C, Barea JM (1992) Interactions between mycorrhizal fungi and other rhizosphere microorganisms. In: Allen MF (ed) Mycorrhizal functioning: an integrative plant-fungal process. Chapmann & Hall, New York, pp 163–198

Bacilio-Jiménez M, Aguilar-Flores S, del Valle MV, Pérez A, Zapeda A, Zenteno E (2001) Endophytic bacteria in rice seeds inhibit early colonization of roots by *Azospirillum brasilense*. Soil Biol Biochem 33:167–172

Bai Y, D'aoust F, Smith DL, Driscoll BT (2002) Isolation of plant-growth-promoting *Bacillus* strains from soybean root nodules. Can J Microbiol 48:230–238

Bell CR, Dickie GA, Harvey WLG, Chan JWYF (1995) Endophytic bacteria in grapevine. Can J Microbiol 41:46–53

Berg G, Roskot N, Steidle A, Eberl L, Zock A, Smalla K (2002) Plant-dependent genotypic and phenotypic diversity of antagonistic rhizobacteria isolated from different *Verticillium* host plants. Appl Environ Microbiol 68:3328–3338

Berg G, Krechel A, Ditz M, Sikora RA, Ulrich A, Hallmann J (2005) Endophytic and ectophytic potato-associated bacterial communities differ in structure and antagonistic function against plant pathogenic fungi. FEMS Microbiol Ecol 51:215–229

Berg G, Eberl L, Hartmann A (2006) The rhizosphere as a reservoir for opportunistic human pathogenic bacteria. Environ. Microbiol. 71:4203–4213

Bird LS, Leverman C, Thaxaton R, Percy RG (1980) Evidence that microorganisms in and on tissues have a role in a mechanism of multi-adversity resistance in cotton. In: Brown JM (ed) Proceedings of Beltville Cotton Production Research Conferences. Nationals Cotton Council, Memphis, TN, pp 283–285

Cavalcante VA, Döbereiner J (1988) A new acid-tolerant nitrogen fixing bacterium associated with sugarcane. Plant Soil 108:23–31

Chen C, Bauske EM, Musson G, Rodríguez-Kábana R, Kloepper JW (1995) Biological control of *Fusarium* wilt on cotton by use of endophytic bacteria. Biol Control 5:83–91

Cho JC, Tiedje JM (2000) Biogeography and degree of endemicity of fluorescent *Pseudomonas* strains in soil. Appl Environ Microbiol 66:5448–5456

Conn, VM, Franco CMM (2004) Analysis of the endophytic actinobacterial population in the roots of wheat (*Triticum aestivum* L.) by terminal restriction fragment length polymorphism and sequencing on 16S rRNA clones. Appl Envrion Microbiol 70:1787–1794

Conn KL, Nowak J, Lazorovita G (1997) A gnotobiotic bioassay for studying interactions between potatoes and plant growth-promoting rhizobacteria. Can J Microbiol 43:801–808

Coombs JT, Franco CMM (2003) Isolation and identification of actinobacteria from surface-sterilized wheat roots. Appl Environ Microbiol 69:5603–5608

Dimock MB, Beach RM, Carlson PS (1988) Endophytic bacteria for delivery of crop protection agents. In: Roberts DW, Granados RR (eds) Biotechnology, biological pesticides and novel plant-pest resistance for pest management. Boyce Thompson Institute for Plant Research, Ithaca, NY, pp 88–92

Dong Y, Iniguez AL, Ahmer BM, Triplett EW (2003) Kinetics and strain specificity of rhizosphere and endophytic colonization by enteric bacteria on seedlings of *Medicago sativa* and *Medicago truncatula*. Appl Environ Microbiol 69:1783–1790

Eberl L (1999) N-acyl homoserinelactone-mediated gene regulation in Gram-negative bacteria. Syst Appl Microbiol 22:493–506

Frommel MI, Nowak J, Lazarovits G (1991) Growth enhancement and developmental modifications of in vitro grown potato (*Solanum tuberosum* ssp. *tuberosum*) as affected by a nonfluorescent *Pseudomonas* sp. Plant Physiol 96:928–936

Fuentes-Ramírez LE, Caballero-Melado J, Sepúlveda J, Martínez-Romero E (1999) Colonization of sugarcane by *Acetobacter diazotrophicus* is inhibited by high N-fertilization. FEMS Microbiol Ecol 29:117–128

Gagné S, Richard C, Rousseau H, Antoun H (1987) Xylem-residing bacteria in alfalfa roots. Can J Microbiol 33:996–1000

Gandhi MS, Golding S, Yaron S, Matthews KR (2001) Use of green fluorescent protein expressing *Salmonella* Stanley to investigate survival, spatial location, and control on alfalfa sprouts. J Food Prot 64:1891–1898

Gardner JM, Feldman AW, Zablotowicz RM (1982) Identity and behavior of xylem-residing bacteria in rough lemon roots of Florida citrus trees. Appl Environ Microbiol 43:1335–1342

Germida JJ, Siciliano SD (2001) Taxonomic diversity of bacteria associated with the roots of modern, recent and ancient wheat cultivars. Biol Fertil Soils 33:410–415

Germida JJ, Siciliano SD, Freitas JR de, Seib AM (1998) Diversity of root-associated bacteria associated with field-grown canola (*Brassica napus* L.) and wheat (*Tricicum aestivum* L.). FEMS Microbiol Ecol 26:43–50

Graner G, Persson P, Meijer J, Alstrom S (2003) A study on microbial diversity in different cultivars of *Brassica napus* in relation to its wilt pathogen, *Verticillium longisporum*. FEMS Microbiol Lett 29:269–276

Grimault V, Prior P (1994) Invasiveness of *Pseudomonas solanacearum* in tomato, eggplant, and pepper: a comparative study. Eur J Plant Pathol 100:259–267

Gyaneshwar P, James EK, Mathan N, Reddy PM, Reinhold-Hurek B, Ladha JK (2001) Endophytic colonization of rice by a diazotrophic strain of *Serratia marcescens*. J Bacteriol 183:2634–2645

Hallmann J (2001) Plant interactions with endophytic bacteria. In: Jeger MJ, Spence NJ (eds) Biotic interactions in plant-pathogen associations. CABI, Wallingford, UK, pp 87–119

Hallmann J (2003) Biologische Bekämpfung pflanzenparasitärer Nematoden mit antagonistischen Bakterien, vol 392. Mitt Biol Bundesanst Land Forstwirtsch, Berlin,

Hallmann J, Quadt-Hallmann A, Mahaffee WF, Kloepper JW (1997a) Bacterial endophytes in agricultural crops. Can J Microbiol 43:895–914

Hallmann J, Kloepper JW, Rodríguez-Kábana R (1997b) Application of the Scholander pressure bomb to studies on endophytic bacteria of plants. Can J Microbiol 43:411–416

Hallmann J, Quadt-Hallmann A, Rodríguez-Kábana R, Kloepper JW (1998) Interactions between *Meloidogyne incognita* and endophytic bacteria in cotton and cucumber. Soil Biol Biochem 30:925–937

Hallmann J, Rodríguez-Kábana R, Kloepper JW (1999) Chitin-mediated changes in bacterial communities of the soil, rhizosphere and within roots of cotton in relation to nematode control. Soil Biol Biochem 31:551–560

Hurek T, Reinhold-Hurek B, van Montagu M, Kellenberger E (1994) Root colonization and systemic spreading of *Azoarcus* sp. Strain BH72 in grasses. J Bacteriol 176:1913–1923

Jacobs MJ, Bugbee WM, Gabrielson DA (1985) Enumeration, location, and characterization of endophytic bacteria within sugar beet roots. Can J Bot 63:1362–1365

Krechel A, Faupel A, Hallmann J, Ulrich A, Berg G (2002) Potato-associated bacteria and their antagonistic potential towards plant-pathogenic fungi and the plant-parasitic nematode *Meloidogyne incognita* (Kofoid & White) Chitwood. Can J Microbiol 48:772–786

Lalande R, Bissonnette N, Coutlée D, Antoun H (1989) Identification of rhizobacteria from maize and determination of their plant-growth promoting potential. Plant Soil 115:7–11
LiPuma JJ (2003) *Burkholderia cepacia* complex as human pathogens. J Nematol 35:213–217
Ludwig JA, Reynolds JF (1988) Statistical ecology. Wiley, New York
Mahaffee WF, Kloepper JW (1997a) Temporal changes in the bacterial communities of soil, rhizosphere, and endorhiza associated with field-grown cucumber (*Cucumis sativus* L.). Microb Ecol 34:210–223
Mahaffee WF, Kloepper JW (1997b) Bacterial communities of the rhizosphere and endorhiza associated with field-grown cucumber plants inoculated with a plant growth-promoting rhizobacterium or its genetically modified derivative. Can J Microbiol 43:34–35
McInroy JA, Kloepper JW (1994) Studies on indigenous endophytic bacteria of sweetcorn and cotton. In: O'Gara F, Dowling DN, Boesten B (eds) Molecular ecology of rhizosphere microorganisms. VCH, Weinheim, pp 19–28
McInroy JA, Kloepper JW (1995) Survey of indigenous bacterial endophytes from cotton and sweet corn. Plant Soil 173:337–342
Minkwitz A, Berg G (2001) Comparison of antifungal activities and 16S ribosomal DNA sequences of clinical and environmental isolates of *Stenotrophomonas maltophilia*. J Clin Microbiol 39:139–145
Misaghi IJ, Donndelinger CR (1990) Endophytic bacteria in symptom-free cotton plants. Phytopathology 80:808–811
Misko AL, Germida JJ (2002) Taxonomic and functional diversity of pseudomonads isolated from the roots of field-grown canola. FEMS Microbiol Ecol 42:399–407
Munif A (2001) Studies on the importance of endophytic bacteria for the biological control of the root-knot nematode *Meloidogyne incognita* on tomato. PhD thesis, University of Bonn, Germany
Petratis PS, Latham RE, Niesenbaum RA (1989) The maintenance of species diversity by disturbance. Q Rev Biol 64:393–418
Philipson MN, Blair ID (1957) Bacteria in clover root tissue. Can J Microbiol 3:135–139
Preston GM, Bertrand N, Rainey PB (2001) Type III secretion in plant growth-promoting *Pseudomonas fluorescens* SBW25. Mol Microbiol 41:999–1014
Reis Junior dos FB, Silva da LG, Reis VM, Dobereiner J (2000) Occurrence of diazotrophic bacteria in different sugar cane genotypes. Pesqui Agropecu Bras 35:985–994
Reiter B, Pfeifer U, Schwab H, Sessitsch A (2002) Response of endophytic bacterial communities in potato plants to infection with *Erwinia carotovora* subsp. *atroseptica*. Appl Environ Microbiol 68:2261–2268
Sardi P, Sarachhi M, Quaroni S, Petrolini B, Borgonovi GE, Merli S (1992) Isolation of endophytic *Streptomyces* strains from surface-sterilized roots. Appl Environ Microbiol 58:2691–2693
Sasser M (1990) Identification of bacteria through fatty acid analysis. In: Klement Z, Rudolph K, Sands D (eds) Methods in phytobacteriology. Akademiai Kiado, Budapest, pp 199–204
Seghers D, Wittebolle L, Top EM, Verstraete W, Siciliano SD (2004) Impact of agricultural practices on the *Zea mays* L. endophytic community. Appl Environ Microbiol 70:1475–1482
Sessitsch A, Reiter B, Berg G (2004) Endophytic bacterial communities of field-grown potato plants and their plant growth-promoting abilities. Can J Microbiol 50:239–249
Shishido M, Loeb BM, Chanway CP (1995) External and internal root colonization of lodgepole pine seedlings by two growth-promoting *Bacillus* strains originated from different root microsites. Can J Microbiol 41:707–713

Smalla K, Wieland G, Buchner A, Zock A, Parzy J, Roskot N, Heuer H, Berg G (2001) Bulk and rhizosphere soil bacterial communities studied by denaturing gradient gel electrophoresis: plant dependent enrichment and seasonal shifts. Appl Envrion Microbiol 67:4742–4751

Sturz AV (1995) The role of endophytic bacteria during seed piece decay and potato tuberization. Plant Soil 175:257–263

Sturz AV, Christie BR, Matheson BG, Nowak J (1997) Biodiversity of endophytic bacteria which colonize red clover nodules, roots, stems and foliage and their influence on host growth. Biol Fertil Soils 25:13–19

Sturz AV, Christie BR, Matheson BG, Arsenault WJ, Buchanan NA (1999) Endophytic bacterial communities in the periderm of potato tubers and their potential to improve resistance to soil-borne plant pathogens. Plant Pathol 48:360–369

Suckstorff I, Berg G (2003) Evidence of dose-dependent effects on plant growth by *Stenotrophomonas* strains from different origins. J Appl Microbiol 95:656–663

Surette MA, Sturz AV, Lada RR, Nowak J (2003) Bacterial endophytes in processing carrots (*Daucus carota* L. var. *sativus*): their localization, population density, biodiversity and their effects on plant growth. Plant Soil 253:381–390

Trevet IW, Hollis JP (1948) Bacteria in storage organs of healthy plants. Phytopathology 38:960–967

Tsiantos J, Stevens WA (1986) The population dynamics of *Corynebacterium michiganense* pv. *michiganensis* and other selected bacteria in tomato leaves. Phytopathol Mediterr 25:160–162

Van Beneden CA, Keene WE, Strang RA, Werker DH, King AS, Mahon B, Hedberg K, Bell A, Kelly MT, Balan VK, MacKenzie WR, Fleming D (1999) Multinational outbreak of *Salmonella enterica* serotype *Newport* infections due to contaminated alfalfa sprouts. JAMA 281:158–162

Vandamme P, Mahenthiralingam E (2003) Strains from the *Burkholderia cepacia* complex: Relationship to opportunistic pathogens. J Nematol 35:208–211

Wolf A, Fritze A, Hagemann M, Berg G (2002) *Stenotrophomonas rhizophila* sp. nov., a novel plant-associated bacterium with antifungal properties. Int J Evol Syst Microbiol 52:1937–1944

3 Bacterial Endophytes as Elicitors of Induced Systemic Resistance

Joseph W. Kloepper, Choong-Min Ryu

3.1 Introduction and Terminology

As indicated elsewhere in this book (e.g. Chap. 1 by Schulz and Boyle), the question of what are endophytes can be answered in different ways. For the purposes of this chapter, only those endophytes that could be isolated from surface-sterilized plant tissue or extracted from within the plant, as proposed by Hallmann et al. (1997), will be discussed. All of the rhizobacteria discussed here were isolated by grinding tissues of surface-sterilized plants, while maintaining sterility controls. It was subsequently discovered that some of these bacterial strains elicited systemic protection against pathogens when the bacteria were inoculated onto seeds or into the potting mix.

Application to crops of many plant-associated bacteria, including some endophytic bacteria, results in a reduction in the incidence or severity of diseases. This phenomenon is referred to as biological control. The most commonly reported mechanism of biological control is antagonism, where the bacterium causes a reduction in the pathogen population or its disease-producing potential. Antagonism includes the more specific mechanisms of predation, competition, and antibiosis. Antagonism is discussed in detail in Chap. 4 by Berg and Hallmann.

An alternative mechanism for biological control is that bacterial metabolites affect the plant in such a way as to increase the plant's resistance to pathogens, a process termed induced systemic resistance (ISR). Resistance can also be elicited in plants by the application of chemicals or necrosis-producing pathogens, and this process is termed systemic acquired resistance (SAR). Pieterse et al. (1998) proposed that ISR and SAR can be differentiated not only by the elicitor but also by the signal transduction pathways that are elicited within the plant. Accordingly, ISR is elicited by rhizobac-

Joseph W. Kloepper: Department of Entomology and Plant Pathology, Auburn University, Auburn, AL 36849, USA, E-mail: kloepjw@auburn.edu

Choong-Min Ryu: Plant Biology Division, The Samuel Roberts Noble Foundation, 2510 Sam Noble Parkway, Ardmore, OK 73401, USA

Soil Biology, Volume 9
Microbial Root Endophytes
B. Schulz, C. Boyle, T. N. Sieber (Eds.)

teria or other nonpathogenic microorganisms, while SAR is elicited by pathogens or chemical compounds. Further, the signal transduction pathway of ISR is independent of salicylic acid but dependent on jasmonate and ethylene, while the pathway of SAR is dependent on salicylic acid and shows variable dependency on jasmonate and ethylene. Recent discoveries show that some rhizobacteria, including some of the endophytic strains discussed in this chapter, elicit systemic protection that may be dependent on salicylic acid and independent of jasmonate or ethylene. Hence, ISR cannot be separated from SAR based on signal transduction pathways. In this chapter, ISR is used to describe the phenomenon whereby application of bacteria to one part of the plant results in a significant reduction in the severity or incidence of a disease following inoculation of a pathogen to another part of the plant.

3.2 Scope of Endophytes that Elicit Induced Resistance and Pathosystems Affected

The first indication that endophytic bacteria could elicit ISR dates to 1991 (Wei et al. 1991). *Pseudomonas fluorescens* strain G8-4, which was later designated 89B-61 and found to colonize plants internally, elicited systemic protection against cucumber anthracnose following application to cucumber seeds.

In efforts to find other strains of endophytic bacteria that elicited ISR, the research group at Auburn University performed isolations from cucumber plants in the field or from cucumber seeds. *Bacillus pumilus* strain INR7 was isolated from a surface-sterilized stem of a surviving cucumber plant in a field heavily infested with cucurbit wilt disease, caused by *Erwinia tracheiphila*. In two field trials, treatment with INR7 resulted in significant growth promotion relative to the nontreated control (Wei et al. 1996). In addition, the severity of angular leaf spot, following inoculation with *Pseudomonas syringae* pv. lachrymans, and the severity of naturally occurring anthracnose were significantly reduced by INR7. The cumulative yield of marketable cucumber fruit was also significantly enhanced by INR7 in both field trials. In the same study, strain 89B-61 also increased plant growth and yield and reduced the incidence of both angular leaf spot and anthracnose. In a subsequent field trial, INR7 reduced the severity of cucurbit wilt (Zehnder et al. 2001).

Erwinia tracheiphila is completely dependent on the striped cucumber beetle and the spotted cucumber beetles for survival and transmission. The finding that cucumber treated with strain INR7 exhibited reduced severity of cucurbit wilt in the field led to investigations aimed at determining if

ISR changed beetle feeding activity. In a 2-year field study, the number of beetles feeding on cucumber was significantly reduced following treatment with strain INR7 (Zehnder et al. 1997b). In both years of the study, the season-long average number of cucumber beetles per plant was significantly lower on plants treated with INR7 than on nonbacterized plants. Reduced beetle feeding on plants treated with strain INR7 was confirmed in subsequent greenhouse studies (Zehnder et al. 1997a). Beetle preference for nontreated plants was evident within the first 24 h of releasing the beetles into cages containing cucumber plants. After feeding for 17 days, beetle damage remained significantly lower on cotyledons and stems of plants treated with INR7 than on nontreated plants.

Elicitation of altered beetle behavior and feeding preferences in the field and greenhouse following seed treatment with INR7 was unexpected. As summarized by Zehnder et al. (1997a), cucumber beetle feeding behavior is influenced by a group of secondary plant metabolites called cucurbitacins, which are bitter compounds toxic to most insects. Cucumber beetles consume cucurbitacins without toxicity, apparently as an evolutionary adaptation that protects the cucumber beetles from predation. The beetles seek out cucurbitacins, and concentrations of 1 ng cause cucumber beetles to demonstrate arrested feeding behavior, whereby the beetles feed intensely on a single plant without moving from plant to plant. Hence, as an explanation for reduced beetle feeding on plants treated with INR7, Zehnder et al. (1997a) reasoned that elicitation of ISR by INR7 might be accompanied by reduced production of cucurbitacins by cucumber plants. Support for this hypothesis was found in a study (Zehnder et al. 1997a) in which treatment of cucumber with strain INR7 resulted in significantly reduced production of cucurbitacin C. Collectively, the results from studies on cucumber beetles demonstrate that specific endophytic bacteria can elicit unexpected yet important physiological changes in plants.

Serratia marcescens strain 90-166 colonizes roots internally (Press et al. 2001) and has been shown to elicit ISR against various diseases of cucumber. Elicitation of ISR against Fusarium wilt was demonstrated using a split-root system (Liu et al. 1995a). Root systems of seedlings were mechanically separated into two halves, with each half then being placed into a separate pot. Strain 90-166 was applied to one pot and the pathogen to the other pot. Numbers of dead plants and severity of the disease were significantly reduced by strain 90-166 over a 6-week experimental period. In another study (Liu et al. 1995b), strain 90-166 was found to elicit ISR against angular leaf spot. Treatment of seeds or cotyledons with strain 90-166 resulted in significant reductions in numbers and size of angular leaf spot lesions when the pathogen was inoculated 3 weeks after planting. ISR against angular leaf spot was also elicited when 90-166 was injected into cotyledons 1 week before pathogen inoculation. When 90-166 was injected into cotyledons, there was a reduction of 1.8 log units in the population of the pathogen inside leaves.

The capacity of strain 90-166 to elicit protection against cucumber anthracnose over a 5-week period was evaluated by Liu et al. (1995c). Strain 90-166 was applied to seeds at the time of planting, and *Colletotrichum orbiculare* was inoculated onto the first, second, third, fourth, or fifth leaf. There was approximately 1 week between each leaf stage. Treatment with strain 90-166 resulted in a significant reduction in the mean total lesion diameter when the pathogen was inoculated onto the fifth leaf, indicating that ISR elicited by the strain persists for at least 5 weeks on cucumber. Elicitation of ISR by strain 90-166 in tobacco against wildfire, caused by *P. syringae* pv. tabaci, was also demonstrated by Press et al. (1997). Stem injection of tobacco with strain 90-166 resulted in a significant decrease in disease severity when the pathogen was sprayed onto leaves 10 days after bacterial treatment. Raupach et al. (1996) reported that strain 90-166 also elicited ISR against *Cucumber mosaic virus* (CMV) on cucumber and tomato. On cucumber, seed treatment with 90-166 completely prevented development of CMV symptoms when the virus was inoculated onto cotyledons. On tomato, the effect of 90-166 was to delay symptom development over time. The area under the disease progress curve (AUDPC) was significantly reduced by strain 90-166.

Elicitation of ISR against viruses has also been reported for other strains of endophytic bacteria. Zehnder et al. (2000) conducted a greenhouse screen of PGPR (plant growth promoting rhizobacteria) for the potential to elicit ISR against CMV on tomato. PGPR were applied as seed treatments and as drenches upon transplanting 2 weeks after seeding. CMV was rub-inoculated with carborundum onto leaves 1 week after transplanting. From among 26 tested strains, three strains of endophytes were selected (*Bacillus subtilis* strain IN937b, *Bacillus pumilus* strain SE34, and *Bacillus amyloliquefaciens* strain IN937a). All of the selected strains significantly reduced disease incidence in each of five experiments. In the same study, Zehnder et al. (2000) conducted two field trials to evaluate the effects of strains IN937b, SE34, and IN937a on CMV. Treatment with all three endophytic bacterial strains resulted in significant reductions in the AUDPC compared to the nonbacterized control in both years of testing.

In another study with CMV in tomato, Murphy et al. (2003) used various two-strain combinations, where one strain was *B. subtilis* strain GB03, which is not reported to be an endophyte, and various endophytic bacteria, including strains IN937a, IN937b, SE34, and INR7. Spores of the bacteria were formulated on chitosan as a carrier and this preparation was mixed into potting mix. All of the bacterial treatments significantly reduced disease severity based on symptoms, decreased disease incidence based on enzyme-linked immunosorbent assay (ELISA), and decreased virus accumulation, compared to controls.

Three field trials were conducted with various formulations of the endophytic strains IN937a, IN937b, and SE34 (Murphy et al. 2000) to determine their capacity to elicit ISR against *Tomato mottle virus* (ToMoV), which is vectored by the silver whitefly (*Bremisia argentifolii*). The plots were inoculated with ToMoV by natural movement of viruliferous whitefly adults from adjacent plantings of ToMoV-resistant tomato germplasm that was inoculated prior to transplanting into the field. The incidence and severity of ToMoV were significantly reduced by one or more of the formulations of each endophyte. The number of whitefly nymphs detected on plants was significantly reduced by strains IN937a and IN937b. Hence, as in the case with cucurbit wilt of cucumber, some endophytic bacteria can elicit ISR against both a plant disease and its insect vectors.

ISR elicited by the endophytes *B. pumilus* strain SE34, *S. marcescens* strain 90-166, and *Pseudomonas fluorescens* strain 89B-61 has been shown to reduce the severity of blue mold of tobacco, caused by *Peronospora tabacina* (Zhang et al. 2002a, 2002b, 2004). In one study (Zhang et al. 2002b), strains SE34, 90-166, and 89B-61 elicited ISR in detached leaf and microtiter plate bioassays as well as in pot trials in the greenhouse. In the pot assay, application of all three strains as a soil drench to 4-week-old plants of three tobacco cultivars resulted in significant reductions in the mean percentage of leaf area with lesions caused by *P. tabacina* inoculated onto leaves 1 week after bacterial treatment. Sporulation of the pathogen on lesions was significantly decreased by treatment with the three strains in pot trials. Strains SE34, 90-166, and 89B-61 also significantly reduced disease severity in the detached leaf (injection of a bacterial suspension into petioles) and microtiter plate bioassays (application of bacterial suspensions to roots). Sporulation of the pathogen was significantly reduced by both strains in the detached leaf bioassay.

In another study (Zhang et al. 2004), strains SE34 and 90-166 were used to explore the relationship between elicitation of plant growth promotion and ISR. Application of the endophytes as a seed treatment alone elicited significantly enhanced tobacco plant growth but not disease protection. When the strains were applied as seed treatments followed by a soil drench, both plant growth promotion and ISR were elicited. Overall, the results from this study indicated that while there was a relationship between growth promotion and ISR, elicitation of growth promotion can occur without elicitation of ISR; however, when ISR was elicited, growth promotion was also elicited with the bacterial strains used in the study.

Endophytic bacteria have also elicited ISR against tomato late blight, caused by *Phytophthora infestans* (Yan et al. 2002). Application of strains SE34 and 89B-61 by incorporation into the potting medium at the time of planting elicited significant reductions in disease severity when *P. infestans* was inoculated onto leaves 5 weeks after planting.

Greenhouse screening of endophytic *Bacillus* spp. that have elicited ISR on some crops was conducted in Thailand (Jetiyanon and Kloepper 2002) as a first step toward employment of endophyte-elicited ISR in tropical agriculture. This study used four different host/pathogen systems: tomato and *Ralstonia solanacearum*, long cayenne pepper (*Capsicum annuum* var. *acuminatum*) and *Colletotrichum gloeosporioides*, green kuang futsoi (*Brassica chinensis* var. *parachinensis*) and *Rhizoctonia solani*, and cucumber and CMV. The goal of the study was to find mixtures of endophytic spore-forming bacteria that elicited ISR in all four host/pathogen systems. Seven individual strains and 11 combinations of 2 strains were tested. One strain (*B. amyloliquefaciens* IN937a) and four mixtures (IN937a + *B. subtilis* IN937b; IN937b + *B. pumilus* SE34; IN937b + *B. pumilus* SE49; and IN937b + *B. pumilus* INR7) significantly reduced incidence or severity of all four diseases. The results are noteworthy for two reasons. First, they indicate that ISR elicited by specific endophytic bacterial strains can protect hosts under tropical conditions. Second, the results show that mixtures of two bacterial strains are superior to individual strains for eliciting significant protection in multiple hosts against different pathogens.

Further evidence for the concept of using strain mixtures of *Bacillus* spp. to increase the repeatability of plant growth promotion or elicitation of ISR by bacteria was reported in a follow-up field investigation (Jetiyanon et al. 2003). Field tests were conducted in Thailand to find mixtures of bacteria that could protect several different hosts against the multiple diseases that are typical under the multi- or inter-cropping agricultural conditions predominant in Thailand. In tests conducted during the rainy and dry seasons, some two-strain mixtures of endophytes more consistently protected against disease than did a single strain. In each season, the mixture of strains IN937a and IN937b significantly protected against all the tested diseases (southern blight of tomato, CMV on cucumber, and anthracnose of long cayenne pepper). The same mixture of endophytes also resulted in significant yield increases of all crops during the rainy season.

Endophytic bacteria have also been shown to elicit ISR in conifers. Enebak and Carey (2000) tested the potential elicitation of ISR on loblolly pine (*Pinus taeda*) by strains *B. sphaericus* SE56 and *B. pumilus* strains INR7, SE34, SE49, and SE52 against *Cronartium quercuum* f. sp. *fusiforme*, which causes fusiform rust. Bacteria were applied at seeding, and suspensions of *C. quercuum* basidiospores from field-collected telia on water oak (*Quercus nigra*) were sprayed onto the pine seedlings at five different times. Six months after the final application of basidiospores, the incidence of fusiform rust was determined by noting the presence or absence of the typical symptoms of main-stem swellings or galls. The experiment was conducted annually for 2 years. All strains except SE49 resulted in significant reductions in disease incidence in either one of the 2 years or in the pooled data from both years.

Before concluding this discussion of case studies of endophytic bacteria that have been shown to elicit ISR, a note should be made about *Azospirillum* spp. *Azospirillum brasilense* is a well characterized endophyte, and nearly all strains of this species have been shown to promote growth of many crop species (Bashan and de-Bashan 2002). Because many of the strains of *Bacillus* spp. cited above that elicit ISR also elicit plant growth promotion, one might expect that *A. brasilense* would also elicit ISR. However, this is not the case. Bashan and de-Bashan (2002) investigated the potential of *A. brasilense* to elicit ISR in tomato against bacterial speck, caused by *P. syringae* pv. Tomato, and concluded that this endophyte does not elicit ISR.

3.3 Internal Colonization of Endophytes that Elicit Induced Resistance

In most of the studies discussed in the previous section, extensive microbial ecology studies to determine the extent of internal colonization of plant tissues by the applied endophytes were not carried out. Typically, isolations are performed near the location where the pathogen was applied. Such isolation is done to test one of the suggested tenants of ISR: that there is physical separation of the pathogen and the inducing agent. According to this tenant, physical separation is required to differentiate ISR from antagonism as a mechanism for protection against pathogens. While testing for physical separation has validity, it also creates an inherent problem with endophytic bacteria that exhibit systemic colonization of plants. An endophyte that can move within the plant and colonize petioles could, theoretically, still elicit ISR at a level sufficient to reduce disease severity of a foliar pathogen. However, based on the tenant of physical separation, one could not state that ISR was the operable mechanism by which disease severity was reduced. Hence, some endophytic bacteria might actually elicit plant defense although they are not spatially separated from the pathogen.

An example illustrating the limited internal colonization of well characterized endophytes that elicit ISR is *P. fluorescens* strain 89B-61, which was initially designated as strain G8-4. Because this strain has reactions in biochemical tests that are intermediate between *P. putida* and *P. fluorescens*, some publications before 1997 refer to 89B-61 as *P. putida*. Later publications designate the strain as *P. fluorescens* based on repeated fatty acid analyses. Chen et al. (1995) found that strain 89B-61 significantly reduced the severity of Fusarium wilt of cotton. In this system, 89B-61 was stab-inoculated into seedling stems 13 days prior to inoculation with *Fusarium oxysporum* f. sp. *vasinfectum* at a point 1.5 cm above the point of bacterial inoculation. Using a rifampicin-resistant mutant of 89B-61, no movement up the stem from the point of bacterial inoculation was detected. In a separate greenhouse study, Kloepper et al. (1992) reported that seed treatment of cucumber with strain G8-4 (89B-61) significantly reduced lesion numbers and size of anthracnose following challenge inoculation of leaves with *C. orbiculare*. This induced resistance was associated with colonization of internal root tissues at log 4.0 cfu/g at 14 days after emergence; however, the bacteria were not isolated from stems or leaves. Elicitation of induced resistance in cucumber by 89B-61 was confirmed in field studies (Wei et al. 1996), where application of the bacterium as a seed treatment resulted in significant reductions in severity of anthracnose and angular leaf spot.

Quadt-Hallmann et al. (1997) used isolation, ELISA, and immunogold labeling with *P. fluorescens* 89B-61-specific polyclonal antibodies to investigate the pattern of internal colonization of cotton by the ISR-eliciting strain 89B-61. Results from isolation studies indicated that, following treatment of seeds, 89B-61 colonized roots internally at a mean population of 1.1×10^3 cfu/g, while the bacterium was not recovered from stems, cotyledons, or leaves. With ELISA, strain 89B-61 was detected outside and inside roots but not inside stems, cotyledons, or leaves. Electron microscopy with immunogold labeling revealed that internal colonization of roots by 89B-61 was restricted mainly to intercellular spaces of the epidermis. Interestingly, the colonization pattern was quite distinct from that of another bacterium (*Enterobacter asburiae* strain JM22) that does not elicit ISR. JM22 colonized throughout the root cortex, including inside the vascular stele, in intercellular spaces close to the conducting elements, as was previously found in other plant species (Quadt-Hallmann and Kloepper 1996). It was suggested that the internal colonization of cotton by 89B-61 and JM22 could be considered as representative of two fundamental options for how endophytic bacteria colonize plants after application to seeds or soils. The first pattern is that of 89B-61 and consists of limited internal colonization of roots. The second pattern, demonstrated by JM22, consists of extensive internal root colonization and ultimately in some vascular colonization.

Internal colonization of roots by *S. marcescens* strain 90-166 was demonstrated by Press et al. (2001) in an investigation into the role of iron in elicitation of ISR. Cucumber root colonization by the wild-type strain was compared to that by a mutant deficient in siderophore production. While the total root population sizes (external and internal colonization) were statistically equivalent, the internal population size was significantly greater with the wild-type than with the siderophore-negative mutant. Because the mutant failed to elicit ISR against anthracnose, while ISR was elicited by the wild-type strain, it was concluded that capacity to elicit ISR was related to the population size of the bacterium inside roots.

3.4 Plant Responses to Endophytic Elicitors

Investigations aimed at determining how plants respond to inoculation with endophytes that elicit ISR is one approach to studying mechanisms of ISR by such bacteria. During the previously discussed study on cotton colonization by strain 89B-61 (Quadt-Hallmann et al. 1997), the epidermal cell walls of plant cells adjacent to cells of 89B-61 in the intercellular space developed electron-opaque appositions of an amorphous matrix.

Two cytological studies were conducted by Benhamou et al. (1996, 1998) with *B. pumilus* strain SE34. In the first study (Benhamou et al. 1996), colonization of pea roots by *Fusarium oxysporum* f. sp. *pisi* was restricted to the epidermis and outer cortex of roots treated with SE34, while in nonbacterized roots, the pathogen colonized the cortex, endodermis, and the paratracheal parenchyma cells. This reduction in fungal colonization by SE34 was associated with strengthening of the epidermal and cortical cell walls. In addition, roots treated with SE34 exhibited newly formed barriers beyond the site of fungal infection. These barriers were cell wall appositions that contained large amounts of callose and were infiltrated with phenolic compounds. Phenolic compounds were detected in transmission electron microscopy using gold-complexed laccase and were found to accumulate in host cell walls, in intercellular spaces, and on the surface of and inside the invading pathogen hyphae.

In another study (Benhamou et al. 1998), the effect of SE34 alone or in combination with chitin on structural and cytochemical changes of tomato infected with *F. oxysporum* f. sp. *radicis-lycopersici* was investigated. Treatment with SE34 reduced the severity of typical symptoms, including wilting of seedlings and numbers of brown lesions on lateral roots. This disease protection by strain SE34 was associated with more limited fungal colonization of roots and with marked changes in host physiology. Physiological changes elicited by strain SE34 included an increase in host cell wall density, the accumulation of polymorphic deposits at sites of potential pathogen penetration, and the occlusion of epidermal cells and intercellular spaces with an osmophilic, amorphous material that appeared to trap the invading fungal hyphae. The extent and magnitude of the physiological changes in the host elicited by SE34 were enhanced by the addition of chitosan. Interestingly, the overall chitin component of the pathogen was structurally preserved in roots treated with SE34 with or without chitosan at the time when hyphal degradation was apparent. This suggests that synthesis of chitinase in bacteria-treated roots is not an early event in the cascade of physiological steps in signal transduction that lead to induced resistance. Benhamou et al. (1998) concluded: "According to our cytological observations, the induction of resistance triggered by *B. pumilus* strain SE34 involves a sequence of events including first the elaboration of structural barriers and the production of toxic substances such as phenolics and phytoalexins, and second the synthesis and accumulation of other molecules including chitinases and other hydrolytic enzymes such as β-1,3-glucanases which probably contribute to the release of oligosaccharides that, in turn, can stimulate other defense reactions."

Jeun et al. (2004) conducted a cytological comparison of cucumber plants in which systemic resistance had been elicited by bacteria or by chemicals. In this study, the endophytes 89B-61 and 90-166 were used to elicit ISR against *C. orbiculare*. Significantly fewer numbers of anthracnose lesions developed on plants treated with 89B-61 and 90-166 than on the control (chemical treatment). Cytological studies using fluorescent microscopy revealed a higher frequency of autofluorescent epidermal cells, which are related to accumulation of phenolic compounds, at the sites of fungal penetration in plants treated with either strain and inoculated with *C. orbiculare*. In addition, callose-like structures (β-1,3-glucan polymers) were frequently deposited at the site of fungal penetration of the leaves of plants treated with either strain.

Investigations on plant responses to elicitation of ISR can also examine the signal transduction pathway of plants to determine general biochemical pathways in plants during ISR. As previously discussed, according to the model pathway for signal transduction (Pieterse et al. 1998), ISR pathways are independent of salicylic acid, but dependent on ethylene, jasmonic acid, and the regulatory gene *npr-1*. Further, according to the model, ISR elicited by bacteria does not result in the accumulation of pathogenesis-related (PR) proteins. PR proteins are accumulated during SAR elicited by pathogens and chemicals, and SAR is dependent upon salicylic acid.

A few studies on signal pathways have been reported with endophytic bacteria that elicit ISR. In the tomato late blight system, ISR was elicited by *B. pumilus* strain SE34 on NahG lines, which breakdown endogenous salicylic acid, but not in the ethylene-insensitive *NR/NR* line or in the jasmonic acid-insensitive *df1/df1* line (Yan et al. 2002). These results are consistent with the model of Pieterse et al. (1998). Similar results were reported by Zhang et al. (2002a). In the tobacco blue mold system, *B. pumilus* strain SE34, as well as two strains of Gram-negative bacteria, elicited ISR on both wild-type and NahG transgenic tobacco lines, as evidenced by significant reductions in the severity of blue mold on bacterized plants compared to nonbacterized plants.

Different results were found with strain 89B-61 (Park and Kloepper 2000), which elicits ISR in tobacco against wildfire caused by *P. syringae* pv. tabaci. In this system, a transgenic line of tobacco with a β-glucuronidase (GUS) reporter gene fused to the *PR-1a* promoter had significantly reduced severity of wildfire compared to nonbacterized controls. Elicitation of ISR by strain 89B-61 was associated with a significant increase in GUS activity in microtiter plate and whole plant bioassays. Hence, with strain 89B-61, elicitation of ISR results in activation of the *PR-1a* gene, which is activated during SAR but not during bacterial-induced ISR according to the model of Pieterse et al. (1998).

Signal pathways in ISR elicited by *P. fluorescens* strain CHA0 in *Arabidopsis* against *Peronospora parasitica* were investigated by Iavicoli et al. (2003) using various transgenic and mutant plant lines: NahG (for degradation of salicylic acid), *sid2-1* (lacks production of salicylic acid), *npr1-1* (nonexpressor of *PR* genes), *jar1-1* (insensitive to jasmonic acid), *ein2-1* (insensitive to ethylene), *eir1-1* (insensitive to ethylene/auxin), and *pad2-1* (phytoalexin-deficient). ISR was elicited by strain CHA0 in all lines except *jar1-1*, *eir1-1*, and *npr1-1*.

In another study of signaling pathways, Ryu et al. (2003b) used endophytic bacterial strains in *Arabidopsis* to elicit ISR against two different pathovars of *P. syringae* (pvs. tomato and maculicola). Strains SE34, 90-166, and 89B-61 elicited ISR against both pathogens. Strain SE34 elicited a salicylic acid-independent pathway against one pathovar and salicylic acid-dependent pathway against a different pathovar. Additional tests of strains 89B-61 and SE34 on various mutant lines of *Arabidopsis* (Ryu et al. 2003b) revealed that, in agreement with the model of Pieterse et al. (1998), ISR elicited by both strains was dependent on NPR1 and ISR elicited by SE34 was dependent on jasmonic acid and ethylene. However, in contrast to the model, ISR elicited by strain 89B-61 was independent of ethylene and jasmonic acid, and ISR by strain 90-166 was dependent on jasmonic acid but independent of ethylene.

Endophyte strains 90-166 and SE34 also elicited ISR against CMV in *Arabidopsis* (Ryu et al. 2004b). Strains 90-166 and SE34 reduced disease severity in NahG plants, indicating that ISR elicited by strains 90-166 and SE34 was independent of salicylic acid. Further investigation on the signal pathway of ISR against CMV elicited by strain 90-166 with lines NahG, *npr1*, and *fad3-2 fad7-2 fad8* (insensitive to jasmonic acid) indicated that ISR against CMV by strain 90-166 is independent of salicylic acid and NPR1, but is dependent on jasmonic acid (Ryu et al. 2004b).

Collectively, the results on signaling pathways of ISR elicited by endophytic bacteria indicate that different pathways are elicited by various strains. Further, the specific signal transduction pathway that is activated during ISR depends on the host plant and, at least in one case, on the pathogen used on a given host.

A new approach to investigations on plant responses to endophytic bacteria was recently opened by the finding that volatile organic compounds produced by the endophyte *B. amyloliquefaciens* IN937a elicit plant growth promotion (Ryu et al. 2003a) and ISR (Ryu et al. 2004a). Significant growth promotion of *Arabidopsis* by IN937a was observed in I-plates, which have a raised plastic divider separating agar on each half of the dish, thus preventing movement of soluble compounds. When IN937a was placed on one side of an I-plate, *Arabidopsis* plants growing on the other side exhibited enhanced growth, presumably as a result of volatiles produced by the bacteria. Characterization of the volatile organic compounds (VOCs) produced by IN937a, coupled with bioassays of fractions of VOCs, revealed that 2,3-butanediol and acetoin elicited plant growth promotion. In a separate study (Ryu et al. 2004a), exposure of *Arabidopsis* to VOCs from strain IN937a resulted in significantly less disease caused by *Erwinia carotovora* subsp. carotovora. Tests with various mutant lines of *Arabidopsis* revealed that elicitation of ISR by VOCs of IN937a is independent of jasmonic acid, ethylene, salicylic acid, and *npr1*. Such a pattern of signal pathway has not been reported with ISR elicited by bacteria and, therefore, it is likely that VOCs of IN937a elicit a distinct, and as yet uncharacterized, pathway in *Arabidopsis*.

3.5 Implementation in Production Agriculture: Two Case Studies

The principle that endophytic bacteria can elicit ISR or plant growth promotion has been extended to use in production agriculture and horticulture through the development of two products. These two products are discussed, not as endorsements of the products, but as case studies indicating that our growing scientific knowledge of endophytic bacteria can be put to practical use.

In the first case study, an agricultural product has been developed using the capacity of a single endophytic strain of *Bacillus* spp. to elicit both ISR and plant growth promotion. The product is Yield Shield, which is produced by Gustafson, LLC. Yield Shield consists of a spore preparation of the *B. pumilus* strain listed in this review as INR7 (Table 3.1) and designated by Gustafson as GB34 (http://www.gustafson.com/Labels/yield_shield_label.pdf). The product received registration from the United States Enivironmental Protection Agency (EPA) in 2003 for use on soybeans to protect against *Rhizoctonia solani* and *Fusarium* spp. Seed treatment of soybean with Yield Shield and strain INR7 results in significant seedling growth promotion and in ISR, which is apparent both by a significant de-

Table 3.1. Endophytic bacterial strains that have been reported to elicit induced systemic resistance (ISR) in at least two publications

Strain no. and identification	Effects reported and systems used[a]	Reference
IN937a *Bacillus amyloliquefaciens*	Reduced incidence or severity of vegetable diseases (caused by *Cucumber mosaic virus* (CMV), *Sclerotium rolfsii*, *Ralstonia solanacearum*, *Colletotrichum gloeosporioides*, and *Rhizoctonia solani*) in greenhouse and field trials in Thailand	Jetiyanon et al. 2002; Jetiyanon and Kloepper 2003
	Reduced incidence of CMV on tomato in the greenhouse	Zehnder et al. 2000
	When applied to tomato with the non-endophyte *Bacillus subtilis* GB03, reduced severity of CMV in the greenhouse	Murphy et al. 2003
	Reduced incidence and severity of tomato mottle virus in the field. Also reduced numbers of the white fly vector feeding on plants	Murphy et al. 2000
	Volatile organic compounds of the strain elicit growth promotion of *Arabidopsis* and ISR against *Erwinia carotovora* subsp. carotovora	Ryu et al. 2003a, 2004
	Component of the product, BioYield (Gustafson LLC, http://www.gustafson.com)	Kloepper et al. 2004
IN937b *Bacillus subtilis*	Reduced incidence of CMV on tomato in the greenhouse	Zehnder et al. 2000
	When applied to tomato with the non-endophyte *B. subtilis* GB03, reduced severity of CMV in the greenhouse	Murphy et al. 2003
	Reduced incidence and severity of tomato mottle virus in the field. Also reduced numbers of the white fly vector feeding on plants	Murphy et al. 2000
90-166 *Serratia marcescens*	Reduced incidence and delayed development of symptoms of Fusarium wilt of cucumber, caused by *Fusarium oxysporum* f. sp. *cucumerinum*	Liu et al. 1995a
	Reduced severity of bacterial angular leaf spot of cucumber, caused by *Pseudomonas syringae* pv. lachrymans	Liu et al. 1995b
	Reduced severity of anthracnose, caused by *Colletotrichum orbiculare*, in two cucumber cultivars over a 5-week period after bacterial treatment	Liu et al. 1995c
	Reduced incidence of CMV on cucumber and tomato and reduced the area under the disease progress curve (AUDPC)	Raupach et al. 1996
	Decreased severity of tobacco wildfire, caused by *Pseudomonas syringae* pv. tabaci	Press et al. 1997

Table 3.1. (continued)

Strain no. and identification	Effects reported and systems used[a]	Reference
	Wild-type strain reduced severity of cucumber anthracnose. A siderophore-negative mutant did not elicit ISR and colonized roots internally at lower populations than the wild-type	Press et al. 2001
	Decreased severity of tobacco blue mold, caused by *Peronospora tabacina*, in NahG transgenic tobacco lines that degrade salicylic acid	Zhang et al. 2002a
	Decreased severity of tobacco blue mold in microtiter plate assays and detached leaf assays. Reduced pathogen sporulation	Zhang et al. 2002b
	Reduced symptoms of *Pseudomonas syringae* pvs. tomato and maculicola on *Arabidopsis*	Ryu et al. 2003b
	Decreased severity of tobacco blue mold, caused by *Peronospora tabacina*	Zhang et al. 2004
SE34 *Bacillus pumilus*	Reduced incidence or severity of vegetable diseases (caused by cucumber mosaic virus, *Sclerotium rolfsii*, *Ralstonia solanacearum*, *Colletotrichum gloeosporioides*, and *Rhizoctonia solani*) in greenhouse and field trials in Thailand	Jetiyanon et. al. 2002; Jetiyanon and Kloepper 2003
	Decreased severity of tobacco blue mold, caused by *Peronospora tabacina*, in NahG transgenic tobacco lines that degrade salicylic acid	Zhang et al. 2002a
	Decreased severity of tobacco blue mold in microtiter plate assays and detached leaf assays. Reduced pathogen sporulation	Zhang et al. 2002b
	Decreased severity of tobacco blue mold when applied as both seed treatment and drench in the greenhouse. Elicitation of ISR was associated with growth promotion	Zhang et al. 2004
	Decreased severity of tomato late blight, caused by *Phytophthora infestans*, and decreased germination of sporangia and zoospores of the pathogen	Yan et al. 2002
	Reduced incidence of CMV on tomato in the greenhouse	Zehnder et al. 2000
	When applied to tomato with the non-endophyte *B. subtilis* GB03, reduced severity of CMV in the greenhouse	Murphy et al. 2003
	Reduced incidence and severity of tomato mottle virus in the field	Murphy et al. 2000
	Reduced incidence of Fusiform rust, caused by *Cronartium quercuum* f. sp. *fusiforme*, on loblolly pine	Enebak and Carey 2000

Table 3.1. (continued)

Strain no. and identification	Effects reported and systems used[a]	Reference
	Restricted colonization of pea roots by *F. oxysporum* f. sp. *pisi* and induced formation of structural barriers in the plant	Benhamou et al. 1996
	Reduced damage of tomato roots to *F. oxysporum* f. sp. *radicis-lycopersici* and induced structural and cytochemical barriers in the plant	Benhamou et al. 1998
	Reduced symptoms of *Pseudomonas syringae* pvs. tomato and maculicola on *Arabidopsis*	Ryu et al. 2003b
INR7 *Bacillus pumilus* (available as the product Yield Shield; Gustafson LLC)	Reduced incidence or severity of vegetable diseases (caused by CMV, *Sclerotium rolfsii*, *Ralstonia solanacearum*, *Colletotrichum gloeosporioides*, and *Rhizoctonia solani*) in greenhouse and field trials in Thailand	Jetiyanon et. al. 2002; Jetiyanon and Kloepper 2003
	Decreased severity of anthracnose (caused by *Colletotrichum orbiculare*) and angular leaf spot (caused by *Pseudomonas syringae* pv. lachrymans) on cucumber in field trials	Wei et al. 1996
	Decreased the incidence of cucurbit wilt disease, caused by *Erwinia tracheiphila*, in field trials	Zehnder et al. 2001
	Decreased numbers of cucumber beetles on plants in the field	Zehnder et al. 1997b
	Decreased beetle feeding activity and transmission of *E. tracheiphila* on cucumber in cages where beetles had a choice between bacterial-treated and nontreated plants	Zehnder et al. 1997a
	When applied to tomato with the non-endophyte *B. subtilis* GB03, reduced severity of CMV in the greenhouse	Murphy et al. 2003
	Reduced incidence of Fusiform rust, caused by *Cronartium quercuum* f. sp. *fusiforme*, on loblolly pine	Enebak and Carey 2000
	Decreased severity of anthracnose (caused by *Colletotrichum orbiculare*) and angular leaf spot (caused by *Pseudomonas syringae* pv. lachrymans) on cucumber in field trials	Wei et al. 1996
89B-61 *Pseudomonas fluorescens* (earlier referred to as G8-4)	Decreased severity of tobacco blue mold, caused by *Peronospora tabacina*	Zhang et al. 2004
	Decreased severity of Fusarium wilt of cotton, caused by *Fusarium oxysporum* f. sp. *vasinfectum*	Chen et al. 1995
	Following seed treatment of cotton, 89B-61 reached an internal root population of 1.1×10^3 cfu/g. Bacteria were not detected inside cotyledons, stems, or leaves	Quadt-Hallmann et al. 1997

Table 3.1. (continued)

Strain no. and identification	Effects reported and systems used[a]	Reference
	Decreased severity of anthracnose (caused by *Colletotrichum orbiculare*) and angular leaf spot (caused by *Pseudomonas syringae* pv. lachrymans) on cucumber	Wei et al. 1996
	Decreased severity of tomato late blight, caused by *Phytophthora infestans*, and decreased germination of sporangia and zoospores of the pathogen when applied to seeds	Yan et al. 2002
	Decreased severity of tobacco blue mold, caused by *Peronospora tabacina*, in NahG transgenic tobacco lines that degrade salicylic acid	Zhang et al. 2002a
	Decreased severity of tobacco blue mold in microtiter plate assays and detached leaf assays. Reduced pathogen sporulation	Zhang et al. 2002b
	Decreased severity of tobacco wildfire, caused by *Pseudomonas syringae* pv. tabaci. Activated the promoter for PR1a [a pathogenesis-related (PR) protein]	Park and Kloepper 2000
	Reduced symptoms of *Pseudomonas syringae* pvs. tomato and maculicola on *Arabidopsis*	Ryu et al. 2003b
	Reduced number of lesions of *Colletotrichum orbiculare* on cucumber and increased deposition of callose-like polymers on leaf cells at the site of pathogen penetration	Jeun et al. 2004
	Reduced mean numbers and size of anthracnose lesions on cucumber	Wei et al. 1991
	Colonized roots internally at log 4.0 cfu/g at 2 weeks after emergence when applied as seed treatments. Bacteria were not detected in leaves or stems	Kloepper et al. 1992
CHA0 *Pseudomonas fluorescens*	Reduced severity of *Tobacco necrosis virus* (TNV) on tobacco.	Maurhofer et al. 1994
	Severity of TNV on tobacco was reduced equivalently by the wild-type strain and a transgenic strain carrying *pchAB* genes for synthesis of salicylic acid.	Maurhofer et al. 1998
	Reduced sporulation of *Peronospora parasitica* on *Arabidopsis*	Iavicoli et al. 2003

[a] In all cases, the stated reductions in disease incidence or severity and the effects on insects are statistically significant at $P \leq 0.05$

crease in incidence and severity of *R. solani* inoculated onto stems at a point where INR7 does not colonize, and by a systemic increase in lignification of plant cell walls (C.-M. Ryu and C.-H. Hu, unpublished). It should be emphasized that Yield Shield is a unique case for a rhizobacterium that elicits ISR in that economically significant efficacy sufficient to warrant the costs of product development and EPA registration was shown for a single bacterial strain.

In the second case study, the product consists of a two-strain mixture of *Bacillus* spp., where one strain (IN937a) is an endophyte that elicits ISR (Table 3.1) and the other strain (GB03) is a non-endophyte. The product is BioYield and is also produced by Gustafson. The development of BioYield was recently reviewed (Kloepper et al. 2004). The underlying concept was to develop a biological formulation consisting of components known to exert different mechanisms for control of diseases. The selected components and their mechanisms were chitosan (as a carrier) for nematode control via promotion of indigenous soil predators and antagonists to root-knot nematodes, *B. subtilis* strain GB03 for control of soil-borne pathogens via production of the antibiotic iturin, and one of several tested endophytic *Bacillus* spp. that elicit ISR. The most unexpected finding was that the three-component combination (chitosan plus two bacterial strains) exhibited more consistent, and a greater magnitude of, growth promotion and systemic protection against pathogens than did any of the individual components (Kloepper et al. 2004). Based on the results, the two-strain combination of *B. amyloliquefaciens* strain IN937a and *B. subtilis* strain GB03 was selected for product development.

3.6 Conclusions

As discussed in this review, selected strains of nonpathogenic endophytic bacteria can elicit ISR in plants, leading to reductions in severity of various diseases. Research on such endophytes has concentrated both on delineating the pathosystems where protection results and in understanding plant responses that occur during the signal transduction pathways that culminate in disease protection. In many cases, elicitation of ISR by endophytic bacilli is associated with increased plant growth, and the relationship between ISR and growth promotion should be further investigated. Elucidation of specific bacterial determinants that account for elicitation of ISR is just beginning, and further work is needed to understand why one strain of a given bacterial species can elicit ISR while another strain of the same species cannot. It is encouraging that implementation of ISR by endophytic bacilli is beginning, even while some basic questions remain to be answered.

References

Bashan Y, de-Bashan LE (2002) Reduction of bacterial speck (*Pseudomonas syringae* pv. *tomato*) of tomato by combined treatments of plant growth-promoting bacterium, *Azospirillum brasilense*, streptomycin sulfate, and chemo-thermal seed treatment. Eur J Plant Pathol 108:821–829

Benhamou N, Kloepper JW, Quadt-Hallmann A, Tuzun S (1996) Induction of defense-related ultrastructural modifications in pea root tissues inoculated with endophytic bacteria. Plant Physiol 112:919–929

Benhamou N, Kloepper JW, Tuzun S (1998) Induction of resistance against *Fusarium* wilt of tomato by combination of chitosan with an endophytic bacterial strain: ultrastructure and cytochemistry of the host response. Planta 204:153–168

Chen C, Bauske EM, Musson G, Rodríguez-Kábana R, Kloepper JW (1995) Biological control of Fusarium wilt of cotton by use of endophytic bacteria. Biol Control 5:83–91

Enebak SA, Carey WA (2000) Evidence for induced systemic protection to fusiform rust in loblolly pine by plant growth-promoting rhizobacteria. Plant Dis 84:306–308

Hallman J, Quadt-Hallmann A, Mahaffee WF, Kloepper JW (1997) Bacterial endophytes in agricultural crops. Can J Microbiol 43:895–914

Iavicoli A, Boutel E, Buchala A, Métraux J-P (2003) Induced systemic resistance in *Arabidopsis thaliana* in response to root inoculation with *Pseudomonas fluorescens* CHA0. Mol Plant Microb Interact 16:851–858

Jeun YC, Park KS, Kim CH, Fowler WD, Kloepper JW (2004) Cytological observations of cucumber plants during induced resistance elicited by rhizobacteria. Biol Control 29:34–42

Jetiyanon K, Kloepper JW (2002) Mixtures of plant growth-promoting rhizobacteria for induction of systemic resistance against multiple plant diseases. Biol Control 24:285–291

Jetiyanon K, Fowler WD, Kloepper JW (2003) Broad-spectrum protection against several pathogens by PGPR mixtures under field conditions. Plant Dis 87:1390–1394

Kloepper JW, Wei G, Tuzun S (1992) Rhizosphere population dynamics and internal colonization of cucumber by plant growth-promoting rhizobacteria which induce systemic resistance to *Colletotrichum orbiculare* In: Tjamos ES, Papavizas GC, Cook RJ (eds) Biological control of plant diseases, Plenum, New York, pp 185–191

Kloepper JW, Reddy MS, Rodríguez-Kabana R, Kenney DS, Kokalis-Burelle N, Martinez-Ochoa N (2004) Application for rhizobacteria in transplant production and yield enhancement. Acta Hortic 631:219–229

Liu L, Kloepper JW, Tuzun S (1995a) Induction of systemic resistance in cucumber against Fusarium wilt by plant growth-promoting rhizobacteria. Phytopathology 85:695–698

Liu L, Kloepper JW, Tuzun S (1995b) Induction of systemic resistance in cucumber against bacterial angular leaf spot by plant growth-promoting rhizobacteria. Phytopathology 85:843–847

Liu L, Kloepper JW, Tuzun S (1995c) Induction of systemic resistance in cucumber by plant growth-promoting rhizobacteria: duration of protection and effect of host resistance on protection and root colonization. Phytopathology 85:1064–1068

Maurhofer M, Hase C, Meuwly P, Métraux J-P, Defago G (1994) Induction of systemic resistance of tobacco to tobacco necrosis virus by the root-colonizing *Pseudomonas fluorescens* strain CHA0: influence of the *gacA*-gene and of pyoverdine production. Phytopathology 84:139–146

Maurhofer M, Reimmann C, Schmidli-Sachere P, Heeb S, Haas D, Defago G (1998) Salicylic acid biosynthesis genes expressed in *Pseudomonas fluorescens* strain P3 improve the induction of systemic resistance in tobacco against tobacco necrosis virus. Phytopathology 88:678–684

Murphy JF, Zehnder GW, Schuster DJ, Sikora EJ, Polston JE, Kloepper JW (2000) Plant growth-promoting rhizobacterial mediated protection in tomato against tomato mottle virus. Plant Dis 84:779–784

Murphy JF, Reddy MS, Ryu C-M, Kloepper JW, Li R (2003) Rhizobacteria-mediated growth promotion of tomato leads to protection against *Cucumber mosaic virus*. Phytopathology 93:1301–1307

Park KS, Kloepper JW (2000) Activation of PR-1a promoter by rhizobacteria that induce systemic resistance in tobacco against *Pseudomonas syringae* pv. *tabaci*. Biol Control 18:2–9

Pieterse CMJ, van Wees SCM, van Pelt JA, Knoester M, Laan R, Gerrits H, Weisbeek, PJ, van Loon LC (1998) A novel signaling pathway controlling induced systemic resistance in *Arabidopsis*. Plant Cell 10:1571–1580

Press CM, Wilson M, Tuzun S, Kloepper JW (1997) Salicylic acid produced by *Serratia marcescens* 90-166 is not the primary determinant of induced systemic resistance in cucumber or tobacco. Mol Plant Microb Interact 10:761–768

Press CM, Loper JE, Kloepper JW (2001) Role of iron in rhizobacteria-mediated induced systemic resistance. Phytopathology 91:593–598

Quadt-Hallmann A, Kloepper JW (1996) Immunological detection and localization of the cotton endophyte *Enterobacter asburiae* JM22 in different plant species. Can J Microbiol 42:1144–1154

Quadt-Hallmann A, Hallmann J, Kloepper JW (1997) Bacterial endophytes in cotton: location and interaction with other plant-associated bacteria. Can J Microbiol 43:254–259

Raupach GS, Liu L, Murphy JF, Tuzun S, Kloepper JW (1996) Induced systemic resistance in cucumber and tomato against cucumber mosaic cucumovirus using plant growth-promoting rhizobacteria (PGPR). Plant Dis 80:891–894

Ryu C-M, Farag MA, Hu C-H, Reddy MS, Wei H-X, Paré PW, Kloepper JW (2003a) Bacterial volatiles promote growth in *Arabidopsis*. Proc Natl Acad Sci USA 100:4927–4932

Ryu C-M, Hu C-H, Reddy MS, Kloepper JW (2003b) Different signaling pathways of induced resistance by rhizobacteria in *Arabidopsis thaliana* against two pathovars of *Pseudomonas syringae*. New Phytol 160:413–420

Ryu C-M, Farag MA, Hu C-H, Munagala SR, Kloepper JW, Paré PW (2004a) Bacterial volatiles induce systemic resistance in *Arabidopsis*. Plant Physiol 134:1017–1026

Ryu C-M, Murphy JF, Mysore KS, Kloepper JW (2004b) Plant growth-promoting rhizobacteria systemically protect *Arabidopsis thaliana* against *Cucumber mosaic virus* by a salicylic acid and NPR1-independent and jasmonic acid-dependent signaling pathway. Plant J 39:381–392

Wei G, Kloepper JW, Tuzun S (1991) Induction of systemic resistance of cucumber to *Colletotrichum orbiculare* by select strains of plant growth-promoting rhizobacteria. Phytopathology 81:1508–1512

Wei G, Kloepper JW, Tuzun S (1996) Induced systemic resistance to cucumber diseases and increased plant growth by plant growth-promoting rhizobacteria under field conditions. Phytopathology 86:221–224

Yan Z, Reddy MS, Ryu C-M, McInroy JA, Wilson M, Kloepper JW (2002) Induced systemic resistance against tomato late blight elicited by plant growth-promoting rhizobacteria. Phytopathology 92:1329–1333

Zehnder G, Kloepper J, Tuzun S, Yao C, Wei G, Chambliss O, Shelby R (1997a) Insect feeding on cucumber mediated by rhizobacteria-induced plant resistance. Entomol Exp Appl 83:81–85

Zehnder G, Kloepper J, Yao C, Wei G (1997b) Induction of systemic resistance in cucumber against cucumber beetles (Coleoptera: Chrysomelidae) by plant growth-promoting rhizobacteria. J Econ Entomol 90:391–396

Zehnder GW, Yao C, Murphy JF, Sikora EJ, Kloepper JW (2000) Induction of resistance in tomato against cucumber mosaic cucumovirus by plant growth-promoting rhizobacteria. BioControl 45:127–137

Zehnder GW, Murphy JF, Sikora EJ, Kloepper JW (2001) Application of rhizobacteria for induced resistance. Eur J Plant Pathol 107:39–50

Zhang S, Moyne A-L, Reddy MS, Kloepper JW (2002a) The role of salicylic acid in induced systemic resistance elicited by plant growth-promoting rhizobacteria against blue mold of tobacco. Biol Control 25:288–296

Zhang S, Reddy MS, Kloepper JW (2002b) Development of assays for assessing induced systemic resistance by plant growth-promoting rhizobacteria against blue mold of tobacco. Biol Control 23:79–86

Zhang S, Reddy MS, Kloepper JW (2004) Tobacco growth enhancement and blue mold disease protection by rhizobacteria: Relationship between plant growth promotion and systemic disease protection by PGPR strain 90-166. Plant Soil 262:277–288

4 Control of Plant Pathogenic Fungi with Bacterial Endophytes

Gabriele Berg, Johannes Hallmann

4.1 Introduction

Interest in biological control has increased over the past years, driven by the need for alternatives to chemicals – which have often lost their activity due to the development of resistant pathogen populations – and to public pressure to develop production systems favourable to the environment (Whipps 2001). In this respect antagonistic bacteria provide an environmentally sound alternative to protect plants against attack by fungal pathogens (Whipps 1997; Bloemberg and Lugtenberg 2001). In the past, rhizosphere bacteria have been shown to be effective antagonists against a broad spectrum of fungal pathogens (Weller 1988; Emmert and Handelsman 1999; Kurze et al. 2001). More recent studies have indicated that bacteria colonising the root interior can even improve plant growth and plant health (Frommel et al. 1991; Sturz et al. 1999; see Chap. 3 by Kloepper and Ryu), and seem to be excellent candidates for use as biological control agents (BCAs) (Chen et al. 1995; Sturz et al. 1997; Downing and Thomson 2000; Sturz et al. 2000; Adhikari et al. 2001; Tjamos et al. 2004; see Chap. 3 by Kloepper and Ryu).

Besides induced resistance (see Chap. 3 by Kloepper and Ryu), little is known about other mechanisms used by antagonistic endophytic bacteria towards fungal pathogens, such as antibiosis, competition and lysis. Furthermore, endophytic bacteria are known to promote plant growth by the production of plant hormones, enhanced nutrient availability and nitrogen fixation (Whipps 2001; Hurek and Reinhold-Hurek 2003). For example, plant hormones produced by endophytic bacteria seem to be essential for bryophyte development (Hornschuh et al. 2002).

So far, most information about the community structure of endophytic bacteria with antagonistic properties has been obtained using cultivation-dependent approaches [Chen et al. 1995; Sturz et al. 1999; see Chaps. 2

Gabriele Berg: Graz University of Biotechnology, Department of Environmental Technology, Petersgasse 12, 8010 Graz, Austria, E-mail: gabriele.berg@TUGraz.at

Johannes Hallmann: Biologische Bundesanstalt für Land- und Forstwirtschaft, Institut für Nematologie und Wirbeltierkunde, Toppheideweg 88, 48161 Münster, Germany

Soil Biology, Volume 9
Microbial Root Endophytes
B. Schulz, C. Boyle, T. N. Sieber (Eds.)

(Hallmann and Berg) and 15 (Schulz)]. However, recently developed cultivation-independent methods (reviewed by Smalla 2004) analyse endophytic communities directly from root tissue. Techniques using bacterial DNA provide information on the structural diversity of the entire endophytic community, while techniques using bacterial RNA identify metabolically active bacteria, e.g. those which are of interest after pathogen infection. Primers are available that specifically recognise bacterial taxa with high percentages of antagonistic strains, such as *Pseudomonas*, *Serratia* or *Burkholderia* (Widmer et al. 1998; Salles et al. 2001). Unfortunately, the methodology does not distinguish between antagonistic and non-antagonistic strains. However, genes involved in bacterial antagonism, such as those involved in the expression of antibiotics, siderophores or phytohormones, are now being cloned and in the near future may lead to the development of primers targeting antagonism and specific microarrays (Raaijmakers et al. 1997; De Souza and Raaijmakers 2003; Zhou 2003). Overall, a polyphasic approach combining cultivation-dependent and cultivation-independent methods is recommended to best gain insights into the dynamics of antagonistic endophytic communities as well as plant/endophyte/pathogen interactions (Garbeva et al. 2001; Krechel et al. 2002; Reiter et al. 2002; Sessitsch et al. 2004). Understanding those interactions provides a platform from which to develop endophytic bacteria as biocontrol agents.

However, formulation, application, risk assessment, fruit quality and potential side-effects are further questions that need to be properly answered before endophytic bacteria can be released as BCAs. This chapter discusses four key aspects of biological control of fungal pathogens using endophytic bacteria: (1) the spectrum of indigenous bacterial antagonists in plant roots, (2) modes of action, (3) use of BCAs, and (4) strategies to enhance biocontrol efficiency.

4.2 Spectrum of Indigenous Endophytic Bacteria with Antagonistic Potential Towards Fungal Plant Pathogens

Antagonists are naturally occurring organisms with the potential to interfere with pathogen infection, growth, and survival (Chernin and Chet 2002). A better understanding of the spectrum of indigenous antagonistic bacteria will (1) increase our knowledge of plant/endophyte interactions, (2) facilitate screening efforts for effective biocontrol organisms, (3) allow breeding of cultivars supporting a high level of antagonistic bacteria, and (4) may even lead to management strategies for increasing the antagonistic potential of endophytic bacteria. This chapter focuses on the spectrum of antagonistic endophytic bacteria, and discusses the factors that influence and regulate them.

Antagonistic communities can be studied from a quantitative or qualitative perspective. Commonly used methods for bacterial identification include morphological and physiological characterisation, fatty methyl-ester analysis and molecular techniques based on specific primers and/or (partial) sequencing of 16S rDNA. While the first method is limited to culturable bacteria, the latter method can also identify non-culturable endophytic bacteria, e.g. in a clone library. Antagonistic activity of endophytic bacteria is generally tested by in vitro inhibition of fungal pathogens in dual cultures and then confirmed in bioassays on host plants. Quantitative analysis to detect antagonistic bacteria is time-consuming as all the endophytic bacteria have to be screened for their antagonistic potential. Primers developed for gene sequences involved in antagonistic activity may, in the future, be able to recognise antagonists without cultivation.

Using cultivation-dependent methods, the proportion of antagonistic endophytes can vary between 0%, as shown for the pathosystem *Phytophthora cactorum*-potato (Sessitsch et al. 2004) and 50%, for *Verticillium longisporum*-oilseed rape (Graner et al. 2003) (Table 4.1). Major factors determining the percentage of antagonists are most likely plant species, pathogen infestation, habitat and vegetation period (Sessitsch et al. 2004; Berg et al. 2005). In conclusion, these data confirm that a significant portion of the indigenous endophytic bacteria in plant roots have antagonistic potential towards fungal pathogens.

Table 4.1. Proportion of antagonistic endophytic bacteria in different pathosystems

Plant	Plant pathogens	Proportion of antagonists (%)	Reference
Rice	*Achlya klebsiana*	25	Adhikari et al. (2001)
	Pythium spinosum	75	
Potato	*Verticillium dahliae*	9	Krechel et al. (2002)
	Rhizoctonia solani		
Potato	*Verticillium dahliae*	13	Berg et al. (2004)
	Rhizoctonia solani	9.7	
Potato	*Verticillium dahliae*	2	Sessitsch et al. (2004)
	Rhizoctonia solani	3	
	Phytophthora cactorum	0	
	Streptomyces scabies	43	
	Xanthomonas campestris	29	
Oilseed rape	*Verticillium longisporum*	50	Graner et al. (2003)
Tomato	*Verticillium dahliae*	12	Tjamos et al. (2004)

But what are the main bacterial species conferring antagonism towards fungal plant pathogens? Although a high diversity of antagonistic endophytic bacteria is found in general, some genera harbour more antagonistic strains than others, e.g. *Bacillus*, *Curtobacterium*, *Methylobacterium*, *Paenibacillus*, *Pseudomonas*, *Serratia*, *Stenotrophomonas* and *Streptomyces* (Sturz et al. 1999; Garbeva et al. 2001; Krechel et al. 2002; Reiter et al. 2002; Sessitsch et al. 2004). Antagonistic species have been isolated from a number of different plant species, but most studies have been done on potato. Table 4.2 lists antagonistic species found in potato in five studies. Surprisingly, a total of 51 different species comprising 27 genera were identified!

The proportion and composition of indigenous endophytic bacteria with antagonistic capacity is influenced by a variety of biotic and abiotic factors, with the plant itself being a major factor (Germida et al. 1998). As shown by

Table 4.2. Endophytic bacterial species of potato with antagonistic properties towards plant pathogenic fungi[a]

Species with antagonistic properties
Agrobacterium tumefaciens
Amycolatopsis meditarranei
Arthrobacter ilicis
Bacillus aquamarinus
Bacillus cereus
Bacillus megaterium
Clavibacter michigenensis
Curtobacterium albidum
Curtobacterium flaccumfaciens
Curtobacterium luteum
Erwinia persicinus
Flavobacterium sp.
Frateuria aurantia
Frigoribacterium sp.
Kingella kingae
Kitasatosporia cystargenia
Methylobacterium sp.
Micrococcus varians
Paenibacillus sp.
Pantoaea agglomerans
Pantoaea anantis
Pseudomonas cichorii
Pseudomonas corrugata
Pseudomonas fluorescens
Pseudomonas graminis
Pseudomonas migulae
Pseudomonas orientalis
Pseudomonas putida
Pseudomonas rhodesiae
Pseudomonas reactans
Pseudomonas straminea
Pseudomonas synthaxa
Pseudomonas syringae
Pseudomonas tolaasii
Pseudomonas veronii
Psychrobacter immobilis
Ralstonia paucula
Rhizobium meliloti
Rhizomonas suberifaciens
Sphigobacterium thalophilum
Sphingomonas adhaesiva
Stenotrophomonas maltophilia
Streptomyces turgidiscabies
Streptomyces bottropensis
Streptomyces diastatochromogenes
Streptomyces galilaeus
Streptomyces griseus
Streptomyces lavendulae
Streptomyces setonii
Streptomyces turgidiscabies
Xanthomonas campestris
Xanthomonas oryzae

[a]According to Sturz et al. (1999), Krechel et al. (2002), Reiter et al. (2002), Sessitsch et al. (2004), and A. Krechel et al., unpublished data

Zinniel et al. (2002) for endophytic bacteria in above-ground plant organs, the bacterial spectrum of 4 agronomic crop species and 27 prairie plant species varied greatly. A similar effect of the plant species on the bacterial spectrum can also be expected for antagonistic root endophytes. There is even an effect of the cultivar or plant genotype on the bacterial spectrum (see Chap. 2 by Hallmann and Berg). For example, the oilseed rape cultivar 'Express', which is tolerant to *V. longisporum*, contained a higher proportion of bacteria with proteolytic, cellulolytic and phosphatase activity than the susceptible cultivar 'Libraska' (Graner et al. 2003). Modern wheat cultivars have a more diverse endophytic community than ancient land races, and are more aggressively colonised by endophytic pseudomonads that produce antifungal metabolites (Germida and Siciliano 2001). Breeder selection for higher productivity might have selected plants that support an endophytic microflora antagonistic to fungal pathogens. Finally, the plant tissue itself can also vary in its antagonistic bacteria. Sturz et al. (1999) made the observation that the outer tissue of a potato tuber yielded higher relative densities of endophytic bacteria antagonistic to soilborne fungal pathogens than deeper layers of the potato. Furthermore, the antifungal potential of bacterial endophytes was highest for isolates recovered from the outermost layer of the tuber. The authors assumed that, in certain communities of endophytic bacteria, antagonism against fungal pathogens may be related to bacterial adaptation to location within a host plant, and may be tissue-type and tissue-site specific, indicating that in plant tissue exposed to the soil, antagonistic endophytes become more prominent. Is this a result of coevolution or a process controlled by the plant itself? Such questions still await proper answers.

Besides the plant itself, several biotic factors affect endophytic communities and the proportion of antagonists (Reiter et al. 2002; Sessitsch et al. 2002, 2004). In this context, it was shown that pathogen stress had a greater impact than plant genotype on bacterial diversity. Furthermore, Sessitsch et al. (2004) found a different spectrum of antagonistic bacteria in good and poor growing potatoes in the field.

Fluctuations in antagonistic bacterial communities of plant roots may also be caused by abiotic factors such as temperature, rainfall, cropping practice or soil amendments (Mocali et al. 2003). Plants are able to select specific bacterial genotypes in response to soil conditions (Siciliano et al. 2001). Therefore, the bacterial community in the soil is an important factor affecting the composition of indigenous endophytic bacteria. Soil amendments can modify the bacterial spectrum in the soil as well as in the roots (Hallmann et al. 1999) and enhance the efficacy of biocontrol agents (Ahmed et al. 2003).

In conclusion, the spectrum and diversity of endophytic bacterial communities in plant roots is never a stable scenario, rather it has its own dynamics in response to biotic as well as abiotic factors. This offers new opportunities for managing the indigenous spectrum of antagonistic endophytic bacteria to increase the benefits to the plant by means of plant breeding, soil amendments or application of endophytic BCAs.

4.3 Mode of Action of Antagonistic Bacteria

Modes of action of antagonistic towards fungal pathogens have been intensively studied for plant growth-promoting rhizobacteria, as reviewed by Fravel (1988), Whipps (2001), Lugtenberg et al. (2001), and Bloemberg et al. (2001). Presumably, endophytic bacteria use similar mechanisms for the control of fungal plant pathogens. However, their hidden life within the plant tissue makes it much more difficult to study such mechanisms (see Chap. 18 by Bloemberg and Camacho Carvajal). Furthermore, it is often difficult to distinguish between direct antagonism (such as antibiosis), competition and lysis, and indirect mechanisms such as induced resistance and improved plant growth (see also Chap. 3 by Kloepper and Ryu).

4.3.1 Antibiosis

Antibiosis describes the ability of an endophytic bacterium to inhibit pathogen growth by the production of antibiotics or toxins. Although the vast majority of endophytic bacteria show antibiosis against fungal pathogens in vitro (Krechel et al. 2002; Sturz et al. 1999), very little is known about the significance of antibiosis controlling fungal pathogens within the root tissue. Examples for antifungal substances released by endophytic bacteria include iturin A (produced by *Bacillus subtilis*) and pyrrolnitrin (produced by *Serratia plymuthica*) (Cho et al. 2002). Further support for the hypothesis that these antifungal metabolites represent the underlying mechanisms in situ could be achieved by antibiotic-deficient mutants that fail to express biocontrol activity. Unfortunately, no such studies have yet been carried out.

However, just as microbial antagonists utilise a diverse arsenal of mechanisms to dominate interactions with fungal pathogens, pathogens have surprisingly diverse responses to counteract these antagonisms (Duffy and Défago 1997). These responses include detoxification, antibiotic resistance, active efflux of antibiotics, and repression of biosynthetic genes expressing proteins involved in biocontrol (Duffy et al. 2003). Again, most work in this area has been carried out on rhizosphere bacteria, and almost nothing is known about the regulation of antifungal metabolites expressed by endophytic bacteria. However, since antibiosis seems to be a mechanism of fungal control used by endophytic bacteria, metabolites released by the bacteria into plant tissue must be carefully monitored to ensure that they pose no risk regarding fruit quality and consumer health.

4.3.2 Competition

Competition is considered an important factor in the control of fungal pathogens by endophytic bacteria, since both organisms colonise similar niches and utilise the same nutrients. Conclusive data demonstrating competition as a major control mechanism of endophytic bacteria are still lacking. Work on rhizosphere bacteria has shown that, under iron-limiting conditions, bacteria produce siderophores with high affinity for ferric iron. By binding available iron these bacteria deprive fungal pathogens of iron, thus restricting their growth (O'Sullivan and O'Gara 1992). Siderophores are also commonly produced by endophytic bacteria (Krechel et al. 2002), indicating that similar mechanisms may occur in the endorhiza.

4.3.3 Lysis

Cell wall lysis is another potential mechanism whereby endophytic bacteria can control fungal pathogens. This mechanism is well established in the biocontrol of fungal pathogens by rhizosphere bacteria. Endophytic bacteria isolated from potato roots express high levels of hydrolytic enzymes such as cellulase, chitinase and glucanase (Krechel et al. 2002). Pleban et al. (1997) analysed the importance of lytic enzymes in antagonism of *Bacillus cereus* strain 65 towards the soilborne fungal pathogen *Rhizoctonia solani*. *B. cereus* strain 65, originally isolated from surface-sterilised seeds of *Sinapis arvensis*, was shown to excrete a chitinase of 36 kDa, responsible for the observed protection of cotton seedlings from root rot disease caused by *R. solani*. Additionally, chitinolytic *Bacillus subtilis* strains were able to reduce symptoms of *Verticillium dahliae* in several host plants (Tjamos et al. 2004). An endophytic chitinase-producing isolate of *Actinoplanes missouriensis* and its culture filtrates were shown to suppress *Plectosporium tabacinum* on lupins (El-Tarabily 2003). The importance of hydrolytic enzymes other than chitinases as biocontrol mechanisms is still unknown.

4.3.4
Induction of Plant Defence Mechanisms

Induction of plant defence mechanisms by endophytic bacteria plays a major role in suppression of fungal plant pathogens and therefore is covered in a separate chapter (Chap. 3 by Kloepper and Ryu).

4.3.5
Plant Growth

Plant growth is a factor that is indirectly involved in pathogen defence. Plants with vigorous growth, such as cucumber, can sometimes outgrow disease by fungal pathogens such as powdery mildew. Therefore, plant growth promotion by endophytic bacteria indirectly affects the pathogenicity of fungal pathogens. Nejad and Johnson (2000) described isolates of endophytic bacteria that significantly improved seed germination and plant growth of oilseed rape and tomato. Plant growth promotion mediated by endophytic bacteria may be exerted by several mechanisms, e.g. production of plant growth hormones, synthesis of siderophores, nitrogen fixation, solubilisation of minerals such as phosphorous, or via enzymatic activities, for example suppression of ethylene by 1-aminocyclopropane-1-carboxylate (ACC) deaminase (Chernin and Chet 2002). Strains of *Pseudomonas*, *Enterobacter*, *Staphylococcus*, *Azotobacter* and *Azospirillum* produce plant growth regulators such as ethylene, auxins or cytokinins, which are assumed to promote plant growth (Arshad and Frankenberger 1991; Leifert et al. 1994). However, in the past, most interest has focussed on the fixation of atmospheric nitrogen by free-living endophytic bacteria, especially of diazotrophs (Döbereiner and Pedrosa 1987; Hecht-Buchholz 1998; Estrada et al. 2002; Hurek and Reinhold-Hurek 2003).

Overall, mechanisms of fungal control by endophytic bacteria may act synergistically, and individual endophytes quite often exhibit several modes of action. However, the expression of antifungal mechanisms is strain-specific (Neiendam-Nielson et al. 1998; Berg 2000; Berg et al. 2002) and most likely under the control of several biotic as well as abiotic factors. A better understanding of the underlying mechanisms has significant relevance for the optimisation of biocontrol strategies (see Sect. 4.5).

4.4
Control Potential of Endophytic Bacteria

A broad spectrum of endophytic bacteria has been described to control fungal plant pathogens on different plant species (Table 4.3). The majority of antagonistic endophytic bacteria are Gram-negative and belong to the group of fluorescent pseudomonads, which are effective BCAs (Bloemberg and Lugtenberg 2001; Whipps 2001). Fluorescent pseudomonads are common members of the endorhiza, making them ideal candidates for biological control measures. Antagonistic isolates have been reported to occur in *Pseudomonas chlororaphis*, *Pseudomonas fluorescens*, *Pseudomonas graminis*, *Pseudomonas putida*, *Pseudomonas tolaasii* and *Pseudomonas veronii* (Table 4.3). Control potential in terms of reduced disease severity can approach 80% under greenhouse conditions (Pleban et al. 1995). Besides pseudomonads, other Gram-negative species with biocontrol activity against fungal pathogens are *Phyllobacterium rubiacearum*, *Burkholderia solanacearum* (Chen et al. 1995), *Sphingomonas trueperi* (Adhikari et al. 2001) and *Serratia plymuthica* (Faltin et al. 2004).

Gram-positive bacterial isolates with antagonistic properties are commonly found within the genus *Bacillus* (Emmert and Handelsman 1999). *Bacillus* spp. occur mainly in the soil/rhizosphere but have also been reported as endophytes of several plant species (see Chap. 2 by Hallmann and Berg). As summarized in Table 4.3, endophytic isolates of *Bacillus*, and the closely related genus *Paenibacillus*, have been shown to significantly control many fungal diseases. Gram-positive biocontrol bacteria other than bacilli include actinomycetes such as *Streptomyces*, *Microbispora* or *Nocardioides* (Coombs et al. 2004). For example, the streptomycete *Actinoplanes missouriensis* gave excellent control of *Plectosporium tabacinum*, the causal agent of lupin root rot (El-Tarabily 2003).

Most studies have been carried out under greenhouse conditions. However, a few studies have achieved similar biocontrol effects under field conditions (Tjamos et al. 2004; Faltin et al. 2004). For example, the endophytic isolate *Pseudomonas trivialis* 3Re2-7 significantly reduced disease incidence of *Rhizoctonia solani* on lettuce and sugar beet by 40% and 86%, respectively (Faltin et al. 2004).

These examples illustrate the high potential of endophytic bacteria in fungal pathogen control. However, further fieldwork is required to confirm their control efficacy in different climatic regions and under different growth conditions. Formulations and applications that meet common farming practice still need to be developed. For promising BCAs, strategies that enhance overall control efficacy should be explored.

4.5 Enhancing Biocontrol Efficiency

It is a well-accepted observation that biological control under field conditions is often inconsistent (Weller 1988). Therefore, enhancing the efficacy

Table 4.3. Examples of antagonistic

Pathosystem	BCA	Trial	Application	Results	Reference
Cotton, Bean (*Rhizoctonia solani*)	*Bacillus cereus* *Bacillus subtilis* *Bacillus pumilus*	Greenhouse	Root incubation in bacterial suspension	Reduction of disease incidence approx. 50%	Pleban et al. 1995, 1997
Bean (*Sclerotium rolfsii*)	*Bacillus subtilis* *Bacillus cereus*	Greenhouse	Root incubation in bacterial suspension	Reduction of disease incidence of 70–80%	Pleban et al. 1995
Cotton (*Fusarium oxysporum* f. sp. *vasinfectum*)	*Aureobacterium saperdae* *Bacillus pumilus* *Phyllobacterium rubiacearum* *Pseudomonas putida* *Burkholderia solanacearum*	Pot experiment	Pierced with a needle	Reduction of disease severity and plant growth promotion	Chen et al. 1995
Tomato (*Verticillium dahliae*)	*Pseudomonas*	Growth chamber	Bacterization of tissue plantlets or seedlings	Reduction of disease severity and plant growth promotion	Sharma and Nowak 1998
Bean (*Rhizoctonia solani*)	*Pseudomonas fluorescens* Rif1::tc4 (with *chiA* gene)	Plant growth chamber	Soil drenching 10^7 cfu ml^{-1} Introduction into the plant	Reduction of disease incidence up to 48%, only if introduced as endophytes	Downing and Thomson 2000
Rice (*Achlya klebsiana*, *Pythium spinosum*)	*Pseudomonas fluorescens* *Pseudomonas tolaasii* *Pseudomonas veronii* *Sphingomonas trueperi*	Pot experiments	Rice seeds were soaked for 1 h in a bacterial suspension 10^8 cfu ml^{-1}	Reduction of disease incidence by 50–73%; plant growth promotion (plant height and dry weight)	Adhikari et al. 2001
Potato [*Rhizoctonia solani*, *Verticillium dahliae*, *Meloidogyne incognita* (nematode)]	*Streptomyces turgidiscabies* *Kitasatosporia cystargenia* *Streptomyces galilaeus* *Streptomyces griseus* *Pseudomonas graminis*	Growth chamber	Seed treatment	Reduction of fungal growth in vitro, of nematode infestation ad planta; plant growth promotion up to 78%	Krechel et al. 2002

Table 4.3. (continued)

Pathosystem	BCA	Trial	Application	Results	Reference
Balloon flower (*Rhizoctonia solani*)	*Bacillus* sp.	Pot experiments	Soil inoculation	Reduction of disease incidence	Cho et al. 2003
Pepper (*Rhizoctonia solani*, *Phytophthora capsici*)	*Bacillus subtilis* *Bacillus licheniformis*	Greenhouse	Seed treatments Root drenching	Reduction of disease incidence of *R. solani* up to 55% and of *P. capsici* up to 55%	Ahmed et al. 2003
Eggplant, potato (*Verticillium dahliae*)	*Bacillus* spp. *Paenibacillus alvei* *Bacillus amiloliquefaciens*	Greenhouse Field	Root dipping Soil drenching 10^7 cfu ml^{-1} talc-gum xanthan formulation	Reduction of disease severity by 40–70%; yield increase 25%	Tjamos et al. 2004
Potato (*Rhizoctonia solani*)	*Serratia plymuthica* *Pseudomonas reactans*	Greenhouse Field	Soil drenching Seed bacterization 10^8 cfu ml^{-1}	Reduction of disease severity up to 76%	Faltin et al. 2004
Wheat (*Gaeumannomyces graminis* var.*tritici*)	*Streptomyces* spp. *Microbispora* spp. *Nocardioides*	Greenhouse	Seed treatments 10^9 to 10^{10} cfu ml^{-1}	Reduction of black lesions up to 71%	Coombs et al. 2004

and consistency of control by endophytic bacteria is a major factor in determining their future success as BCAs. Current strategies to enhance biocontrol efficiency include (1) optimised formulation and application technologies, (2) integrated biocontrol strategies, (3) management of the indigenous endophytic microflora, and (4) genetic engineering.

The formulation of BCAs has undergone major progress in recent years (Burghes 1998). Most work has been done on rhizosphere bacteria and the acquired know-how now needs to be transferred to endophytic bacteria. Basically, two types of formulation seem to be preferable for endophytic bacteria: dry products and liquid suspensions. For Gram-positive endophytes, dry formulations with a long shelf life are already standard (Emmert and Handelsmann 1999; Whipps 1997) whereas stable formulations for Gram-negative bacteria are more difficult, due mainly to the lack of resting spores that can withstand the formulation process. Optimum formulations should ensure bacterial survival, but also promote bacterial activity after application. Microcapsules that protect the incorporated bacteria, while also providing nutrient sources, are an interesting alternative (Burghes 1998).

The delivery of BCAs into the host plant is a prerequisite for successful biocontrol. Musson et al. (1995) studied eight methods for delivering endophytic bacteria into cotton stems and roots: stab-inoculation of bacteria into stems, soaking seed in bacterial suspensions, soil drench, methyl cellulose seed coating, foliar spray, bacteria-impregnated granules applied in-furrow, vacuum infiltration, and the pruned-root dip. Whereas each of the methods effectively established most of the endophytic bacteria based on re-isolation studies, none of the methods successfully delivered all 15 strains tested, indicating that the optimum method is strain-specific. For practical reasons, soil drench, root dipping, and seed coating are the most promising methods. Standard technology can be used and the bacterium is delivered directly into the root system where it can immediately colonise the root and protect the plant against invasion by fungal pathogens.

Fahey et al. (1991) described a seed inoculation technique in which pressure is applied to infuse the bacterial suspension into imbibed seeds, followed by redrying of the seeds. The inoculated seeds met the requested shelf-life requirements of several months, and application of commonly used fungicides had no adverse impact on bacterial survival or efficacy of the bacterial inoculation. This leads to another interesting option for biocontrol enhancement: the combination of endophytic bacteria with other BCAs or even chemicals. Integrated biological control strategies against fungal pathogens using a combination of antagonistic endophytes with complementary modes-of-action and/or colonisation sites, or of endophytes with synthetic control agents take advantage of synergies and should be exploited further.

The naturally occurring bacterial antagonistic potential within a plant is influenced by several biotic and abiotic factors (see Chap. 2 by Hallmann and Berg). By changing these factors, the antagonistic potential can be managed. For example, chitin application not only increased total numbers of antagonistic *Burkholderia cepacia* (Hallmann et al. 1999) but, as shown by Ahmed et al. (2003), also improved control of root rot disease in pepper by endophytic *Bacillus subtilis* and *B. licheniformis*. Furthermore, the chitin derivate chitosan was successfully used in combination with the endophytic bacterium *Bacillus pumilus* to enhance plant resistance towards Fusarium wilt of tomato (Benhamou et al. 1998). It is hypothesised that chitin as well as chitosan stimulate the establishment of a chitinolytic microflora, which then decomposes fungal cell wall chitin.

Biocontrol efficiency might also be improved by breeding plant genotypes to support a high level of antagonistic endophytes. For example, plants expressing high levels of N-acylhomoserine lactones, or which are capable of degrading this important signal molecule of bacterial communication, have been shown to also influence plant-associated bacteria (Fray 2002). On the other hand, endophytic bacteria might also be promoted by transgenic means. Promising biocontrol targets for genetic engineering are gene-regulated factors of endophytic bacteria involved in modes-of-action, colonisation potential, survival, fitness, and adaptation to environmental conditions. For example, the endophytic bacterium *Pseudomonas fluorescens*, originally isolated from micropropagated apple plantlets, was genetically modified by Downing and Thomson (2000) to harbour a gene encoding the major chitinase of *Serratia marcescens*. The gene *chi*A was cloned under the control of the tac promoter in the broad-host-range plasmid pKT240 and the integration vector pJFF350. *P. fluorescens* carrying *tac-chi*A either on the plasmid or integrated into the chromosome significantly controlled *Rhizoctonia solani* on beans. Although a promising approach, genetically modified BCAs still face many restrictions, making broad-scale application in the near future improbable and difficult.

4.6 Conclusions

Most plants are colonised by a broad spectrum of endophytic bacteria that are potentially antagonistic towards fungal plant pathogens. This enormous potential needs to be further explored for its use in modern plant disease control strategies. This requires not only a better understanding of the underlying mechanisms and their regulation in response to environmental factors, but also a more comprehensive picture of what triggers endophytic colonisation as well as of the population dynamics of antagonistic bacterial endophytes within the plant. Continuing research in this area will hopefully lead to new and innovative concepts for biological control of fungal pathogens.

References

Adhikari TB, Joseph CM, Yang G, Phillips DA, Nelson LM (2001) Evaluation of bacteria isolated from rice for plant growth promotion and biological control of seedling disease of rice. Can J Microbiol 47:916–924

Ahmed AS, Ezziyyani M, Pérez Sánchez C, Candela ME (2003) Effect of chitin on biological control activity of *Bacillus* spp. and *Trichoderma harzianum* against root rot disease in pepper (*Capsicum annuum*) plants. Eur J Plant Pathol 109:633–637

Arshad M, Frankenberger WT (1991) Microbial production of plant hormones. In: Keister DL, Cregan PB (eds) The rhizosphere and plant growth. Kluwer, Dordrecht, pp 327–334

Benhamou N, Kloepper JW, Tuzun S (1998) Induction of resistance against *Fusarium* wilt of tomato by combination of chitosan with an endophytic bacterial strain. Planta 204:153–168

Berg G (2000) Diversity of antifungal and plant-associated *Serratia* strains. J Appl Microbiol 88:952–960

Berg G, Roskot N, Steidle A, Eberl L, Zock A, Smalla K (2002) Plant-dependent genotypic and phenotypic diversity of antagonistic rhizobacteria isolated from different *Verticillium* host plants. Appl Environ Microbiol 68:3328–3338

Berg G, Krechel A, Ditz M, Faupel A, Sikora RA, Ulrich A, Hallmann J (2005) Endophytic and ectophytic potato-associated bacterial communities differ in structure and antagonistic function against plant pathogenic fungi. FEMS Microbiol Ecol 51:215–229

Bloemberg GV, Lugtenberg BJJ (2001) Molecular basis of plant growth promotion and biocontrol by rhizobacteria. Curr Opin Plant Biol 4:343–350

Burghes HD (1998) Formulation of biopesticides. Kluwer, Dordrecht

Chen C, Bauske EM, Musson G, Rodríguez-Kábana R, Kloepper JW (1995) Biological control of Fusarium wilt on cotton by use of endophytic bacteria. Biol Control 5:83–91

Chernin, L, Chet I (2002) Microbial enzymes in biocontrol of plant pathogens and pests. In: Burns R, Dick R (eds) Enzymes in the environment: activity, ecology, and applications. Dekker, New York, pp 171–225

Cho SJ, Lim WJ, Hong SY, Park SR, Yun HD (2002) Endophytic colonization of balloon flower by antifungal strain *Bacillus* sp. CY22. Biosci Biotech Biochem 67:2132–2138

Coombs JT, Michelson PP, Franco CMM (2004) Evaluation of endophytic actinobacteria as antagonists of *Gaeumannomyces graminis* var. *tritici* in wheat. Biol Control 29:359–366

De Souza JT, Raaijmakers JM (2003) Polymorphisms within the *prnD* and *pltC* genes from pyrrolnitrin and pyoluteorin-producing *Pseudomonas* and *Burkholderia* spp. FEMS Microbiol Ecol 43:21–34

Döbereiner J, Pedrosa FO (1987) Nitrogen-fixing bacteria in non-leguminous crop plants. Science Tech, Madison, WI

Downing KJ, Thomson JA (2000) Introduction of the *Serratia marcescens chi*A gene into an endophytic *Pseudomonas fluorescensfluorescens* for the biocontrol of phytopathogenic fungi. Can J Microbiol 46:363–369

Duffy BK, Défago G (1997) *Fusarium* pathogenicity factor blocks antibiotic biosynthesis by *Pseudomonas* biocontrol strains. IOBC WPRS Bull 21:145–148

Duffy B, Schouten A, Raaijmakers JM (2003) Pathogen self-defense: mechanisms to counteract microbial antagonism. Annu Rev Phytopathol 41:501–538

El-Tarabily KA (2003) An endophytic chitinase-producing isolate of *Actinoplanes missouriences*, with potential for biological control of root rot of lupin caused by *Plectosporium tabacinum*. Aust J Bot 51:257–266

Emmert EAB, Handelsman J (1999) Biocontrol of plant disease: a (Gram$^+$) positive perspective. FEMS Microbiol Lett 171:1–9

Estrada P, Mavingui P, Cournoyer B, Fontaine F, Balandreau J, Caballero-Mellado J (2002) A N_2-fixing endophytic *Burkholderia* sp. associated with maize plants cultivated in Mexico. Can J Microbiol 48:285–294

Fahey JW, Dimock MB, Tomasino SF, Taylor JM, Carlson, PS (1991) Genetically engineered endophytes as biocontrol agents: a case study from industry. In: Andrews JH, Hirano SS (eds) Microbial ecology of leaves. Springer, Berlin Heidelberg New York, pp 401–411

Faltin F, Lottmann J, Grosch R, Berg G (2004) Strategy to select and assess antagonistic bacteria for biological control of *Rhizoctonia solani* Kühn. Can J Microbiol 50:811–820

Fray RG (2002) Altering plant-microbe interaction through artificially manipulating bacterial quorum sensing. Ann Bot 89:245–253

Fravel DR (1988) Role of antibiosis in the biocontrol of plant diseases. Annu Rev Phytopathol 26:75–91

Frommel MI, Nowak J, Lazarovits J(1991) Growth enhancement and developmental modifications of in vitro grown potato (*Solanum tuberosum* ssp. *tuberosum*) as affected by a nonfluorescent *Pseudomonas* sp. Plant Physiol 96:928–936

Garbeva P, van Overbeek LS, van Vuurde JWL, van Elsas JD (2001) Analysis of endophytic bacterial communities of potato by plating and denaturing gradient gel electrophoresis (DGGE) of 16S rDNA based PCR fragments. Microb Ecol 413:69–383

Germida JJ, Siciliano SD (2001) Taxonomic diversity of bacteria associated with the roots of modern, recent and ancient wheat cultivars. Biol Fertil Soils 33:410–415

Germida JJ, Siciliano SD, Freitas JR de, Seib AM (1998) Diversity of root-associated bacteria associated with field-grown canola (*Brassica napus* L.) and wheat (*Triticum aestivum* L.). FEMS Microbiol Ecol 26:43–50

Graner G, Persson P, Meijer J, Alstrom S (2003) A study on microbial diversity in different cultivars of *Brassica napus* in relation to its wilt pathogen, *Verticillium longisporum*. FEMS Microbiol Lett 29:269–276

Hallmann J, Rodríguez-Kábana R, Kloepper JW (1999) Chitin-mediated changes in bacterial communities of the soil, rhizosphere and within roots of cotton in relation to nematode control. Soil Biol Biochem 31:551–560

Hecht-Buchholz C (1998) The apoplast-habitat of endophytic nitrogen-fixing bacteria and their significance for the nitrogen nutrition on nonleguminous plants. Z Pflanzenernähr Bodenkd 161:509–520

Hornschuh M, Grotha R, Kutschera U (2002) Epiphytic bacteria associated with the bryophyte *Funaria hygrometrica*: effects of *Methylobacterium* strains on protonema development. Plant Biol 4:682–687

Hurek T, Reinhold-Hurek B (2003) *Azoarcus* sp. strain BH72 as a model for nitrogen-fixing grass endophytes. J Biotechnol 106:169–178

Krechel A, Faupel A, Hallmann J, Ulrich A, Berg G (2002) Potato-associated bacteria and their antagonistic potential towards plant-pathogenic fungi and the plant-parasitic nematode *Meloidogyne incognita* (Kofoid & White) Chitwood. Can J Microbiol 48:772–786

Kurze S, Dahl R, Bahl H, Berg G (2001) Biological control of fungal strawberry diseases by *Serratia* HRO-C48. Plant Dis 85:529–534

Leifert C, Morris CE, Waites WM (1994) Ecology of microbial saprophytes and pathogens in tissue culture and field-grown plants: reasons for contamination problems in vitro. Crit Rev Plant Sci 13:139–183
Lugtenberg BJJ, Dekkers L, Bloemberg GV (2001) Molecular determinants of rhizosphere colonization by *Pseudomonas*. Annu Rev Phytopathol 39:461–490
Mocali S, Bertelli E, Di Cello F, Mengoni A, Sfalanga A, Viliani F, Caciotti A, Tegli S, Surico G, Fani R (2003) Fluctuation of bacteria isolated from elm tissues during different seasons and from different plant organs. Res Microbiol 154:105–114
Musson G, McInroy JA, Kloepper JW (1995) Development of delivery systems for introducing endophytic bacteria into cotton. Biocontrol Sci Technol 5:407–416
Neiendam-Nielson M, Sörensen J, Fels J, Pedersen HC (1998) Secondary metabolite- and endochitinase-dependent antagonism toward plant-pathogenic microfungi of *Pseudomonas fluorescens* isolates from sugar beet rhizosphere. Appl Environ Microbiol 64:3563–3569
Nejad P, Johnson PA (2000) Endophytic bacteria induce growth promotion and wilt disease suppression in oilseed rape and tomato. Biol Control 18:208–215
O'Sullivan DJ, O'Gara F (1992) Traits of fluorescent *Pseudomonas* spp. involved in suppression of plant root pathogens. Microbiol Rev 56:662–676
Pleban S, Ingel F, Chet I (1995) Control of *Rhizoctonia solani* and *Sclerotium rolfsii* in the greenhouse using endophytic *Bacillus* spp. Eur J Plant Pathol 101:665–672
Pleban S, Chernin L, Chet I (1997) Chitinolytic activity of an endophytic strain of *Bacillus*. Lett Appl Microbiol 25:284–288
Raaijmakers JM, Weller DM, Thomoshow LS (1997) Frequency of antibiotic-producing *Pseudomonas* spp. in natural environments. Appl Environ Microbiol 63:881–887
Reiter B, Pfeifer U, Schwab H, Sessitsch A (2002) Response of endophytic bacterial communities in potato plants to infection with *Erwinia carotovora* subsp. *atroseptica*. Appl Environ Microbiol 68:2261–2268
Salles JF, De Souza FA, van Elsas JD (2001) Molecular method to assess the diversity of *Burkholderia* species in environmental samples. Appl Environ Microbiol 68:1595–1603
Sessitsch A, Reiter B, Pfeifer U, Wilhelm E (2002) Cultivation-independent population analysis of bacterial endophytes in three potato varieties based on eubacterial and *Actinomycetes*-specific PCR of 16S rRNA genes. FEMS Microbiol Ecol 39:23–32
Sessitsch A, Reiter B, Berg G (2004) Endophytic bacterial communities of field-grown potato plants and their plant growth-promoting abilities. Can J Microbiol 50:239–249
Sharma VK, Nowak J (1998) Enhancement of Verticillium wilt resistance in tomato transplants by in vitro co-culture of seedlings with a plant growth promoting rhizobacterium (*Pseudomonas* sp.) strain PsJN. Can J Microbiol 44:528–536
Siciliano DS, Forin N, Mihoc A, Wisse G, Labell S, Beaumier D, Ouellette D, Roy R, Whyte LG, Banks MK, Schwab P, Lee K, Greer CW (2001) Selection of specific endophytic bacterial genotypes by plants in response to soil contamination. Appl Environ Microbiol 67:2469–2475
Smalla K (2004) Culture-independent microbiology. In: Bull AT (ed) Microbial diversity and bioprospecting. ASM, Washington DC, pp 88–99
Sturz AV, Christie BR, Matheson BG (1997) Associations of bacterial endophyte populations from red clover and potato crops with potential for beneficial allelopathy. Can J Microbiol 44:162–167
Sturz AV, Christie BR, Matheson BG, Arsenault WJ, Buchanan NA (1999) Endophytic bacterial communities in the periderm of potato tubers and their potential to improve resistance to soilborne plant pathogens. Plant Pathol 48:360–369
Sturz AV, Christie BR, Nowak J (2000) Bacterial endophytes: potential role in developing sustainable systems of crop production. Crit Rev Plant Sci 19:1–30

Tjamos EC, Tsitsigiannis DI, Tjamos SE, Antoniou P, Katinakis P (2004) Selection and screening of endorhizosphere bacteria from solarised soils as biocontrol agents against *Verticillium dahliae* of solanaceous hosts. Eur J Plant Pathol 110:35–44

Weller DM (1988) Biological control of soilborne plant pathogens in the rhizosphere with bacteria. Annu Rev Phytopathol 26:379–407

Whipps JM (1997) Ecological considerations involved in commercial development of biological control agents for soil-borne diseases. In: Van Elsas JD, Trevors JT, Wellington EMH (eds) Modern soil microbiology. Dekker, New York, pp 525–545

Whipps J (2001) Microbial interactions and biocontrol in the rhizosphere. J Exp Bot 52:487–511

Widmer F, Seidler RJ, Gillevet PM, Watrud LS, Di Giovanni GD (1998) A highly selective PCR protocol for detecting 16S rRNA genes of the genus *Pseudomonas* (sensu stricto) in environmental samples. Appl Environ Microbiol 64:2545–2553

Zhou J (2003) Microarrays for bacterial detection and microbial community analysis. Curr Opin Microbiol 6:288–294

Zinniel DK, Lambrecht P, Harris NB, Feng Z, Kuczmarski D, Higley P, Ishimaru CA, Arunakumari A, Barletta RG, Vidaver AK (2002) Isolation and characterization of endophytic colonizing bacteria from agronomic crops and prairie plants. Appl Environ Microbiol 68:2198–2208

5 Role of Proteins Secreted by Rhizobia in Symbiotic Interactions with Leguminous Roots

Maged M. Saad, William J. Broughton, William J. Deakin

5.1 Introduction

Rhizobia are Gram-negative soil inhabitants with the ability to induce the formation of highly specialised organs called nodules on the roots or stems of leguminous plants. Some rhizobial species provoke nodule formation on a limited number of legume genera and are said to have narrow host ranges, e.g. *Rhizobium meliloti*, which nodulates only three genera of legumes. Other (broad host-range) rhizobia provoke the formation of nodules on many different legumes, e.g. *Rhizobium* sp. NGR234 (hereafter called NGR234), which nodulates more than 112 genera of legumes as well as the non-legume *Parasponia andersonii* (Pueppke and Broughton 1999; Trinick 1980).

To form root nodules, legume roots undergo several new developmental changes. Initially, rhizobia attach to root hairs, causing deformation and then curling of the root hair. Rhizobia invade the root through newly formed tubular structures, called infection threads, which grow toward the root cortex. During invasion, rhizobia cause the induction of division of cortical cells, thus forming nodule primordia (Relić et al. 1994). Infection threads travel inter- and intra-cellularly toward the primordia. Wall-degrading enzymes help the passage of infection threads from cell to cell (van Spronsen et al. 1994). Rhizobia are released from the infection threads into the cytoplasm of host cells by a process resembling endocytosis (Stacey et al. 1991). Extensive cell division in the primordia leads to functional nodules containing rhizobia, in which the bacteria differentiate into their endosymbiotic form, known as the bacteroids (Franssen et al. 1992). Bacteroids, together with the surrounding plant-derived peribacteroid membrane (PBM) are

Maged M. Saad: Université de Genève, Laboratoire de Biologie Moléculaire des Plantes Supérieures, Sciences III, 30 Quai Ernest-Ansermet, 1211 Genève 4, Switzerland

William J. Broughton: Université de Genève, Laboratoire de Biologie Moléculaire des Plantes Supérieures, Sciences III, 30 Quai Ernest-Ansermet, 1211 Genève 4, Switzerland, E-mail: william.broughton@bioveg.unige.ch

William J. Deakin: Université de Genève, Laboratoire de Biologie Moléculaire des Plantes Supérieures, Sciences III, 30 Quai Ernest-Ansermet, 1211 Genève 4, Switzerland

Soil Biology, Volume 9
Microbial Root Endophytes
B. Schulz, C. Boyle, T. N. Sieber (Eds.)

called symbiosomes. At this stage, the bacteria synthesise nitrogenase, an enzyme complex that catalyses the reduction of atmospheric nitrogen to ammonia, which is subsequently assimilated into amino acids. In return, the plant reduces carbon dioxide to sugars during photosynthesis and translocates these compounds to the roots, where the bacteria use them as an energy source (Atkins et al. 1982; Fisher and Long 1992). Depending upon the plant species, at least two types of nodules are formed. Indeterminate nodules have a persistent meristem that grows continuously giving the nodules an elongated shape. Determinant nodules lack the persistent meristem, and are round in shape as the meristematic activity is limited to the early stages of nodule development.

Coordination of this complex developmental programme requires the exchange of many signals between the two symbiotic partners and it is the response to (and synchronisation of) these signals that controls nodule formation. Amongst the first signal molecules are phenolic compounds, mainly flavonoids that are secreted by roots into the rhizosphere. The rhizobial protein NodD functions first as an environmental sensor of these phenolics, and later as a transcriptional activator of a series of rhizobial genes that encode proteins responsible for the production of rhizobial signals. NodD proteins belong to the LysR family of transcriptional regulators, which have the ability to bind to specific, highly conserved DNA sequences (*nod*-boxes) present in the promoter regions of many nodulation genes/loci (Perret et al. 2000). The number of *nod*-boxes varies in different rhizobia. For example, in NGR234 there are at least 19 *nod*-boxes that help regulate transcription of genes involved in a range of different signalling compounds (Kobayashi et al. 2004). The first rhizobial signal molecules to be synthesised and secreted are encoded by the nodulation genes (*nod*, *noe*, and *nol*), which are responsible for the synthesis of host-specific lipo-chito-oligosaccharide molecules called Nod-factors. Nod-factors provoke deformation and curling of the root hair and allow rhizobia to enter roots through infection-threads (Relić et al. 1993, 1994; D'Haeze et al. 1998). All invasive rhizobia produce Nod-factors, and the addition of purified Nod-factors alone causes root-hair deformation and cortical-cell division (Downie 1998; Perret et al. 2000). Nod-factors consist of a backbone of three to six β-1.4-linked *N*-acetyl-D-glucosamine residues. A fatty-acid chain of variable length and structure (depending on the *Rhizobium* species) is attached at the C-2 position of the non-reducing sugar residue. Synthesis of the backbone is brought about by the products of the *nodABC* genes, known as the core enzymes, as they are found in all rhizobia. NodB and NodC are responsible for the synthesis of the backbone while NodA is an acyl-transferase, which adds the fatty-acid side chain. Nod-factors are further modified by the action of other Nod enzymes that add various chemical groups to the backbone (Hanin et al. 1999; Broughton et al. 2000).

The development of the infection threads needs other sets of rhizobial signals, including surface polysaccharides (SPS) as well as secreted proteins. SPSs include extracellular polysaccharides (EPS), capsular polysaccharides (CPS), lipo-polysaccharides (LPS) as well as cyclic β-glucans. Most of these polysaccharides function during infection thread development, where they possibly help suppress plant defence reactions. SPSs have highly diverse structures and may contribute to rhizobial host-range (for reviews, see Broughton et al. 2000; Perret et al. 2000). In this chapter we will concentrate on protein secretion by rhizobia, which are thought to play a role in the infection process leading to nodule formation and, thus, in successful nitrogen fixation.

5.2 Bacterial Protein Secretion Systems

Gram-negative bacteria possess at least four systems (type I–type IV) to secrete proteins into the external environment (see Fig. 5.1) (Thanassi and Hultgren 2000). Examples of all four systems have been found in rhizobia. Type II secretion is said to be *sec*-dependent as it requires the *sec* system to export proteins into the periplasm prior to secretion across the outer membrane. Proteins secreted by the type II system thus possess a classical amino-terminal hydrophobic signal sequence, which is cleaved during *sec*-export. In contrast, type I and type III secretion systems are *sec*-independent; proteins are secreted across the bacterial inner- and outer-membranes in a single step process. Such proteins do not possess cleavable amino-terminal signal sequences. A number of proteins secreted by type I or type III secretion systems are able to directly affect nodulation in a variety of legume-rhizobia associations.

5.2.1 Type I Secretion Systems

Many Gram-negative bacteria utilise type I [or ATP-binding cassette (ABC)] secretion systems (T1SS). Generally, the substrates of ABC transporters are toxins, proteases or lipases. All T1SS secreted proteins possess a carboxy-terminal secretion signal of approximately 60 amino acids, which is not cleaved during export. The secretion machine consists of multimers of three proteins: an inner membrane exporter, an outer membrane pore, and an inner-membrane anchored protein that spans the periplasm linking the proteins found in the inner and outer membranes. Proteins can thus be directly secreted from the cytoplasm to the external environment without

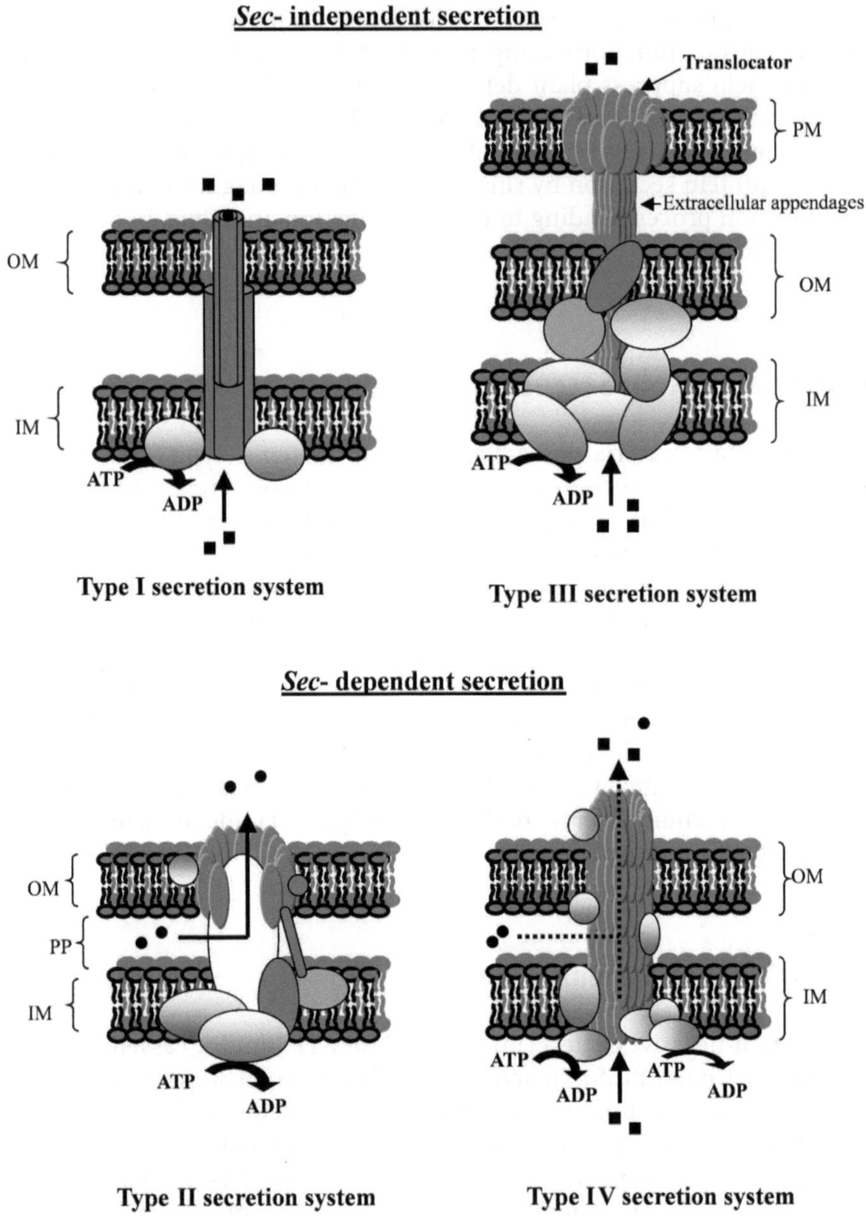
Sec- independent secretion
Translocator
PM
Extracellular appendages
OM
IM
ATP
ADP
Type I secretion system
Type III secretion system
Sec- dependent secretion
PP
Type II secretion system
Type IV secretion system

▶ **Fig. 5.1.** Diagrammatic scheme describing four different types of protein secretion systems in Gram-negative bacteria. *OM* Bacterial outer membrane, *IM* bacterial inner membrane, *PP* periplasmic space, *PM* host plasma membrane. A number of proteins are involved in the assembling of the different secretion apparatus (indicated by *spheres* and *ellipsoids*). T2SS (type II secretion system) and T4SS (type IV secretion system) are *Sec*-dependent, thus proteins to be exported from the bacterial cell are first transported to the periplasm by the *Sec* system. The direction of this two-step secretion is shown by the *black arrow*, from the periplasm to the external environment. T4SS can also export other protein substrates directly from the cytosol, which does not require the Sec system. T1SS (type I secretion system) and T3SS (type III secretion system) are *Sec*-independent, transporting proteins directly from the bacterial cytoplasm to the external environment or into eukaryotic cells (for T3SS). Extracellular appendages are known to be components of several T3SS and T4SS. Secreted proteins are represented by *black circles* if they are exported from the periplasm (*Sec*-dependent) or *black squares* if they originate in the bacterial cytoplasm (*Sec*-independent)

the formation of periplasmic intermediates (Hueck 1998; Thanassi and Hultgren 2000).

T1SSs Involved in Symbiosis

T1SSs have been found in a number of rhizobia, including *Rhizobium leguminosarum* bv. *viciae* and *Rhizobium* sp. BR816 (van Rhijn et al. 1996). NodO was first identified as a secreted protein in *R. leguminosarum* bv. *viciae* by de Maagd et al. (1989). The *nodO* gene is flavonoid inducible in a *nod*-box- and *nodD*-dependent manner (de Maagd et al. 1989; van Rhijn et al. 1996). NodO exists as a dimer in its native form, with an approximate molecular mass of 67 kDa (Sutton et al. 1994, 1996). The amino acid sequence of NodO contains putative repeated Ca^{2+}-binding domains in the amino-terminal region, and has homology to a number of pore-forming bacterial toxins known as RTX proteins (Economou et al. 1990). Studies in vitro have shown that purified NodO forms cation-selective pores in plasma-membrane lipid bilayers (Sutton et al. 1994). By analogy, NodO may thus form a pore in the root cell membrane, causing an influx of Ca^{2+}, which could act as a second messenger thereby stimulating the cytoskeletal changes required for infection thread growth.

Mutation of the *nodO* gene has little effect on nodulation (Downie and Surin 1990), although double mutants of *nodO* and those involved in Nod-factor synthesis display clear Nod^- phenotypes. This was unexpected as NodO is not involved in the biosynthesis or export of Nod-factors (Spaink et al. 1991; Sutton et al. 1994). Yet on *Pisum sativum* and *Vicia sativa* for example, NodO and a functional *nodE* are necessary for nodulation (Downie and Surin 1990; Economou et al. 1994). As infection threads appear to abort in the *nodO/nodE* double mutant, it is possible that NodE (an

α-keto-acyl-synthase) is also involved in acylation of some components of the infection thread. Furthermore, mobilising *nodO* into different rhizobia extends the host range of the trans-conjugant (e.g. *nodO* into a *nodE* mutant of *R. leguminosarum* bv. *trifolii* allows transconjugants to nodulate *V. sativa*; Economou et al. 1994), thus perhaps pointing to a host-specific role for NodO. Another example of *nodO* complementing a defect in Nod-factor synthesis was shown when a *nodO* homologue of *Rhizobium* sp. strain BR816 was used to complement a *nodS* mutant of NGR234 for the nodulation of *Leucaena leucocephala* (van Rhijn et al. 1996). Again, NodO and NodS have distinct biochemical functions: NodS is a N-methyl transferase that methylates Nod-factors (Geelen et al. 1995; Jabbouri et al. 1995). The BR816 *nodO* gene was also shown to suppress the nodulation defect of the *nodU* mutants of NGR234 and *R. tropici* CIAT899 on *L. leucocephala* and of the *nodE* mutant of *R. leguminosarum* bv. *trifolii* ANU842 on white clover (Vlassak et al. 1998).

Based on the observation that over-expression of *nodO* rescues nodulation by the multiple mutant *nodFEMNTLO* (of *R. leguminosarum* bv. *viciae*), Walker and Downie (2000) proposed a role for NodO in the complementation of Nod-factor defects. This mutant produces Nod-factors that are devoid of decorations and results in abortion of infection-thread development in *Vicia sativa*. They suggested that NodO stimulates ion flow across the cell membrane, thereby amplifying a weaker-than-normal signal transmitted by the undecorated version of Nod-factor.

NodO is not the only T1SS Protein Involved in Symbiosis

Although the phenotype of a *R. leguminosarum nodO* mutant was Fix^+ on *P. sativum*, inactivation of the type I system that secretes NodO results in Fix^- nodules (Economou et al. 1994). It is thus possible that the T1SS is capable of secreting other proteins that play important roles in nodulation. Two such proteins were identified as PlyA and PlyB, two similar enzymes that function as extra-cellular glycanases, which are involved in processing rhizobial EPS (Finnie et al. 1998). ExpEI is secreted in a type I-dependent fashion by *Rhizobium meliloti*. Like the PlyAB proteins of *R. leguminosarum*, ExpEI is thought to be involved in the extra-cellular processing of EPS (Becker et al. 1997; Moreira et al. 2000). Thus it seems that proteins secreted via T1SSs in rhizobia play indirect roles in symbiosis, perhaps by amplifying or modifying other signal molecules. In this way they could increase ion flux following pore formation in plasma membranes, as proposed for NodO, or by augmenting the amounts of the active forms of EPS, as is possibly the case with ExpE1, and the PlyAB enzymes.

5.2.2
Type II Secretion Systems

A wide variety of Gram-negative bacteria utilise type II secretion systems (T2SSs) as a stepwise process to export proteins from the periplasm across the outer membrane. The amino-terminal signal peptides of the secreted proteins are first recognised and then translocated by a *sec*-dependent mechanism through the inner membrane. The signal peptide is cleaved, releasing the protein into the periplasm (Pugsley 1993; Sandkvist 2001). The T2SS is also called the general secretory pathway (GSP), and is responsible for the secretion of a large variety of degradative enzymes and toxins. T2SSs are composed of a core of between 12 and 15 proteins (Fig. 5.1), not all of which are present in every T2SS, as some appear to be dispensable for secretion (Sandkvist 2001). The core proteins are thought to form a multiprotein complex, spanning the periplasmic compartment that is specifically required for the translocation of any secreted proteins across the outer membrane (Sandkvist 2001; Peabody et al. 2003). In rhizobia, there is no clear evidence that any T2SSs play a role in symbiosis or nodule formation.

Interestingly, the type II secretion machine shares many features with the type IV pilus biogenesis system found in many *Rhizobium* species, e.g. *Bradyrhizobium japonicum* USDA110, *Mesorhizobium loti* MAFF303099, *R. meliloti* and NGR234 (Kaneko et al. 2000, 2002; Galibert et al. 2001; Streit et al. 2004). Type IV pili are found on the surface of many Gram-negative bacteria, where they play an important role in bacterial adhesion to host cells, bio-film formation and conjugative DNA transfer (Wolfgang et al. 2000). Nitrogen fixing bacteria of the genus *Azoarcus* utilise type IV pili to colonise grasses (Dörr et al. 1998). It remains to be seen whether rhizobia use type IV pili during the symbiotic interaction with legumes.

5.2.3
Type III Secretion Systems

Type III secretion systems (T3SSs) are characteristic of pathogenic Gram-negative bacteria, where their function is to inject proteins into the cytoplasm of eukaryotic cells, so facilitating the onset of disease. The T3SS is composed of a complex of about 20 proteins that spans both bacterial membranes (Fig. 5.1). Ten of these proteins are highly conserved in all T3SS-possessing bacteria and even show similarities to components of the flagella assembly apparatus, from which the pathogenic T3SS are thought to have evolved (Hueck 1998). Proteins that are secreted by T3SSs can be separated into four classes based on their functions. Some of them polymerise into extra-cellular components of the secretion apparatus forming pili. There are also effector proteins that are actually injected into the cytosol of host cells, which then modulate cellular functions of the host by interfering with signalling cascades or disrupting the cytoskeleton. The third class of secreted proteins is termed translocators, and they polymerise to form a pore in the membrane of eukaryotic cells that allows the effectors to pass into the cells (Hueck 1998; Feldman and Cornelis 2003). Finally, in certain T3SS-possessing bacteria, secreted regulatory proteins that control cell contact-dependent secretion have also been identified (He 1998; Hueck 1998). Proteins secreted by a T3SS do not require the *sec* system for their transit from the bacterial cytoplasm to the eukaryotic cell, although the *sec* pathway might be required for assembly of the type III secretion apparatus within the bacterial membranes. (Several components of the apparatus carry *sec*-characteristic amino-terminal signal sequences; Hueck 1998).

Identification and Function of T3SSs in Rhizobia

Given their importance in pathogenicity, it was a surprise to find T3SSs in symbiotic rhizobia. A complete T3SS was first identified in *Rhizobium* sp. NGR234 (Freiberg et al. 1997). All ten genes encoding the conserved components of T3SSs were found and named *rhc* (***Rhizobium conserved***) but the same final letter as used to describe pathogenic T3SS genes was maintained (Viprey et al. 1998). Sequencing other large replicons of *B. japonicum* USDA110 (Göttfert et al. 2001; Kaneko et al. 2002) and *M. loti* MAFF303099 (Kaneko et al. 2000) also revealed the existence of loci encoding T3SSs. T3SSs are not ubiquitous in rhizobia, however, as no such system was found within the completed genome of *R. meliloti* (Galibert et al. 2001). Random mutagenesis also identified mutants of *R. fredii* strains that were affected in nodulation. Subsequent analysis of the insertion sites showed them to be within T3SSs (Marie et al. 2001; Krishnan et al. 2003). In fact, mutagenesis of T3SSs of rhizobia proved that although they are not absolutely essential for nodulation of all legumes, they have cultivar-specific effects and thus can be viewed as determinants of rhizobial host-range (Marie et al. 2001). This is exemplified by studies on the T3SS of NGR234 (Viprey et al. 1998). In NGR234, the T3SS genes are grouped within a 30-kb region of the symbiotic plasmid pNGR234*a* (Freiberg et al. 1997). Knockout mutants in the T3SS machine of NGR234 cause three host-dependent effects. As compared to the wild-type rhizobia, there can be a dramatic impairment of nodule development e.g. on *Tephrosia vogelii*, resulting in the formation of a majority of non-fixing pseudo-nodules. Second, a dramatic enhancement of nodulation is often seen, as with *Crotalaria juncea* and *Pachyrhizus tuberosus*, while a third group of legumes seems to be unaffected by the presence/absence of a functional T3SS (e.g. *Lotus japonicus* and *Vigna unguiculata*; Marie et al. 2003; Viprey et al. 1998).

Regulation of Rhizobial T3SSs

In rhizobia, transcription of the T3SS-related genes requires the presence of flavonoids and two bacterial regulatory elements: NodD1 and TtsI (Krishnan et al. 1995; Viprey et al. 1998; Krause et al. 2002). TtsI shares homology with transcriptional activators of the two-component sensor-regulator family (Viprey et al. 1998). The gene encoding TtsI is found within the T3SS loci of rhizobia and is preceded by a *nod*-box. It has been shown that, after flavonoid activation, NodD1 induces *ttsI* transcription, which in turn activates genes within T3SS loci (Viprey et al. 1998; Kobayashi et al. 2004). Induction of the T3SS genes occurs after induction of genes involved in Nod-factor synthesis, implying that the T3SS functions after Nod-factors in the symbiosis. A conserved promoter motif called the *tts*-box has been identified upstream of most T3SS regulons (Krause et al. 2002). Although it has not been demonstrated experimentally, it is thought that TtsI may bind to *tts*-boxes to induce transcription of the downstream genes. Transcriptional studies suggest that the T3SS of NGR234 does not function throughout the symbiosis, as the majority of T3SS gene transcripts could not be detected in nodules of *V. unguiculata* and *Cajanus cajan* (Perret et al. 1999).

Functions of Proteins Secreted via Rhizobial T3SSs

Protein secretion by T3SSs of rhizobia has been demonstrated in vitro for *R. fredii* USDA257 and NGR234. Proteins secreted in a T3SS-dependent manner are called Nops (nodulation outer proteins) (Marie et al. 2001). Five Nops are known to be secreted by USDA257 (Krishnan and Pueppke 1993) and at least eight by NGR234 (Marie et al. 2003). Several of these Nops have been identified and subsequently characterised. Functional studies have shown that rhizobial Nops can be placed into three of the general classes of type-III-secreted proteins. Figure 5.2 summarises the role of the Nops within a rhizobial T3SS.

External Components of the Machinery

Some type-III-secreted proteins of phyto-pathogenic bacteria have been shown to polymerise and form pili. These pili serve to connect the bacterium to the plant cell and allow proteins to pass through the hollow pili. As protein secretion is abolished in mutants of pili-genes, their phenotype resembles that produced by knock-outs of the T3SS machine (He and Jin 2003). In the presence of flavonoids, USDA257 produces extra-cellular appendages (pili), and requires a functional T3SS (Krishnan et al. 2003). Preliminary results indicate that NGR234 also produces pilus-like structures on its surface in a flavonoid and T3SS-dependent manner. These appendages were purified and shown to be composed predominantly of NopA

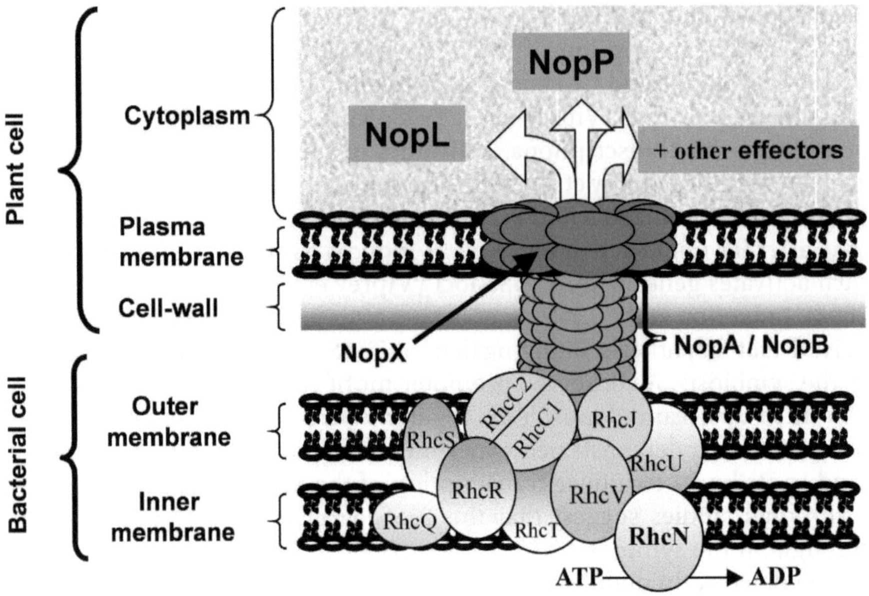

Fig. 5.2. Hypothetical structure of the T3SS of *Rhizobium* sp. NGR234, adapted from Viprey et al. (1998) and Bartsev et al. (2004b). The conserved components (Rhc proteins) of the T3SS form a channel through the bacterial inner and outer membranes. The known roles of the Nops are also illustrated. NopA and NopB are thought to be the major components of a T3SS-dependent pilus that links the rhizobial cell to the plant cell. Nops are secreted through the pilus and can thus cross the plant cell wall. NopX may polymerise to form a pore in the plant root cell plasma membrane and, finally, NopL and NopP are possible effector proteins that function within the plant root cell. NGR234 probably secretes many other effector proteins

(W.J. Deakin, unpublished data), the smallest protein secreted by NGR234 (Marie et al. 2003). Furthermore, homologues of *nopA* have been identified in all T3SS-possessing rhizobia, suggesting that NopA is an essential component of the secretion machinery. Mutation of *nopA* also blocks the secretion of all the other Nops. This phenotype resembles that of mutations in other genes that encode external components of the T3SS. Although NopA is the major component of the rhizobial T3SS-pili, mutations in other genes that encode Nops also block Nop secretion, suggestion that other, minor components of the external secretion apparatus might exist (unpublished data).

Translocators

NopX, one of the first Nops to be identified in NGR234 (Viprey et al. 1998), has significant homology to a number of proteins secreted by T3SSs of phytopathogens. Perhaps the best studied of these is HrpF of *Xanthomonas campestris* pv. *vesicatoria* (Huguet and Bonas 1997). HrpF is secreted in a T3SS-dependent manner and may function as a translocator of the effectors proteins into host cells (Rossier et al. 2000). Indeed, HrpF has been shown to form pores in lipid bilayers (Büttner et al. 2002) We have thus proposed that NopX could perform the role of translocator for rhizobial T3SSs (Marie et al. 2003). NopX of USDA257 has been localised to the infection threads, where it could play a role in the infection thread growth (Krishnan 2002). Although *nopX* homologues are present in most T3SS-possessing rhizobia, there does not appear to be a homologue in *B. japonicum* USDA110 (Krause et al. 2002). This is extremely puzzling as the translocon is thought to be essential for the transport of T3SS proteins into eukaryotic cells.

Effectors

This group of secreted proteins are thought to function within the root cells. So far, only one example of a rhizobial T3SSs effector has been identified (NopL), although it is suspected that there could be many more. Homologues of *nopL* are found only in T3SS-possessing rhizobia, although *M. loti* MAFF303099 does not appear to have a copy. Mutations in genes encoding effector proteins do not affect the secretion of any other T3SS proteins, and this was shown to be the case for a mutation in *nopL* of NGR234 (Marie et al. 2003). A *nopL* mutant has a similar nodulation phenotype to NGR234 on the majority of legumes tested. This is another characteristic of effector proteins of phytopathogens, for there are many of them and they are thought to be redundant in function. NopL is important for the efficient nodulation of *Flemingia congesta* (Marie et al. 2003), suggesting that it is a rhizobial "virulence factor" for this plant. Functional characterisation of NopL revealed that it can be phosphorylated by plant kinases (Bartsev et al. 2003). Furthermore, expression of *nopL* in *Nicotiana tabacum* inhibited this plant's ability to accumulate pathogenesis-related (PR)-proteins in response to pathogen attack. We thus suggest that NopL could suppress root-cell defence responses by disrupting the intracellular signalling cascades required for activation of PR-genes (Bartsev et al. 2004a).

Sequence analyses of NGR234 and USDA110 revealed the presence of genes homologous to secreted effector proteins from other T3SS-possessing pathogenic bacteria. The proteins encoded by these rhizobial genes are thus good candidates for secretion in a T3SS-dependent manner, and possibly function as effectors within legume root cells (Marie et al. 2001, Krause et al. 2002). These proteins have been studied extensively in pathogenic bacteria, where they act to suppress host defence responses, and it will be interesting to determine whether their rhizobial counterparts function in a similar manner.

The Role of Rhizobial T3SSs

At this stage, explanations for the roles of T3SSs in rhizobia can only be hypothetical. It is thought that the T3SS functions during infection thread development and perhaps while the rhizobia are being released from infection threads into cortical cells. Similarities between attempts by rhizobia to colonise roots and the attack by pathogens of other plant cells are obvious. Undoubtedly, legumes mount defences against rhizobia. NGR234 and similar rhizobia may have acquired a T3SS to suppress such defence responses, and effectors like NopL are the inter-cellular messengers in this process. Legumes, like most other plants have probably evolved sophisticated methods for detecting T3SS-containing pathogens. To some plants, NGR234 (and similar rhizobia) would reveal themselves as "pathogens" provoking defence-responses that block nodulation (e.g. *P. tuberosus*). In this scenario, other legumes would have evolved immunity to T3SS proteins (*L. japonicus*, *V. unguiculata*), while still others would positively welcome them (e.g. *T. vogeli*). Furthermore, it is possible that rather than responding in one of three ways to T3SS-proteins, a continuum of responses exists, some too slight to be detected. Other possibilities include that the T3SS is for some reason not activated by certain plants, or that some plants possess other signalling systems that over-ride or complement the T3SS.

5.2.4
Type IV Secretion Systems

Type IV secretion systems (T4SSs) were initially defined on the basis of the homologies between components of three different macromolecular complexes: the *Agrobacterium tumefaciens* T-DNA transfer system that is required for exporting oncogenic T-DNA to susceptible plant cells; the conjugal transfer (Tra) system of the conjugative IncN plasmid pKM101; and the *Bordetella pertussis* toxin exporter (Ptl) machine (Winans et al. 1996; Christie 1997). Like T2SSs, T4SSs use a stepwise process to translocate macromolecular substrates first across the inner membrane, prior to transport across the cell envelope (Christie 2001). Some symbiotic nitrogen-fixing bacteria also possess genes that could encode a T4SS e.g. *R. etli*, *M. loti* strain R7A and *R. meliloti* (Galibert et al. 2001; Sullivan et al. 2002; Gonzalez et al. 2003). It is interesting to note that in *M. loti* R7A the location of the genes that may encode a T4SS is exactly at the T3SS locus of *M. loti* MAFF303099 strain, although each strain possesses only one type of secretion machine. The role of T4SSs in symbiosis is not known, but there is a suggestion that it could affect the nodulation process, as a putative *nod*-box is located in the promoter region of one of the genes of the R7A T4SS (Sullivan et al. 2002).

5.3 Conclusions

Successful symbiotic associations between rhizobia and legumes require the exchange of many signal molecules. Both partners secrete these signals and it is the timing of their emission and perception as well as the quantity that are probably important. Symbiotic harmony depends on the precise meshing of these signals. Plant flavonoids are the first important group of signalling molecules and they act as inducers of nodulation genes (*nod*, *noe* and *nol*) (Broughton et al. 2000; Perret et al. 2000). The regulatory networks of flavonoids, the NodD family of transcriptional activators, and their conserved promoter sequences (*nod*-boxes) guarantee the timing of expression of downstream genes that are responsible for the synthesis of diffusible lipo-chito-oligosaccharidic Nod-factors - early symbiotic "master keys". Once the legume "doors" have been opened to allow rhizobia in, different morphological and cytological changes in the roots occur. Nod-factors play only secondary roles in the later steps of invasion, at which time other signal molecules occupy centre stage. Bacterial SPSs and secreted proteins contribute to the infection process, where they assist in infection thread development within the root hair, and help modify host-defence mechanisms.

Acknowledgements. We thank D. Gerber for general support and encouragement. Research in LBMPS is financed by the Founds National Suisse de la recherché Scientifique (Project 31-63893.00) and the Université de Genève.

References

Atkins CA, Ritchie A, Rowe PB, McCairns E, Sauer D (1982) De novo purine synthesis in nitrogen-fixing nodules of cowpea (*Vigna unguiculata* [L.] Walp.) and soybean (*Glycine max* [L.] Merr.). Plant Physiol 70:55–60

Bartsev AV, Boukli NM, Deakin WJ, Staehelin C, Broughton WJ (2003) Purification and phosphorylation of the effector protein NopL from *Rhizobium* sp. NGR234. FEBS Lett 554:271–274

Bartsev AV, Deakin WJ, Boukli NM, McAlvin CB, Stacey G, Malnöe P, Broughton WJ, Staehelin C (2004a) NopL, an effector protein of *Rhizobium* sp. NGR234, thwarts activation of plant defense reactions. Plant Physiol 134:871–879

Bartsev AV, Kobayashi H, Broughton WJ (2004b) Rhizobial signals convert pathogens to symbionts at the legume interface. In: Gillings M, Holmes A (eds) Plant microbiology, Bios Scientific, Oxfordshire, UK, pp 19–28

Becker A, Ruberg S, Kuster H, Roxlau AA, Keller M, Ivashina T, Cheng HP, Walker GC, Pühler A (1997) The 32-kilobase *exp* gene cluster of *Rhizobium meliloti* directing the biosynthesis of galactoglucan: genetic organization and properties of the encoded gene products. J Bacteriol 179:1375–1384

Broughton WJ, Jabbouri S, Perret X (2000) Keys to symbiotic harmony. J Bacteriol 182:5641–5652

Büttner D, Nennstiel D, Klusener B, Bonas U (2002) Functional analysis of HrpF, a putative type III translocon protein from *Xanthomonas campestris* pv. *vesicatoria*. J Bacteriol 184:2389–2398

Christie PJ (1997) *Agrobacterium tumefaciens* T-complex transport apparatus: a paradigm for a new family of multifunctional transporters in eubacteria. J Bacteriol 179:3085–3094

Christie PJ (2001) Type IV secretion: intercellular transfer of macromolecules by systems ancestrally related to conjugation machines. Mol Microbiol 40:294–305

De Maagd RA, Wijfjes AH, Spaink HP, Ruiz-Sainz JE, Wijffelman CA, Okker RJ, Lugtenberg BJ (1989) *nodO*, a new *nod* gene of the *Rhizobium leguminosarum* biovar *viciae* sym plasmid pRL1JI, encodes a secreted protein. J Bacteriol 171:6764–6770

D'Haeze W, Gao M-S, De Rycke R, Van Montagu M, Engler G, Holsters M (1998) Roles for Azorhizobial Nod factors and surface polysaccharides in intercellular invasion and nodule penetration, respectively. Mol Plant Microbe Interact 11:999–1008

Downie JA (1998) Functions of rhizobial nodulation genes. In: Spaink HP, Kondorosi A, Hooykaas JP (eds) The *Rhizobiaceae*. Kluwer, Dordrecht, pp 387–402

Dörr J, Hurek T, Reinhold-Hurek B (1998) Type IV pili are involved in plant-microbe and fungus-microbe interactions. Mol Microbiol 30:7–17

Downie JA, Surin BP (1990) Either of two *nod* gene loci can complement the nodulation defect of a *nod* deletion mutant of *Rhizobium leguminosarum* bv *viciae*. Mol Gen Genet 222:81–86

Economou A, Hamilton WD, Johnston AW, Downie JA (1990) The Rhizobium nodulation gene *nodO* encodes a Ca^{2+} binding protein that is exported without N-terminal cleavage and is homologous to haemolysin and related proteins. EMBO J 9:349–354

Economou A, Davies AE, Johannes E, Downie JA (1994) The *Rhizobium leguminosarum* biovar *vicia nodO* gene can enable a *nodE* mutant of *Rhizobium leguminosarum* biover *trifolii* to nodulate vetch. Microbiology 140:2341–2347

Feldman MF, Cornelis GR (2003) The multitalented type III chaperones: all you can do with 15 kDa. FEMS Microbiol Lett 219:151–158

Finnie C, Zorreguieta A, Hartley NM, Downie JA (1998) Characterization of *Rhizobium leguminosarum* exopolysaccharide glycanases that are secreted via a type I exporter and have a novel heptapeptide repeat motif. J Bacteriol 180:1691–1699

Fisher RF, Long SR (1992) *Rhizobium*-plant signal exchange. Nature 357:655–660

Franssen HJ, Vijn I, Yang WC, Bisseling T (1992) Developmental aspects of the *Rhizobium*-legume symbiosis. Plant Mol Biol 19:89–107

Freiberg C, Fellay R, Bairoch A, Broughton WJ, Rosenthal A, Perret X (1997) Molecular basis of symbiosis between *Rhizobium* and legumes. Nature 387:394–401

Galibert F, Finan TM, Long SR, Puhler A, Abola P, Ampe F, Barloy-Hubler F, Barnett MJ, Becker A, Boistard P, Bothe G, Boutry M, Bowser L, Buhrmester J, Cadieu E, Capela D, Chain P, Cowie A, Davis RW, Dreano S, Federspiel NA, Fisher RF, Gloux S, Godrie T, Goffeau A, Golding B, Gouzy J, Gurjal M, Hernandez-Lucas I, Hong A, Huizar L, Hyman RW, Jones T, Kahn D, Kahn ML, Kalman S, Keating DH, Kiss E, Komp C, Lelaure V, Masuy D, Palm C, Peck MC, Pohl TM, Portetelle D, Purnelle B, Ramsperger U, Surzycki R,

Thebault P, Vandenbol M, Vorholter FJ, Weidner S, Wells DH, Wong K, Yeh KC, Batut J (2001) The composite genome of the legume symbiont *Sinorhizobium meliloti*. Science 293:668–672

Geelen D, Goethals K, Van Montagu M, Holsters M (1995) The *nodD* locus from *Azorhizobium caulinodans* is flanked by two repetitive elements. Gene 164:107–111

Gonzalez V, Bustos P, Ramirez-Romero MA, Medrano-Soto A, Salgado H, Hernandez-Gonzalez I, Hernandez-Celis JC, Quintero V, Moreno-Hagelsieb G, Girard L, Rodriguez O, Flores M, Cevallos MA, Collado-Vides J, Romero D, Davila G (2003) The mosaic structure of the symbiotic plasmid of *Rhizobium etil* CFN42 and its relation to other symbiotic genome compartments. Genome Biol 4:R36

Göttfert M, Röthlisberger S, Kündig C, Beck C, Marty R, Hennecke H (2001) Potential symbiosis-specific genes uncovered by sequencing a 410-kilobase DNA region of the *Bradyrhizobium japonicum* chromosome. J Bacteriol 183:1405–1412

Hanin M, Jabbouri S, Broughton WJ, Fellay R, Quesada-Vincens D (1999) Molecular aspects of host-specific nodulation. In: Stacey G, Keen NT (eds) Plant-microbe interaction. American Phytopathological Society, St Paul, MN, pp 1–37

He SY (1998) Type III protein secretion systems in bacterial pathogenic bacteria. Annu Rev Phytopathol 36:363–392

He SY, Jin Q (2003) The Hrp pilus: learning from flagella. Curr Opin Microbiol 6:15–19

Hueck CJ (1998) Type III protein secretion systems in bacterial pathogens of animals and plants. Microbiol Mol Biol Rev 62:379–433

Huguet E, Bonas U (1997) *hrpF* of *Xanthomonas campestris* pv. *vesicatoria* encodes an 87-kDa protein with homology to NoIX of *Rhizobium fredii*. Mol Plant Microbe Interact 10:488–498

Jabbouri S, Fellay R, Talmont F, Kamalaprija P, Burger U, Relić B, Prome JC, Broughton WJ (1995) Involvement of *nodS* in N-methylation and *nodU* in 6-O-carbamoylation of *Rhizobium* sp. NGR234 Nod factors. J Biol Chem 270:22968–22973

Kaneko T, Nakamura Y, Sato S, Asamizu E, Kato T, Sasamoto S, Watanabe A, Idesawa K, Ishikawa A, Kawashima K, Kimura T, Kishida Y, Kiyokawa C, Kohara M, Matsumoto M, Matsuno A, Mochizuki Y, Nakayama S, Nakazaki N, Shimpo S, Sugimoto M, Takeuchi C, Yamada M, Tabata S (2000) Complete genome structure of the nitrogen-fixing symbiotic bacterium *Mesorhizobium loti*. DNA Res 7:331–338

Kaneko T, Nakamura Y, Sato S, Minamisawa K, Uchiumi T, Sasamoto S, Watanabe A, Idesawa K, Iriguchi M, Kawashima K, Kohara M, Matsumoto M, Shimpo S, Tsuruoka H, Wada T, Yamada M, Tabata S (2002) Complete genomic sequence of nitrogen-fixing symbiotic bacterium *Bradyrhizobium japonicum* USDA110. DNA Res 9:189–197

Kobayashi H, Naciri-Graven Y, Broughton W, Perret X (2004) Flavonoids induce temporal shifts in gene expression of *nod*-box controlled loci in *Rhizobium* sp. NGR234. Mol Microbiol 51:335–347

Krause A, Doerfel A, Göttfert M (2002) Mutational and transcriptional analysis of the type III secretion system of *Bradyrhizobium japonicum*. Mol Plant Microbe Interact 15:1228–1235

Krishnan HB (2002) NolX of *Sinorhizobium fredii* USDA257, a type III-secreted protein involved in host range determination, Is localized in the infection threads of cowpea (*Vigna unguiculata* [L.] Walp) and soybean (*Glycine max* [L.] Merr.) nodules. J Bacteriol 184:831–839

Krishnan HB, Pueppke SG (1993) Flavonoid inducers of nodulation genes stimulate *Rhizobium fredii* USDA257 to export proteins into the environment. Mol Plant Microbe Interact 6:107–113

Krishnan HB, Kuo C-I, Pueppke SG (1995) Elaboration of flavonoid-induced proteins by the nitrogen-fixing soybean symbiont *Rhizobium friedii* is regulated by both *nodD1* and *nodD2*, and is dependent on the cultivar-specificity locus, *nolXWBTUV*. Microbiology 141:2245–2251

Krishnan HB, Lorio J, Kim WS, Jiang G, Kim KY, DeBoer M, Pueppke SG (2003) Extracellular proteins involved in soybean cultivar-specific nodulation are associated with pilus-like surface appendages and exported by a type III protein secretion system in *Sinorhizobium fredii* USDA257. Mol Plant Microbe Interact 16:617–625

Marie C, Broughton WJ, Deakin WJ (2001) Rhizobium type III secretion systems: legume charmers or alarmers? Curr Opin Plant Biol 4:336–342

Marie C, Deakin WJ, Viprey V, Kopcinska J, Golinowski W, Krishnan HB, Perret X, Broughton WJ (2003) Characterization of Nops, nodulation outer proteins, secreted via the type III secretion system of NGR234. Mol Plant Microbe Interact 16:743–751

Moreira LM, Becker JD, Puhler A, Becker A (2000) The *Sinorhizobium meliloti* ExpE1 protein secreted by a type I secretion system involving ExpD1 and ExpD2 is required for biosynthesis or secretion of the exopolysaccharide galactoglucan. Microbiology 146:2237–2248

Peabody CR, Chung YJ, Yen MR, Vidal-Ingigliardi D, Pugsley AP, Saier MH (2003) Type II protein secretion and its relationship to bacterial type IV pili and archaeal flagella. Microbiology 149:3051–3072

Perret X, Freiberg C, Rosenthal A, Broughton WJ, Fellay R (1999) High-resolution transcriptional analysis of the symbiotic plasmid of *Rhizobium* sp. NGR234. Mol Microbiol 32:415–425

Perret X, Staehelin C, Broughton WJ (2000) Molecular basis of symbiotic promiscuity. Microbiol Mol Biol Rev 64:180–201

Pueppke SG, Broughton WJ (1999) *Rhizobium* sp. strain NGR234 and *R. fredii* USDA257 share exceptionally broad, nested host ranges. Mol Plant Microbe Interact 12:293–318

Pugsley AP (1993) The complete general secretory pathway in Gram-negative bacteria. Microbiol Rev 57:50–108

Relić B, Talmont F, Kopcinska J, Golinowski W, Prome JC, Broughton WJ. (1993) Biological activity of *Rhizobium* sp. NGR234 Nod-factors on *Macroptilium atropurpureum*. Mol Plant Microbe Interact 6:764–774

Relić B, Perret X, Estrada-Garcia MT, Kopcinska J, Golinowski W, Krishnan HB, Pueppke SG, Broughton WJ (1994) Nod factors of *Rhizobium* are a key to the legume door. Mol Microbiol 13:171–178

Rossier O, Van den Ackerveken G, Bonas U (2000) HrpB2 and HrpF from *Xanthomonas* are type III-secreted proteins and essential for pathogenicity and recognition by the host plant. Mol Microbiol 38:828–838

Sandkvist M (2001) Biology of type II secretion. Mol Microbiol 40:271–283

Spaink HP, Sheeley DM, van Brussel AA, Glushka J, York WS, Tak T, Geiger O, Kennedy EP, Reinhold VN, Lugtenberg BJ (1991) A novel highly unsaturated fatty acid moiety of lipo-oligosaccharide signals determines host specificity of *Rhizobium* Nature 354:125–130

Stacey G, So JS, Roth LE, Bhagya Lakshmi SK, Carlson RW (1991) A lipopolysaccharide mutant of *Bradyrhizobium japonicum* that uncouples plant from bacterial differentiation. Mol Plant Microbe Interact 4:332–340

Streit WR, Schmitz RA, Perret X, Staehelin C, Deakin WJ, Raasch C, Liesegang H, Broughton WJ (2004) An evolutionary hot spot: the pNGR234b replicon of *Rhizobium* sp. Strain NGR234. J Bacteriol 186:535–542

Sullivan JT, Trzebiatowski JR, Cruickshank RW, Gouzy J, Brown ST, Elliot RM, Fleetwood DJ, McCallum NG, Rossbach U, Stuart GS, Weaver JE, Webby RJ, de Bruijn FJ, Ronson cw (2002) Comparative sequence analysis of the symbiosis island of *Mesorhizobium loti* StrainR7A. J Bacteriol 184:3086–3095

Sutton JM, Lea EJ, Downie JA (1994) The nodulation-signalling protein NodO from *Rhizobium leguminosarum* biovar *viciae* forms ion channels in membranes. Proc Natl Acad Sci USA 91:9990–9994

Sutton JM, Peart J, Dean G, Downie JA (1996) Analysis of the C-terminal secretion signal of the *Rhizobium leguminosarum* nodulation protein NodO; a potential system for the secretion of heterologous proteins during nodule invasion. Mol Plant Microbe Interact 9:671–680

Thanassi DG, Hultgren SJ (2000) Multiple pathways allow protein secretion across the bacterial outer membrane. Curr Opin Cell Biol 12:420–430

Trinick MJ (1980) Relationships amongst the fast-growing rhizobia of *Lablab purpureus*, *Leucaena leucocephala*, Mimosa spp., *Acacia farnesiana* and *Sesbania grandiflora* and their affinities with other rhizobial groups. J Appl Bacteriol 49:39–53

Van Rhijn P, Luyten E, Vlassak K, Vanderleyden J (1996) Isolation and characterization of a *pSym* locus of Rhizobium sp. BR816 that extends nodulation ability of narrow host range *Phaseolus vulgaris* symbionts to *Leucaena leucocephala* Mol Plant Microbe Interact 9:74–77

Van Spronsen PC, Bakhuizen R, van Brussel AA, Kijne JW (1994) Cell wall degradation during infection thread formation by the root nodule bacterium *Rhizobium leguminosarum* is a two-step process. Eur J Cell Biol 64:88–94

Viprey V, Del Greco A, Golinowski W, Broughton WJ, Perret X (1998) Symbiotic implications of type III protein secretion machinery in *Rhizobium*. Mol Microbiol 28:1381–1389

Vlassak KM, de Wilde P, Snoeck C, Luyten E, van Rhijn P, Vanderleyden J (1998) The *Rhizobium* sp. BR816 *nodD3* gene is regulated by a transcriptional regulator of the AraC/XylS family. Mol Gen Genet 258:558–561

Walker SA, Downie JA (2000) Entry of *Rhizobium leguminosarum* bv. *viciae* into root hairs requires minimal Nod factor specificity, but subsequent infection thread growth requires *nodO* or *nodE* Mol Plant Microbe Interact 13:754–762

Winans SC, Burns DL, Christie PJ (1996) Adaptation of a conjugal transfer system for the export of pathogenic macromolecules. Trends Microbiol 4:64–68

Wolfgang M, van Putten JP, Hayes SF, Dorward D, Koomey M (2000) Components and dynamics of fibres formation define a ubiquitous biogenesis pathway for bacterial pili. EMBO J 19:6408–6418

6 Research on Endophytic Bacteria: Recent Advances with Forest Trees

Richa Anand, Leslie Paul, Chris Chanway

6.1 Introduction

Plants can be considered as complex microecosystems that provide different habitats to a variety of microorganisms. These habitats are represented by the plant external surfaces as well as internal tissues (McInroy and Kloepper 1994). Whereas the importance of microbial colonisation of plant surfaces in plant growth promotion has been well understood for a long time, interior tissue colonisation was, until recently, largely perceived as being related only to the perpetuation of systemic diseases. It is now well known that tissues of healthy plants are also colonised internally by various microorganisms that establish neutral or, more interestingly, beneficial interactions with their host plants. The term "endophyte" is commonly used to describe such microorganisms.

Although a variety of definitions have been applied to the term "endophyte", it refers mainly to bacteria and fungi that live inside plant tissues without causing disease (Wilson 1995; see Chap. 1 by Schulz and Boyle). Whether or not latent pathogens can be considered endophytes has been a major topic of debate in the general acceptance of this definition (Misaghi and Donndelinger 1990; James and Olivares 1997; see Chap. 1 by Schulz and Boyle).

The best-characterised microbial endophytes are fungi of the Balansiaceae, for which the most compelling evidence of plant–microbe mutualism has been provided (Clay 1988; Schardl et al. 2004). Some of the non-balansiaceous endophytic fungi are also mutualistic with their hosts (Carroll 1988; see Chap. 15 by Schulz), and produce compounds that render plant tissues less attractive to herbivores, while other strains may increase

Chris Chanway: Faculty of Land and Food Systems, Faculty of Forestry, University of British Columbia, Vancouver, British Columbia, Canada V6T 1Z4, E-mail: cchanway@interchg.ubc.ca

Richa Anand, Leslie Paul: Faculty of Land and Food Sciences, Systems, University of British Columbia, Vancouver, British Columbia, Canada V6T 1Z4

Leslie Paul: Department of Forest Mycology and Pathology, Swedish University of Agricultural Sciences (SLU), Box 7026, 750 07 Uppsala, Sweden (Current Address)

Soil Biology, Volume 9
Microbial Root Endophytes
B. Schulz, C. Boyle, T. N. Sieber (Eds.)

host plant drought resistance. In return, fungal endophytes are thought to benefit from the comparatively nutrient rich, buffered environment inside plants.

Apart from fungi, bacteria belonging to various genera have also been shown to exist inside plants without causing apparent disease symptoms. Some of these bacteria are known to impart benefits to their host plants by the same mechanisms as their soil- or rhizosphere-colonising counterparts. The primary mechanisms thought to lead to beneficial effects for the plant are nitrogen fixation (Boddey and Döbereiner 1995) and biocontrol of pathogenic and detrimental microorganisms, either through direct antagonism of pathogens or by inducing systemic resistance to such organisms (Hallman et al. 1997). Other known mechanisms by which beneficial bacteria can have a positive influence on plant performance are the production or stimulation of plant growth hormones and facilitation of nutrient uptake [see Chaps. 3 (Kloepper and Ryu) and 4 (Berg and Hallmann)].

Since their first reported isolation from potato plants (Tervet and Hollis 1948; Hollis 1951), all the information available on endophytic bacteria has been derived from studies on plant species of agricultural and horticultural importance. The endophytic bacteria of rice (Reinhold-Hurek and Hurek 1998), corn (Triplett 1996) and sugarcane (Döbereiner et al. 1995) are by far the best studied so far. In contrast to these crop species, very much less is known about bacterial endophytes of trees. Some trees survive and grow well in very difficult terrain under extreme conditions, for example lodgepole pine (*Pinus contorta* Dougl. var. *latifolia*) in dry interior regions of British Columbia and western Alberta, Canada, as well as *Tecomella undulata* (Bignoniaceae) in the extremely arid deserts of northwestern India. It is possible that endophytic bacteria that enhance host survival and growth in exchange for protection in the relatively buffered environment of internal plant tissues may be involved under such extreme environmental conditions (Law and Lewis 1983).

Although the realisation of this possibility has led to occasional reports of endophytic bacteria in asymptomatic angiosperm and gymnosperm tree species, little is known about their diversity and influence on plant growth. The earliest report of bacterial endophytes from trees was from Gardner et al. (1982), who isolated representatives of 13 genera from xylem fluid of rough lemon rootstock, and found population sizes ranging from 10^2–10^4 colony forming units (cfu) g^{-1} xylem fluid. Only 48 of the 850 isolates turned out to be phytopathogenic, but the role of the other 802 isolates was not determined. Similarly, several strains of *Pseudomonas syringae* were isolated and characterized from inside pear seedlings by Whitesides and Spotts (1991), but their exact role could not be determined.

The procedures of isolation and identification of endophytic bacteria from trees is the same as their isolation from agronomic crops, and suffers from the same limitations observed in crop species, e.g. the difficulty, or perhaps impossibility, of absolute surface sterilisation of external plant tissues as well as our inability to culture many bacteria we know to exist. The impact of these problems can be reduced by the use of standardised protocols and molecular techniques (James 2000; see Chap. 17 by Hallmann et al.). The major difficulty, therefore, lies in the evaluation of the effects of these bacteria on their host trees, owing to the long tree life-cycle and a lack of detailed physiological information on trees, particularly forest trees.

The following sections provide a detailed review of our current knowledge about bacterial endophytes of forest trees, the mechanisms by which they benefit their host plants, and their potential application in the practice of sustainable forestry.

6.2 Bacterial Endophytes of Forest Trees

Although limited, the results of research on endophytic bacteria and their role in growth promotion of forest trees so far are very encouraging and will hopefully draw more attention to this developing area of study. Brooks et al. (1994) conducted an extensive study in which endophytic bacteria were isolated from surviving live oak (*Quercus fusiformis*) in Texas, where oak wilt is epidemic, and evaluated as potential biological control agents for the disease. Of the 889 bacterial isolates tested, 183 showed in vitro inhibition of the pathogen *Ceratocystis fagacearum*. Six isolates were further evaluated for colonisation of containerised Spanish oak (*Quercus texana*) and live oak. Interestingly, in containerised live oaks inoculated with the oak wilt pathogen, preinoculation with 15 isolates of *Pseudomonas denitrificans* reduced the number of diseased trees by 50% and decreased the percentage of crown loss by 17%. In a subsequent trial, no reduction in numbers of diseased trees was observed but preinoculation with the same isolates of *P. denitrificans* or a strain of *Pseudomonas putida* significantly reduced crown loss. These results clearly established the potential of such endophytic bacteria as pre-plantation nursery treatments for wilt control.

Several endophytic aerobic heterotrophic bacteria belonging to the genera *Bacillus*, *Curtobacterium*, *Pseudomonas*, *Stenotrophomonas*, *Sphingomonas*, *Enterobacter*, and *Staphylococcus*, have also been isolated from phloem tissue of roots and branches of elm trees (*Ulmus* spp.: Mocali et al. 2003). An attempt was also made to determine the correlation between the seasonal fluctuations in the structure of the endophytic bacterial community and phytoplasma disease infection of these trees; however, no consequential effect of the bacterial community on phytoplasmosis of elm trees could be demonstrated (Mengoni et al. 2003).

Apart from these studies on bacterial endophytes of deciduous trees, most other reports of endophytic bacteria and their role in tree growth promotion are from studies on various conifer tree species conducted mostly by our research group. Our interest in endophytic bacteria of conifers has been largely inspired by the immense commercial, social, and environmental importance of forestry in Canada and the rest of North America and the fact that conifers are the most dominant trees in the temperate forests of this region.

6.3 Endophytic Bacteria of Conifers

Conifers are members of the plant division Coniferophyta (2 Domain classification), which are characterised by naked seeds borne in specialised sporophylls or cones. Their vascular tissues differ from angiosperms in not having xylem vessels, and companion cells in phloem. The division is comprised of 550 species spread over seven families, each dating back to the Mesozoic era. Distributed throughout the world with extensive latitudinal and longitudinal ranges, conifers are of great commercial and ecological value.

Traditionally, fungi, particularly mycorrhizae, were considered to be the only microorganisms that could exert a positive influence on the growth and survival of forest trees. The continuity of this trend until now is evident from the results of keyword searches for "endophytic bacteria + conifers" in all well known scientific databases.

Although some confirmed reports of conifer tree growth promotion by naturally occurring soil and rhizosphere bacteria were available (Pokojska-Burdziej 1982; Chanway and Holl 1992, 1993, 1994; O'Neill et al. 1992), the mechanisms employed by these bacteria for growth promotion could not be determined. It was generally believed that the primary mechanism of plant growth promotion by these bacteria was only indirect, i.e. by facilitating the establishment and growth of mycorrhizae (Fitter and Garbaye 1994). The focus of research on endophytic microflora of conifers thus remained on fungi, even after the importance of endophytic bacteria had been well established in agronomic crop species.

In an initial study of conifer root-associated bacteria, O'Neill et al. (1992) isolated 22 strains from surface-sterilised roots of naturally regenerating white x Engelmann (*Picea glauca* x *P. engelmannii*) hybrid spruce seedlings. A range of effects on seedling growth in a greenhouse-screening assay using spruce were found: three strains were inhibitory, five strains were stimulatory and the remaining strains had no significant effect on seedling growth (O'Neill et al. 1992). Based on the magnitude and consistency of seedling growth effects, the two best plant growth-promoting endophytes were identified and selected for further study: one isolate was *Pseudomonas putida* and the other belonged to *Staphylococcus*. While the positive effect of both of these strains on plant growth was reproducible in the greenhouse, a field trial with two ecotypes of 1-year-old spruce seedlings planted at three different reforestation sites yielded mixed results (Chanway and Holl 1993). For example, *P. putida* enhanced seedling growth of only one of two spruce ecotypes planted at two of three reforestation sites. In addition, it had inhibitory effects in three of the spruce ecotype x planting site treatment combinations.

Evaluation of gymnosperm bacterial endophytes was only a small part of a larger project designed to characterise gymnosperm root-associated bacterial (i.e. external and internal) colonists (O'Neill et al. 1992; Chanway and Holl 1992, 1994). Therefore, our group undertook a subsequent bacterial isolation and screening program emphasising endophytic bacteria as possible tree seedling growth-promoting agents (Chanway et al. 1994, 1997). As seen in our earlier work (O'Neill et al. 1992), several bacterial strains isolated from surface-sterilised roots of white x Engelmann hybrid spruce seedlings caused reproducible spruce seedling biomass increases of up to 36% 2 months after seed was sown and inoculated in greenhouse trials (Chanway et al. 1994). Three of these strains belonged to *Paenibacillus*, three were actinomycetes, most likely *Streptomyces* spp., and one was a *Phyllobacterium*. An additional strain that performed well in greenhouse assays could not be identified with certainty.

In addition, the seedling growth promotion efficacy of some of these strains was altered significantly when assays were conducted in the presence of a small amount (2% v/v) of forest soil known to contain seedling growth inhibiting organisms (i.e. minor pathogens). One of the endophytic actinomycetes (isolate W2) as well as the *Phyllobacterium* isolate (W3) clearly stimulated spruce seedling growth only in the absence of forest soil. In its presence, seedling growth was inhibited, as it was when forest soil alone was used. These results suggested that growth promotion by W2 and W3 occurred via a mechanism unrelated to biocontrol of minor pathogens, and may have involved one of the direct plant growth promotion mechanisms, possibly production of plant growth regulators (Kloepper 1993; Glick 1995; Chanway 1997). Interestingly, actinomycete isolate N1 and *Bacillus* isolate N4 stimulated seedling growth only in the presence of forest soil, which suggested that these strains acted through a biocontrol mechanism, either by direct antagonism or by inducing systemic resistance in the host plant. Elucidation of these possibilities requires further experimentation.

We have also looked for bacterial endophytes in lodgepole pine. After isolation of several bacterial strains and screening trials for effects on seedling growth, we identified a plant-growth-promoting *Bacillus polymyxa* (now *Paenibacillus*, Ash et al. 1993) strain (Pw2) that originated from internal root tissues of a naturally regenerating 2- to 3-year-old pine seedling (Shishido et al. 1995). Our studies indicate that Pw2 can colonise external and internal pine and spruce root tissues after seed or root inoculation. Colonisation of internal root tissues may depend on lateral root development, and results in endophytic bacterial population sizes approaching 10^6 cfu g^{-1} fresh root tissue (Shishido et al. 1995; Chanway 1997; Shishido 1997). In addition, using a surface-sterilisation, dilution plating assay as well as immunofluorescence microscopy, a rifamycin-resistant derivative of this strain, Pw2-R, was shown to be capable of colonising internal pine and white x Engelmann hybrid spruce stem tissues after soil or root inoculation (Chanway et al. 2000). Five months after root inoculation, internal stem bacterial populations reached 10^5 cfu g^{-1} fresh stem tissue (Shishido 1997).

In order to examine the effects of endophytes on conifer plant growth and to investigate the host specificity of bacterial endophytes in terms of the ability to promote growth of inoculated host plants other than the ones from which they were initially isolated, initial field trials with *P. polymyxa* strain Pw2-R and *Pseudomonas chloroaphis* strain Sm3-RN, another bacterial endophyte capable of stimulating white x Engelmann hybrid seedling growth in the greenhouse (Chanway et al. 1997), were also performed. Two years after bacterial inoculation and planting at nine sites in British Columbia and Alberta, Canada, white x Engelmann hybrid spruce treated with strain Pw2-R (initially isolated from pine) showed mean biomass increases up to 33% above controls at seven of the nine sites, but increases were significant at only one site. In contrast, *Pseudomonas* strain Sm3-RN (isolated from white x Engelmann hybrid spruce) caused significant white x Engelmann hybrid spruce biomass increases of up to 57% at three of the nine sites but a significant decrease in spruce biomass at one site. Site productivity was not correlated with plant growth promotion or inhibition.

Population sizes of Pw2-R and Sm3-RN were generally below the assay detection limit of ca. 10^2 cfu g^{-1} plant tissue, which led us to question how effectively internal plant tissues were colonised at the onset of the experiment. Therefore, we also evaluated seedlings that were inoculated with strains Pw2-R and Sm3-RN and grown in the greenhouse for 4 months before planting at four of the reforestation sites described above (Shishido and Chanway 2000). The period of growth in the greenhouse facilitated internal tissue colonisation by these microorganisms so that mean internal root populations reached ca. 10^3–10^4 cfu g^{-1} tissue in the greenhouse. As expected, mean seedling biomass also increased due to bacterial inoculation in the greenhouse. Because seedling growth responses in the field would be inseparable from those that occurred in the greenhouse, simple measurement of biomass accumulation after a period of growth in the field would yield spurious results. We evaluated plant growth using relative growth rates (RGRs), in which plant growth increments over time are expressed as a proportion of the biomass that existed at some previous time in the plant's life (Hunt 1982).

In general, after the first growing season, RGRs of seedlings containing endophytic bacteria were greater than those of control seedlings at all four planting sites (Shishido and Chanway 2000). In some cases, RGRs of inoculated plants were double the control value. This was particularly interesting in view of results with seedlings that we inoculated and planted immediately at the same sites. At two of the four sites, seedlings inoculated at the time of planting (i.e., with no greenhouse growth period) did not respond to bacterial treatment, and in one case responded negatively. However, shoot and root RGRs of seedlings pretreated in the greenhouse before planting at the same sites were 23–132% greater than controls, and endophytic populations in root tissues of between 10^2 and 4×10^4 cfu g^{-1} plant tissue were detected in seedlings at three of the four sites. Similar effects on establishment and functioning of bacterial endophytes were observed by Brooks et al. (1994) in wilt-infested oak trees. These results suggest that a period of growth under a controlled environment to facilitate establishment of endophytic bacterial populations may be an important step in successful application of plant-growth-promoting bacterial endophytes in forestry.

It has also been demonstrated that the benefits of pre-outplanting inoculation of seedlings with bacterial endophytes can be maximised by careful matching of the inoculant bacterial strain with outplanting sites (Chanway et al. 2000). However, much research into site quality and plant growth responses will be required before reliable recommendations can be made.

In addition, much more research is warranted to answer the many questions regarding the entry and operation of endophytic bacteria in conifers, some of which are being actively pursued by our group.

6.4 Modes and Sites of Entry

Endophytic bacteria have been shown to be able to gain entry in plants through wounded as well as intact tissues (Sprent and James 1995; see Chap. 18 by Bloemberg et al.). In an attempt to understand the modes and sites of entry of endophytic bacteria, Timmusk and Wagner (1999) followed the colonisation of a green fluorescent protein (*gfp*)-tagged endophytic strain of *Paenibacillus polymyxa* in *Arabidopsis thaliana*; they observed a slight degradation of the root tips within 5 h of inoculation. It was found that *P. polymyxa* had two preferred zones of infection. The first is located at the root tip in the zone of elongation, which sometimes results in the loss of the root cap. The other colonisation region was observed in the differentiation zone. Similar colonisation zones have been reported for other endophytes, e.g. *Azoarcus* by Hurek et al. (1994), who suggested that plant cells were destroyed after bacteria had penetrated cell walls. Perhaps this is the reason why most endophytic bacteria are limited to the intercellular spaces inside tissues. However it is not clear how they are stopped from entering cells and causing necrosis.

To determine which microbial characteristic(s) facilitate entrance of bacterial endophytes into plant tissues, we compared the biochemical capabilities of the endophytic *Paenibacillus polymyxa* strain Pw2 with those of another plant-growth promoting, non-endophytic strain, *P. polymyxa* L6-16R. Interestingly, strain L6-16R is unable to enter plant tissues even when co-inoculated with an endophytic microorganism (Shishido et al. 1995; Bent and Chanway 1997). According to Biolog analysis, both strains possessed similar metabolic capabilities with some potentially important exceptions (Shishido et al. 1995). For example, strain Pw2-R was able to metabolise sorbitol, but strain L6-16R was not. Mavingui et al. (1992) found that, in general, *P. polymyxa* strains isolated from the rhizoplane of wheat (*Triticum aestivum* L.) were capable of metabolising sorbitol while rhizosphere and non-rhizosphere isolates were not. They hypothesised that intense competition for oxygen would occur on the root surface due to root respiration, which would result in selection pressure for bacteria capable of anaerobic growth on highly reduced, scarce substrates, such as sorbitol.

In addition, strain Pw2 was able to metabolise D-melezitose, a sugar that has been detected in the sap of conifers (Lehninger 1975). However, the occurrence of sorbitol and D-melezitose in lodgepole pine root tissues and their utilisation by other *Paenibacillus* root endophytes must be demonstrated before a role for these substrates in internal root colonisation by *Paenibacillus* can be postulated with greater confidence.

To facilitate root colonisation, it is logical to suspect that root endophytic bacteria may also possess the ability to metabolise structural components of plant cells. In particular, the ability to metabolise pectin (polygalacturonic acids), a major component of the middle lamellae of plant cell walls, has been proposed to at least partly explain why bacterial root endophytes are often found in the root cortex intercellularly (Balandreau and Knowles 1978; Baldani and Döbereiner 1980). Both strains L6 and Pw2 possessed pectolytic activity in vitro, but only strain Pw2 was able to metabolise D-galacturonic acid (Shishido et al. 1995), the primary monomeric component of pectin (Paul and Clark 1989).

It is not clear whether the capability of strain Pw2 to metabolise monomeric galacturonic acid after breakdown of the pectin polymer was related to its ability to enter root tissues. However, breakdown products of plant cell walls are known to induce systemic disease responses in plants (Brock et al. 1994), which leads to the possibility that Pw2 avoids plant defence mechanisms by metabolising cell wall components before they elicit a defence response by the host plant. This possibility also requires further investigation.

If, in fact, the entry of Pw2 in plant roots is facilitated by its capability to metabolise the primary components of the cell wall, the question as to why it does not cause necrosis of interior tissues, also needs to be answered.

6.5 Mechanisms of Plant Growth Promotion

Unlike symbiotic rhizobia, mechanisms of plant growth promotion by plant growth-promoting rhizobacteria (PGPR) vary greatly, and have been broadly categorised into two groups, direct and indirect (Kloepper et al. 1989; see Chap. 3 by Kloepper and Ryu). Direct plant growth mechanisms may involve nitrogen fixation (Cavalcante and Döbereiner 1988), production of plant growth regulators and antibiotics, or increased availability of plant growth-limiting nutrients. Indirect mechanisms may involve suppression of deleterious microorganisms as well as enhancement of mutualisms between host plants and other symbionts such as mycorrhizae (Kloepper et al. 1989). Similar to other aspects of studies on endophytic bacteria, there is a great deal of information on the mechanisms of plant growth promotion employed by these bacteria in agronomic crops (Lodewyckx et al. 2002).

In the case of conifers, it was generally believed that these plants could derive benefits from bacteria only indirectly through their mycorrhizal symbionts (Fitter and Garbaye 1994). However, growth studies on lodgepole pine seedlings (Chanway and Holl 1991; Shishido et al. 1996) and hybrid spruce (*Picea glauca* x *P. engelmannii*) (Shishido et al. 1996) co-inoculated with PGPR and mycorrhizal fungi have clearly shown that growth promotion of these conifers by PGPR is independent of the mycorrhizal status of the seedlings.

Despite many efforts, determination of the exact mechanisms of conifer growth promotion by PGPR has not been possible. *Paenibacillus polymyxa* strain L6-16R was shown to produce cytokinins (Holl et al. 1988), and this property was advanced as a likely explanation of pine growth promotion mediated by this strain.

A detailed study was also conducted to determine the mechanisms of growth promotion of spruce by six *Paenibacillus* and *Pseudomonas* strains, including the endophyte, *B. polymyxa* Pw2 (Shishido 1999). It could only be concluded that more than one mechanism was responsible for growth promotion. Production of plant growth promoters and enhancement of nutrient uptake were designated as the most likely of these mechanisms.

Strain Pw2 isolated from lodgepole pine (Shishido 1996) possessed diazotrophic properties. This led to the intriguing possibility that lodgepole pine harbours a systemically endophytic nitrogen-fixing bacterial population, similar to that found in sugar cane (Boddey et al. 1995), which would explain its ability to grow, and even thrive, under nitrogen-deficient conditions in the absence of significant rhizospheric nitrogen fixation (Binkley 1995). Indeed, the $^{15}N/^{14}N$ ratio of pine foliage in a central coast forest in British Columbia devoid of nitrogen-fixing species was observed to be low enough to suggest that biological nitrogen fixation supplies plant N (F.B. Holl, personal communication).

However, nitrogen fixation could not be shown to be the primary mechanism of growth promotion by *P. polymyxa* strain Pw2-R, since seedlings inoculated with it failed to support sufficient rhizosphere acetylene reduction activity (ARA) even after 48 h of incubation with acetylene (Shishido 1997). Interestingly, similar limitations were encountered by Rhodes-Roberts (1981) and Achouak et al. (1999) while working with other strains of *P. polymyxa*. However, they were able to measure the nitrogen gains of seedlings by microkjeldahl analysis, which led them to suggest that the acetylene reduction assay is not always able to provide positive results for the nitrogen-fixing ability of *P. polymyxa*. Therefore, conclusions on the occurrence of N_2 fixation in vivo should be drawn from a number of lines of evidence, including a positive nitrogenase activity test (acetylene reduction assay), ^{15}N dilution and detection of conserved *nif* genes in the purported diazotrophic endophyte.

We have also observed nitrogen-fixing bacteria inside what can only be described as a unique and enigmatic type of mycorrhizae on lodgepole pine, first described by Zak (1971) on Douglas-fir (*Pseudotsuga menziesii*) roots. These mycorrhizal structures, often referred to as tuberculate mycorrhizae, look more like leguminous root nodules than mycorrhizae (Fig. 6.1). They are fully enclosed subterranean "nodules" or tubercles attached to the tree root system, with hundreds more typical mycorrhizal root tips crowded inside the outer covering, or peridium (Fig. 6.2). Nitrogen-fixing bacteria have been previously detected on the peridium (Li et al. 1992), but more recently, in our laboratory, a limited number of strains representing four diazotrophic bacterial species have been detected inside the peridium, colonising the fungal hyphae within the tubercle (unpublished data). It is yet to be demonstrated that these endophytic diazotrophs fix N_2 in situ, let alone transfer it to the host plant, but these intriguing possibilities remain to be evaluated.

Fig. 6.1. External morphology of tuberculate ectomycorrhizae on *Pinus contorta* roots. *Bar* 5 mm

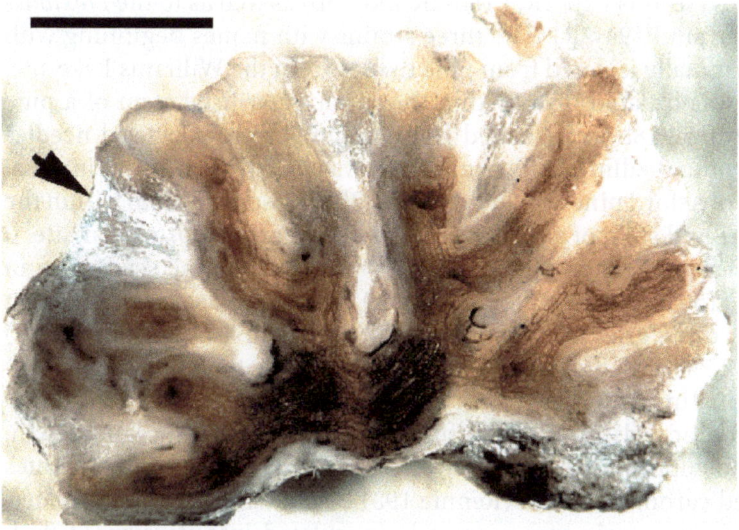

Fig. 6.2. Cross section through a mature tubercle from *P. contorta* revealing mycorrhizal root tips (*brown*) and interstitial hyphae (*arrow*). Note pinnate radiated fan form and dichotomous branching of root tips within the tubercle. *Bar* 2 mm

Recently, we tested the hypothesis that diazotrophic endophytes isolated from internal tissues of immature and mature, naturally regenerated, lodgepole pine produce biologically significant amounts of N through

N-fixation under controlled environmental conditions. Entire lodgepole pine seedlings, as well as root, stem, and needle samples from trees were collected from 40- to 140-year-old stands near Williams Lake, British Columbia (52°05′N, 122°54′W, elevation 1,300 m) and Chilliwack Lake, British Columbia (49°10′N, 121°57′W, elevation 600 m). In addition, western red cedar (*Thuja plicata* Donn ex D. Don) samples were collected from a similar aged stand near Boston Bar, British Columbia (49°50′N, 121°31′W, elevation 600 m). Cedar samples were obtained in the same manner in which pine samples were collected except cedar stem samples from trees were obtained by cutting small wedges from stems using a pruning knife wiped down with 6% NaOCl prior to each sampling.

Stem samples from mature trees were obtained by taking cores with a surface disinfested increment borer after shaving a thin layer of bark from the stem with a sterile scalpel. Root samples were surface-sterilised and triturated (Chanway et al. 2000) before endophytic bacteria were isolated by plating the resulting slurry.

Four diazotrophic endophytes were isolated using this sampling procedure and 16S rDNA sequencing identified them as belonging to the genus *Paenibacillus* (strains P2b-2R, P18b-2R and C3b) as well as to the *Flexibacter* group (strain P19a-2R). The three strains with names beginning with "P" were originally isolated from pine tissues from the Williams Lake site. Strain P2b was isolated from within the surface-sterilised stem of a pine seedling, strain P18b was isolated from within surface-sterilised needles of another pine seedling, and strain P19a was isolated from the internal stem tissue of a third pine seedling. Strain C3b was isolated from within the surface-sterilised stem of a western red cedar tree at the Boston Bar site.

These microorganisms were then used to inoculate pine seed sown in glass tubes (150 mm × 25 mm in diameter) filled with a severely nitrogen deficient seedling growth medium to which a small amount (0.0576 g/l) $Ca(^{15}NO_3)_2$ (5% ^{15}N label) was added to facilitate identification of foliar nitrogen originating from the growth medium versus the atmosphere. Other nutrients were added in amounts sufficient to support healthy plant growth. Bacterial inoculum was prepared by streaking frozen cultures onto plates of combined carbon medium (Rennie 1981). Following growth on plates, a loopful of each strain was separately inoculated into its own 1 l flask containing 500 ml CCM broth. Flasks were then secured on a rotary shaker (150 rpm; room temperature) and agitated for up to 2 days. All bacterial cultures were harvested by centrifugation (10,000 *g* for 30 min), and resuspended in sterile phosphate buffer (SPB). Strains P2b-2R and C3b were resuspended to a density of ca. 10^7 cfu/ml and strains P18b-2R and P19a-2R were resuspended to a density of ca. 10^6 cfu/ml. For inoculation, 5.0 ml of each bacterial suspension was pipetted into separate tubes. Control seeds received 5.0 ml SPB. Tubes were placed in a growth chamber (Conviron

CMP3244, Conviron Products Company, Winnipeg, MB). Photosynthetically active radiation (PAR) at canopy level was ca. 300 μmol s^{-1} m^{-2} during an 18-h photoperiod, and 20°C/14°C day/night temperature cycle.

Of the four diazotrophic endophytes, only inoculation with strain P2b-2R resulted in large, statistically significant increases of 27–66% in pine foliar nitrogen derived from the atmosphere in all three plant growth trials we have conducted to date (Table 6.1). Interestingly, strain P18b-2R, which was found to be phylogenetically very similar to P2b-2R, had no such effect on seedling foliar nitrogen. Clearly, minor genetic differences between endophytic strains can result in profoundly different effects on plant growth and foliar nitrogen derived from the atmosphere (NDFA).

While large, statistically significant amounts of foliar NDFA were detected in all three trials with pine, there was no corresponding positive growth response in the first two experiments. Indeed, pine seedling growth was inhibited by strain P2b-2R compared to noninoculated controls in the first two trials (Bal 2003). That inoculated seedlings may derive nitrogen primarily from the atmosphere is an intriguing phenomenon; poor growth of inoculated seedlings may be an indication of the energetic cost of supporting nitrogen-fixing bacteria in the plant. Seedling growth reduction during the establishment of symbiosis is not uncommon, due to the energy diverted from the plant, and has been reported previously (Chanway and Holl 1991). This idea is supported by results from the third growth trial. By the end of the trial, control seedlings had a restricted growth rate, presumably due to N limitation, compared to the inoculated seedlings. P2b-2R-inoculated seedlings had greater biomass (47%) and total nitrogen (38%) compared to noninoculated controls, and derived 27% of foliar N from the atmosphere. In trials 1 and 2, control seedlings were apparently scavenging the small amount of soil N that was initially provided, and indeed outgrew the inoculated seedlings, which may have been in a "symbiosis development" mode. This tenet is hypothetical at this time and requires further investigation.

Table 6.1. Percentages of nitrogen derived from the atmosphere (NDFA) in pine seedlings after inoculation with *Paenibacillus polymyxa* strain P2b-2R in three separate growth trials

Growth Trial	Duration (months after planting and inoculation)	%NDFA[a]
1	8	30
2	9	66
3	9	27

[a]Calculated according to Rennie et al. (1978), where:

$$\%NDFA = 1 - \left[\frac{\text{atom \% } ^{15}\text{N excess(inoculated plant)}}{\text{atom \% } ^{15}\text{N excess (uninoculated plant)}} \times 100 \right]$$

In addition, strain P2b-2R was detected inside root, stem and needle tissue using a surface-sterilisation-trituration plating technique as well as with confocal laser scanning microscopy after insertion of green fluorescent protein into the bacterium. These results leave us with little doubt that pine can derive biologically significant amounts of nitrogen from endophytic diazotrophs.

Two conclusions can be drawn from our work so far. Nitrogen-fixing bacteria can be isolated from internal root, stem and needle tissues of lodgepole pine seedlings and mature trees, some of which can contribute biologically significant amounts of fixed nitrogen to lodgepole pine seedlings under controlled environments. It is possible that a significant amount of plant nitrogen originates from diazotrophic endophytes in lodgepole pine, but additional research is required to elucidate the role of these bacteria in forest ecosystems.

6.6 Future Work

We are currently involved in detecting in planta expression of strain P2b-2R *nif* genes to demonstrate that internal pine stem, root and needle tissues provide a suitable environment for nitrogen fixation, as we have already confirmed the existence of *nif* genes in this bacterial strain. These experiments will allow localisation of P2b-2R within the plant tissues as well as provide evidence that the observed nitrogen gains were in fact derived from the endophytic P2b-2R. In addition, we need to understand the effect of soil nitrogen availability on growth promotion by strain P2b-2R and the extent to which other mechanisms are responsible for growth promotion. Plant growth studies involving a non-diazotrophic mutant of strain P2b-2R, under various levels of available nitrogen will provide data to answer these questions.

Systemically colonising endophytic bacteria such as *P. polymyxa* strain Pw2-R and P2b-2R could also be used as vectors to deliver specific gene products to plants. Such an approach may be more feasible than attempting to genetically alter the plant host directly. In addition, diazotrophic endophytic bacteria hold great potential for reducing application of fertilisers, especially of mineral N. Inoculation of forest seedlings with effective diazotrophic or plant growth promoting endophytic bacteria could enhance growth and yield of trees significantly, especially at nutrient-poor sites. However, there is much more work to be done if we are to understand these plant/microbe associations to the degree that we can manage them effectively for more efficient and sustainable nursery and field treatments.

More knowledge of the population dynamics and activity of endophytic bacteria in their host plants is required. A considerable research effort is also required to design strategies for the reinoculation of endophytic bacteria. In order to guarantee reproducibility, reliable methods of inoculum delivery should be developed. This is especially the case for the inoculation of trees with endophytic bacteria. Intense testing of different delivery systems has indicated that the efficacy of the application method for introducing endophytic bacteria into plant tissue is strain specific (Musson et al. 1995). The development of successful application technologies would fully depend on improving our understanding of how bacterial endophytes enter and colonise plants. This is just one aspect of the study of bacterial endophytes that needs to be undertaken in order to fully realise their potential use in forestry as in agriculture.

Acknowledgements. The authors acknowledge the excellent help of Ms.S.E. Chanway in reference management and typing this manuscript.

References

Achouak W, Normand P, Heulin T (1999) Comparative phylogeny of *rrs* and *nif*H genes in Bacillaceae. Int J Syst Bacteriol 49:961–967

Ash C, Farrow JAE, Collins MD (1993) Molecular identification of rRNA group 3 bacilli using a PCR probe test: proposal for the creation of a genus *Paenibacillus*. Antonie van Leeuwenhoek 64:253–260

Bal A (2003) Can lodgepole pine derive biologically significant amounts of N from bacterial endophytes? MSc Thesis, The University of British Columbia, Vancouver BC

Balandreau J, Knowles R (1978) The rhizosphere. In: Dommergues YR, Krupa SV (eds) Interactions between non-pathogenic soil microorganisms and plants. Elsevier, Amsterdam, pp 243–268

Baldani VLD, Döbereiner J (1980) Host plant specificity in the infection of cereals with *Azospirillum* ssp. Soil Biol Biochem 12:433–439

Bent E, Chanway CP (1997) PGPR-mediated growth promotion effects on lodgepole pine can be inhibited by the presence of a rhizobacterial competitor. In: Ogoshi A et al. (eds) Plant growth-promoting Rhizobacteria: present status and future prospects. Nakanishi, Sapporo, pp 233–239

Binkley D (1995) The influence of tree species on forest soils: processes and patterns. In: Mead DJ, Cornforth IS (eds) Proceedings of the trees and soil workshop, Lincoln University, 28 February–2 March 1994, Lincoln University Press, New Zealand, pp 1–33

Boddey RM, Döbereiner J (1995) Nitrogen fixation associated with grasses and cereals: recent progress and perspectives for the future. Fertil Res 42:241–250

Brock TD, Madigan MT, Martinko JM, Parker J (1994) Biology of microorganisms. Prentice Hall, New Jersey

Brooks DS, Gonzalez CF, Appel DN, Filer TH (1994) Evaluation of endophytic bacteria as potential biological control agents for oak wilt. Biol Control 4:373–381

Carroll G (1988) Fungal endophytes in stems and leaves: from latent pathogen to mutualistic symbiont. Ecology 69:2–9

Cavalcante VA, Döbereiner J (1988) A new acid-tolerant nitrogen-fixing bacterium associated with sugarcane. Plant Soil 108:23–31

Chanway CP (1997) Inoculation of tree roots with plant growth promoting soil bacteria: an emerging technology for reforestation. For Sci 43:99–112

Chanway CP, Holl FB (1991) Biomass increase and associative nitrogen fixation of mycorrhizal *Pinus contorta* seedlings inoculated with a plant growth promoting *Paenibacillus* strain. Can J Bot 69:507–511

Chanway CP, Holl FB (1992) Influence of soil biota on Douglas-fir (*Pseudotsuga menziesii* (Mirb.) Franco) seedling growth: the role of rhizosphere bacteria. Can J Bot 70:1025–1031

Chanway CP, Holl FB (1993) Field performance of spruce seedlings after inoculation with plant growth promoting rhizobacteria. Can J Microbiol 39:1084–1088

Chanway CP, Holl FB (1994) Growth of outplanted lodgepole pine seedlings one year after inoculation with plant growth promoting rhizobacteria. For Sci 40:238–246

Chanway CP, Shishido M, Holl FB (1994) Root-endophytic and rhizosphere plant growth promoting rhizobacteria for conifer seedlings. In: Ryder MH, Stephens PM, Bowen GD (eds) Improving plant productivity with rhizosphere bacteria. CSIRO Division of Soils 1994:72–74

Chanway CP, Shishido M, Jungwirth S, Nairn J, Markham G, Xiao, Holl FB (1997) Second year growth responses of outplanted conifer seedlings inoculated with PGPR. In: Ogoshi A, Kobayashi K, Homma Y, Kodama F, Kondo N, Akino S (eds) Plant growth-promoting Rhizobacteria present status and future prospects. Nakanishi, Sapporo, pp 172–176

Chanway CP, Shishido M, Nairn J, Jungwirth S, Markham J, Xiao G, Holl FB (2000) Endophytic colonisation and field responses of hybrid spruce seedlings after inoculation with plant growth promoting rhizobacteria. For Ecol Manage 133:81–88

Clay K (1988) Fungal endophytes of grasses: a defensive mutualism between plants and fungi. Ecology 69:10–16

Döbereiner J, Urquiaga S, Boddey RM (1995) Alternatives for nitrogen nutrition of crops in tropical agriculture. Fertil Res 42:339–346

Fitter AH, Garbaye J (1994) Interaction between mycorrhizal fungi and other soil organisms. Plant Soil 159:123–132

Glick BR (1995) The enhancement of plant growth by free-living bacteria. Can J Microbiol 41:109–117

Gardner JM, Feldman AW, Zablotowicz M (1982) Identity and behavior of xylem-residing bacteria in rough lemon roots of Florida citrus trees. Appl Environ Microbiol 43:1335–1342

Hallmann J, Quadt-Hallmann A, Mahafee WF, Kloepper JW (1997) Bacterial endophytes in agricultural crops. Can J Microbiol 43:895–914

Holl FB, Chanway CP, Turkington R, Radley RA (1988) Response of crested wheatgrass (*Agropyron cristatum* L), perennial ryegrass (*Lolium perenne* L) and white clover (*Trifolium repens* L.) to inoculation with *Paenibacillus polymyxa*. Soil Biol Biochem 20:19–24

Hollis JP (1951) Bacteria in healthy potato tissue. Phytopathology 41:320–366

Hunt R (1982) Plant growth curves. University Park Press, Baltimore

Hurek T, Reinhold-Hurek B, Van Montagu M, Kellenberger E (1994) Root colonisation and systemic spreading of *Azoarcus* sp. strain BH72 in grasses. J Bacteriol 176:1913–1923

James EK (2000) Nitrogen fixation in endophytic and associative symbiosis. Field Crops Res 65:197–209

James EK, Olivares FL (1997) Infection and colonisation of sugar cane and other graminaceous plants by endophytic diazotrophs. Crit Rev Plant Sci 17:77–119

Kloepper JW (1993) Plant growth-promoting rhizobacteria as biological control agents. In: Metting FB (ed) Soil microbial ecology applications in agricultural and environmental management. Dekker, New York, pp 255–274

Kloepper JW, Lifshitz R, Zablotowicz RM (1989) Free living bacterial inocula for enhancing crop productivity. Trends Biotechnol 7:39–44

Law R, Lewis DH (1983) Biotic environments and the maintenance of sex – some evidence from mutualistic symbioses. Biol J Linn Soc 20:249–276

Lehninger AL (1975) Biochemistry: the molecular basis of cell structure and function. Worth, New York

Li CY, Massicotte HB, Moore LV (1992) Nitrogen fixing Bacillus sp. associated with Douglas fir tuberculate ectomycorrhizae. Plant Soil 140:35–40

Lodewyckx C, Vangronsveld I, Porteous F, Moore ERB, Taghavi S, Mezgeay M, Vander lella D (2002) Endophytic bacteria and their potential applications. Crit Rev Plant Sci 21:583–606

Mavingui P, Laguerre PG, Berge O, Heulin T (1992) Genotypic and phenotypic variability of *Paenibacillus polymyxa* in soil and in the rhizosphere of wheat. Appl Environ Microbiol 58:1894–1903

McInroy JA, Kloepper JW (1994) Novel bacterial taxa inhabiting internal tissue of sweet corn and cotton. In: Ryder MH, Stephens PM, Bowen GD (eds) Improving plant productivity with rhizosphere bacteria. CSIRO, Melbourne, Australia

Mengoni A, Mocali S, Surico G, Tegli S, Fani R (2003) Fluctuation of endophytic bacteria and phytoplasmosis in elm trees. Microbiol Res 158:363–369

Misaghi IJ, Donndelinger CR (1990) Endophytic bacteria in symptom-free cotton plants. Phytopathology 80:808–811

Mocali S, Bertelli E, Di Cello F, Mengoni A, Sfalanga A, Viliani F, Caciotti A, Tegli S, Surico G, Fani R (2003) Fluctuation of bacteria isolated from elm tissues during different seasons and from different plant organs. Microbiol Res 154:105–114

Musson G, McInroy JA, Kloepper JW (1995) Development of delivery systems for introducing endophytic bacteria into cotton. Biocontrol Sci Technol 5:407–416

O'Neill GA, Chanway CP, Axelrood PE, Radley RA, FB Holl (1992) Growth response specificity of spruce inoculated with coexistent rhizosphere bacteria. Can J Bot 70:2347–2353

Paul EA, Clark FE (1989) Soil microbiology and biochemistry. Academic, New York

Pokojska-Burdziej A (1982) The effect of microorganisms, microbial metabolites and plant growth regulators on the growth of pine seedlings (*Pinus sylvestris* L.). Pol J Soil Sci 15:137–143

Rennie RJ (1981) A single medium for the isolation of acetylene-reducing (dinitrogen-fixing) bacteria from soils. Can J Microbiol 27:8–14

Rennie RJ, Rennie DA, Fried M (1978) Concepts of ^{15}N usage in dinitrogen fixation studies. In: Isotopes in biological dinitrogen fixation. International Atomic Energy Agency, Vienna, pp 107–133

Reinhold-Hurek B, Hurek T (1998) Interactions of Gramineous plants with *Azoarcus* spp. and other diazotrophs: identification, localization, and perspectives to study their function. Crit Rev Plant Sci 17:29–54

Rhodes-Roberts M (1981) The taxonomy of some nitrogen fixing Paenibacillus species with special reference to nitrogen fixation. In: Berkeley RCW, Goodfellow M (eds) The aerobic-endosperm forming bacteria classification and identification. Academic, London, pp 315–335

Schardl CL, Leuchtmann A, Spiering MJ (2004) Symbioses of grasses with seedborne fungal endophytes. Annu Rev Plant Biol 55:315–340

Shishido M (1997) PGPR for interior spruce seedlings. PhD Thesis University of British Columbia, Vancouver BC

Shishido M, Chanway CP (2000) Colonisation and growth promotion of outplanted spruce seedlings pre-inoculated with plant growth-promoting rhizobacteria in the greenhouse. Can J For Res 30:845–854

Shishido M, Loeb BM, Chanway CP (1995) External and internal root colonisation of lodgepole pine seedlings by two growth-promoting *Paenibacillus* strains originated from different root microsites. Can J Microbiol 41:707–713

Shishido M, Massicotte HB, Chanway CP (1996) Effect of plant growth promoting *Paenibacillus* strains on pine and spruce seedling growth and mycorrhizal infection. Ann Bot 77:433–441

Sprent JL, James EK (1995) N_2-fixation by endophytic bacteria: questions of entry and operation. In: Fendrick I (ed) NATO ASI series, Azospirillum VI and related microorganisms, vol G. Springer, Berlin Heidelberg New York, pp 15–30

Tervet IW, Hollis JP (1948) Bacteria in the storage organs of healthy plants. Phytopathology 38:960–967

Timmusk S, Wagner EGH (1999) The plant-growth-promoting rhizobacterium *Paenibacillus polymyxa* induces changes in *Arabidopsis thaliana* gene expression: a possible connection between biotic and abiotic stress responses. Mol Plant-Microbe Interact 12:951–959

Triplett EW (1996) Diazotrophic endophytes: progress and prospects for nitrogen fixation in monocots. Plant Soil 186:29–38

Whitesides SK, Spotts RA (1991) Frequency, distribution, and characteristics of endophytic *Pseudomonas syringae* in pear trees. Phytopathology 81:453–457

Wilson D (1995) Endophyte: the evolution of a term, and clarification of its use and definition. Oikos 73:274–276

Zak B (1971) Characterization and classification of mycorrhizae of Douglas fir. II. *Pseudotsuga menziesii + Rhizopogon vinicolor*. Can J Bot 49:1079–1084

7 Biodiversity of Fungal Root-Endophyte Communities and Populations, in Particular of the Dark Septate Endophyte *Phialocephala fortinii* s. l.

Thomas N. Sieber, Christoph R. Grünig

7.1 Introduction

The peripheral root tissues form a morphologically, physically and chemically complex microcosm that provides different habitats for diverse communities of microorganisms. This microcosm is not stable, and changes over space and time because the boundaries between soil, rhizosphere, and living roots are continually shifted as a result of root growth and the constant modification of nearby soil by root mechanical and metabolic activity (Foster et al. 1983). Microorganisms colonise the rhizoplane, epidermis and outer cortex in a nonrandom patchy manner and contribute to the modification of the soil-rhizosphere-root continuum. Microorganisms affect their plant hosts, and hosts reciprocally affect their symbionts, leading to a feedback that drives changes in both the microbial and plant communities (Bever et al. 1997). Many soil bacteria and fungi are able to colonise epidermal and outer cortical cells of healthy roots inter- and intra-cellularly. A comparatively small number of organisms, e.g. mycorrhizal fungi, endophytic and pathogenic fungi and bacteria, possess, however, the ability to cross the inner boundary of the rhizosphere and to penetrate deeper into the root (Bazin et al. 1990). The interaction of host and endophyte depends on the disposition of host and fungus or bacterium and the environmental conditions, but may be neutral, mutualistic or antagonistic and may change over time. Some endophytic fungi adopt mycorrhizal functions and/or place plants at a competitive advantage against herbivores, insect pests or pathogens (Carroll 1988; Hawksworth 1991). Other endophytes can switch to a pathogenic behaviour when conditions are unfavourable for the host (Schulz et al. 1999). The biodiversity of root endophyte communities varies in relation to environmental factors, type of vegetation,

Thomas N. Sieber: Swiss Federal Institute of Technology, Department of Environmental Sciences, Institute of Integrative Biology, Forest Pathology and Dendrology, 8092 Zürich, Switzerland, E-mail: thomas.sieber@env.ethz.ch

Christoph R. Grünig: Swiss Federal Institute of Technology, Department of Environmental Sciences, Institute of Integrative Biology, Forest Pathology and Dendrology, 8092 Zürich, Switzerland

Soil Biology, Volume 9
Microbial Root Endophytes
B. Schulz, C. Boyle, T. N. Sieber (Eds.)

spatiotemporal patterns of the root microcosm and interactions among microorganisms. There is currently an urgent need to assess biodiversity in pristine ecosystems and to use these data as references to measure the effects of disturbances on diversity and to better enable informed decision-making on the fate of threatened natural habitats (Cannon 1997). Threats may come from a variety of sources, including exploitation by logging, machine-graded soils, urban development, pollution, climate change and input of pesticides and fertilisers. Biodiversity can be explored at several levels, i.e. in terms of communities, species and populations (Hawksworth 1991). Here, we will explore current knowledge on the biodiversity of non-mycorrhizal fungal root endophytes at all levels. The first part of this review will be dedicated to biodiversity at the community level in relation to environmental factors. In the second part, special emphasis will be placed on the diversity of dark septate endophytes (DSE), in particular of *Phialocephala fortinii* s. l.

Readers of this chapter should always bear in mind that the methods of detection are highly selective and, thus, the species list and species diversity derived for any habitat will be incomplete and will be biased in respect to physiological features selected for by the method used [Sieber 2002; Swift 1976; see Chaps. 9 (Bayman and Otero), 18 (Bloemberg and Camacho Carvajal) and 19 (Van Overbeek et al.)].

7.2 Species Diversity of Root Endophyte Communities

"Species diversity" comprises two distinct components: the total number of species, which ecologists refer to as "species richness", and "evenness" or equitability, which refers to how species abundances are distributed among the species present. An ecosystem is said to be more diverse if many species with equal population sizes are present and less diverse if some species are rare and a few are very common. Other helpful terms are "spectrum of species" or "community composition" to describe habitat or ecosystem differences with respect to the species found. The species diversity and the species spectrum of root-endophyte communities are related to various factors, which can tentatively be arranged into four groups: (1) geography and climate, (2) soil, (3) multitrophic interactions, and (4) natural and anthropogenic disturbances. This grouping is rather artificial and does not account for the intricate interplay among factors that often makes it impossible to determine the contribution of each factor. Another aspect obscuring the effects of different factors is that of site history, i.e. the dynamics of plant and endophyte communities. Nevertheless, the above grouping seems to be the most appropriate structure for this section.

7.2.1 Geography and Climate

Fungal species diversity is higher in tropical than in temperate regions owing in part to the great diversity of hosts, but also to the optimal growth conditions for many fungi as a result of the hot and moist climate (Cannon and Hawksworth 1995). Whether this relationship is also valid for fungal root endophytes remains to be tested. Compared to habitats in the temperate or the tropical zones, species diversity is distinctly reduced in arctic-alpine environments, not only because of the lower number of available host species, but also with respect to the number of endophyte species in each host. For example, only seven root endophyte species were detected in *Dryas octopetala* in arctic Spitsbergen (Fisher et al. 1995) [Table 7.1(i)]. Correspondingly, species richness in *Erica carnea* was highest at an altitude of 640 m and lowest at 2,140 m in Switzerland (Oberholzer-Tschütscher 1982). The species spectra differed greatly among sites, as expressed by very low between-site similarities [Table 7.1(ii)]. Evenness was lowest at the lowest altitude where the comparatively species-rich community was dominated by only four to five species.

There is strong evidence for a shift from arbuscular mycorrhizal fungi (AMF) and ectomycorrhizal fungi (ECM) in temperate habitats towards symbioses of uncertain status, especially dark septate endophytes (DSE), in arctic-alpine ecosystems (Bledsoe et al. 1990; Christie and Nicolson 1983; Read and Haselwandter 1981; Väre et al. 1992). Correspondingly, the frequency of roots colonised by *Phialocephala fortinii* s. l., a ubiquitous and dominant DSE in conifer roots (see Sect. 7.3.3), was positively correlated with the altitude in forest ecosystems (Ahlich and Sieber 1996).

Weather and climatic conditions are assumed to have a weaker direct effect on species diversity of endophyte assemblages in root tissues than in aerial plant parts due to the insulating and compensating properties of soils (Fitter et al. 1985). Thus, changes in root endophyte assemblages become manifest only if the climatic conditions deviate from the "normal" over an extended period of time, i.e. if the climate changes. In fact, long-term changes in mean annual temperature, frequency and amount of precipitation, as well as enhanced CO_2 may affect root endophyte diversity through shifts in the quantity and quality of photosynthates and secondary plant metabolites translocated to the roots, the rate of root turnover, and shifts in the competivity of endophytes and other soil microorganisms (Coûteaux et al. 1999; Rillig et al. 1999; Körner 2000). However, nothing is known about the direction and magnitude of effects on root-endophyte diversity.

Table 7.1. Influence of geographical, physical, chemical and biological factors on species diversity and similarity of communities of fungal root endophytes

Hosts and factors	Observed species richness[a]	Adjusted species richness[b]	Number of very abundant species[c]	Evenness index[d]	Total number of isolates	Sample size[e]	Pairwise similarities of communities[f] (2)	(3)	(4)	(5)	Reference
(i) *Dryas octopetala*						R					Fisher et al. 1995
(1) Spitzbergen, site A	4	3.8 ± 0.4	2.4	0.81	42	50	0.73				
(2) Spitzbergen, site B	7	5.8 ± 0.4	3.8	0.83	24	50	–				
(ii) *Erica carnea*						R					Oberholzer-Tschütscher 1982
(1) Fläsch (640 m)	35	26.5 ± 2.2	4.2	0.51	296	333	0.26	0.21	0.30		
(2) Näfels (760 m)	19	17.1 ± 1.2	4.7	0.71	219	120	–	0.20	0.19		
(3) Davos-Wolfgang (1,640 m)	22	21.8 ± 0.4	7.0	0.72	173	298	–	–	0.24		
(4) Davos-Schatzalp (2,140 m)	12	10.6 ± 1.0	2.7	0.68	235	300	–	–	–		
(iii) *Picea abies*						R					Kattner and Schönhar 1990
(1) Soil pH neutral	27	26.9 ± 0.3	11.9	0.70	153	480	0.57				
(2) Soil pH acidic	29	28.8 ± 0.4	12.6	0.72	154	480	–				
(iv) *Alnus glutinosa*						R					Fisher et al. 1991
(1) Submerged roots	45	44.1 ± 0.8	17.4	0.66	114	40	0.37				
(2) Non-submerged roots	31	29.2 ± 1.2	13.8	0.75	126	40	–				
(v) *Rhizophora mucronata*						R					Ananda and Sridhar 2002
(1) Low-tide level	6	5.3 ± 0.7	2.2	0.80	21	30	0.50	0.38			
(2) Mid-tide level	14	9.7 ± 1.2	3.6	0.87	34	30	–	0.58			
(3) High-tide level	10	9.4 ± 0.6	3.2	0.91	17	30	–	–			

Table 7.1. (continued)

Hosts and factors	Observed species richness[a]	Adjusted species richness[b]	Number of very abundant species[c]	Evenness index[d]	Total number of isolates	Sample size[e]	Pairwise similarities of communities[f] (2)	(3)	(4)	(5)	Reference
(vi) *Phragmites australis*											Wirsel et al. 2001
Location 1						R					
(1) Flooded site	10	9.5 ± 0.7	4.6	0.74	63	45	0.52	0.52	–		
(2) Dry site	13	12.9 ± 0.3	7.4	0.80	51	45	–	–	0.75		
Location 2											
(3) Flooded site	13	11.9 ± 0.9	5.6	0.71	47	45			0.58		
(4) Dry site	11	10.0 ± 0.9	4.8	0.72	52	45			–		
(vii) *Triticum aestivum*											Sieber et al. 1988
Development stage:						R					
(1) One leaf – end of tillering	63	61.0 ± 1.3	15.3	0.59	547	5040	0.67				
(2) Inflorescence emerged – caryopsis hard	62	39.5 ± 2.9	6.3	0.53	1857	3360	–				
(viii) *Triticum aestivum*											Sieber et al. 1988
Preceding crop:						R					
(1) Sugar beet	51	44.7 ± 2.1	9.3	0.56	623	2400	0.58	0.57	0.67		
(2) Red clover	49	48.3 ± 3.3	10.9	0.60	413	1200	–	0.60	0.59		
(3) Maize	44	35.9 ± 2.2	4.7	0.45	773	2400	–	–	0.65		
(4) Potatoes	39	33.5 ± 1.9	6.9	0.61	595	2400	–	–	–		

Table 7.1. (continued)

Hosts and factors	Observed species richness[a]	Adjusted species richness[b]	Number of very abundant species[c]	Evenness index[d]	Total number of isolates	Sample size[e]	Pairwise similarities of communities[f] (2)	(3)	(4)	(5)	Reference
(ix) Various vegetables						R					Narisawa et al. 2002
(1) Eggplant	7	5.5 ± 0.9	3.2	0.71	35	45	0.67	0.55	0.67	0.86	
(2) Tomato	5	4.7 ± 0.4	2.7	0.78	19	45	–	0.44	0.60	0.50	
(3) Melon	4	3.9 ± 0.3	2.6	0.83	17	45	–	–	0.44	0.55	
(4) Strawberry	5	4.5 ± 0.6	2.1	0.71	21	45	–	–	–	0.67	
(5) Chinese cabbage	7	5.9 ± 0.8	4.4	0.79	29	45	–	–	–		
(x) *Betula pendula*						T					Görke (1998)
(1) Plantation in cleared windthrow	30	12.5 ± 1.7	10.6	0.59	87	160	0.34	0.29	0.38		
(2) Natural regeneration in untouched windthrow	11	9.3 ± 1.0	5.7	0.73	27	75	–	0.41	0.50		
(3) Natural regeneration in cleared windthrow	18	10.9 ± 1.5	7.4	0.60	48	100	–	–	0.57		
(4) Natural regeneration in low density forest[g]	17	9.3 ± 1.5	6.3	0.70	58	100	–	–	–		
(x) *Pinus sylvestris*						T					Görke (1998)
(1) Plantation in cleared windthrow	16	6.2 ± 1.2	6.8	0.72	56	160	0.42	0.40	0.38		
(2) Natural regeneration in untouched windthrow	8	5.5 ± 1.0	4.7	0.79	20	75	–	0.55	0.53		
(3) Natural regeneration in cleared windthrow	14	7.0 ± 1.2	6.9	0.69	26	100	–	–	0.38		
(4) Natural regeneration in low density forest	7	5.4 ± 0.9	4.8	0.80	19	100	–	–	–		

Table 7.1. (continued)

Hosts and factors	Observed species richness[a]	Adjusted species richness[b]	Number of very abundant species[c]	Evenness index[d]	Total number of isolates	Sample size[e]	Pairwise similarities of communities[f] (2)	(3)	(4)	(5)	Reference
(xi) *Erica carnea*						R					Cevnik et al. 2000
(1) Control (Cd 1.4; Pb 171; Zn 61.8)[h]	9	8.2 ± 0.7	3.9	0.78	104	240	0.50	0.60	0.63		
(2) Low pollution (Cd 6.9; Pb 667; Zn 177)	7	6.9 ± 0.2	3.3	0.81	72	240	–	0.67	0.71		
(3) High pollution (Cd 35.8; Pb 5422; Zn 582)	11	10.5 ± 0.6	8.0	0.87	107	240	–	–	0.67		
(4) Highest pollution (Cd 87.7; Pb 31320; Zn 1330)	10	8.9 ± 0.8	3.8	0.72	142	240	–	–	–		

[a] Diversity index N0 according to Hill (1973); number of species
[b] Mean and standard error of the number of species adjusted to the lowest within-study number of isolates using rarefaction according to Hurlbert (1971)
[c] Diversity index N2 according to Hill (1973)
[d] Evenness index according to Hill (1973); this index converges towards 1 as one species tends to dominate
[e] Number of root segments (R) or trees (T) examined
[f] Soerensen index (Soerensen 1948); column numbers (in brackets) correspond to factor identifiers in the column "Hosts and factors"; $0 \leq$ Soerensen index ≤ 1, the index is 0 if two communities have no species in common, and it is 1 if all species occur in both communities
[g] Low density forest = selectively logged forest stand; aim: increased solar radiation within the stand
[h] Concentrations of heavy metals in micrograms per gram of soil

7.2.2 Soil

Soil and rhizosphere are highly variable habitats. Chemical properties such as pH or the availability of minerals and carbohydrates may vary significantly within a few centimetres of soil (Papritz and Flühler 1991). Similarly, differences in soil texture and water regime contribute to the variability of soils. In addition, roots constantly modify the nearby soil structure by depletion of minerals, ions and water and by the secretion of root exudates. Soils offer habitats for various communities of microorganisms including potential root endophytes. Plant and microbial metabolites may differentially influence the surrounding soil and change some of its properties, thus preparing the soil for the microorganisms of the next successional stage (Van Der Putten 2003).

Physical and Chemical Soil Characteristics

Soil pH had an effect on community composition but not on species diversity of endophytic fungi in Norway spruce roots (*Picea abies*) (Kattner and Schönhar 1990) [Table 7.1(iii)]. The similarity of only 57% of the endophyte communities in roots from neutral and acidic soils reflects either the selectivity of soil pH or the historical presence/absence of certain endophyte species, e.g. endophytes with low dispersion and/or survival rates. For example, *Phialocephala fortinii* preferentially occurs in roots growing in acidic soils (Ahlich et al. 1998).

Species richness was not related to soil texture in wheat roots (*Triticum aestivum*) (Riesen and Sieber 1985; Sieber et al. 1988). However, texture affected the frequency of *Microdochium bolleyi* and *Periconia macrospinosa*. *M. bolleyi* was more frequently isolated from roots originating from silty loam, whereas *P. macrospinosa* was isolated more often from roots growing in pure loam.

Root endophytes differ in their ability to metabolise minerals and carbohydrates, making some endophytes more successful than others in a given habitat. DSE are thought to be excellent metabolisers of phosphorus (P) and to mediate P uptake for their hosts (Jumpponen et al. 1998; Barrow and Osuna 2002). In fact, DSE were more abundant in habitats poor in P (Haselwandter and Read 1982; Ruotsalainen et al. 2002). Similarly, differential utilisation of carbohydrates as well as which carbohydrates were available determined fungal species diversity and endophyte-community composition in the experiments of Hadacek and Kraus (2002).

Water Regime

The water regime in soils and streams has a strong impact on species diversity and especially on the species spectrum of endophytic fungi (see Chap. 10 by Bärlocher). In roots of the same tree, 45 species were isolated from roots submerged in a river as opposed to only 31 species from non-submerged roots (Fisher et al. 1991) [Table 7.1(iv)]. The similarity of the community composition in submerged and non-submerged roots of the same individual black-alder tree was only 37%. Colonisation of submerged roots by aquatic hyphomycetes, together with the absence or scarcity of these specialists in non-submerged roots, emphasise the importance of the milieu in which roots grow in determining the composition and diversity of endophyte communities. For example, high water tables restricted the occurrence of *P. fortinii* in wetlands (Addy et al. 2000). The endophyte species diversity in roots of the mangrove *Rhizophora mucronata* strongly depended on the tidal level at which the roots were collected. Diversity was highest at the mid-tide level, i.e. the zone submerged in seawater approximately half of the time, and roots from the high-tide and the low-tide level had, on average, only 38% of species in common (Ananda and Sridhar 2002) [Table 7.1(v)]. Flooding and site conditions affected endophyte species spectra but not species richness in roots of common reed (*Phragmites australis*) (Wirsel et al. 2001) [Table 7.1(vi)]. In contrast, species spectra in bracken rhizomes (*Pteridium aquilinum*) did not differ among wetland and woodland sites (Petrini et al. 1992).

7.2.3 Multitrophic Interactions

The diversity of soil microorganisms is tremendous; 1 g soil can contain between 5,000 and 10,000 species of microorganisms (Torsvik et al. 1990). However, only 1,200 species of fungi have been isolated from soil (Watanabe 1994), perhaps because, as estimates suggest, only 17% of known fungi can be readily grown in culture (Hawksworth 1991). If this percentage were applied to the 1,200 species as suggested by Watanabe (1994), this would give an estimate of approximately 7,000 species of soil fungi (Bridge and Spooner 2001). The total length of fungal hyphae varies greatly according to soil type and soil biology and has been reported to be as high as 66,900 m in 1 g dry soil (Bååth and Söderström 1979). The high number of species and the high amount of microbial biomass in such small volumes of soil suggest that multitrophic interactions among soil bacteria, soil fungi, soil microfauna and plants are frequent. Interspecific competition may be "the" factor that overrides all others in regulating species abundance of soil fungi (Gochenaur 1984). If a community is dominated by inter- and intra-specific competition, the resources are more likely to be fully exploited. Endophyte species diversity and spectrum will then depend on the range of available resources, including host tissues, the extent to which species are specialists, antagonism among competitors, their ability to overcome host defences and the permitted extent of habitat overlap.

Microdochium bolleyi is a frequent and successful endophyte in cereal roots, where it functions as an effective antagonist of various root pathogens. For example, its presence in wheat roots was negatively correlated with the presence of *Septoria nodorum*, the causal agent of glume blotch disease of wheat (Riesen and Sieber 1985; Sieber et al. 1988). Similarly, *M. bolleyi* inhibited various *Fusarium* species and *Gaeumannomyces graminis* var. *tritici* (Kirk and Deacon 1987; Reinecke 1978). Whether *M. bolleyi* interacts with these pathogens indirectly by inducing systemic resistance in the host plant, or directly by either parasitising pathogens or producing inhibitory metabolites, remains to be examined.

The phenological state of the roots and/or the season may influence endophyte species diversity by affecting the probability of interactions among endophytic thalli. For example, the number of dominant species was higher in young than in mature winter wheat, presumably because freshly established thalli were small. Growth was reduced due to the cold temperatures in winter, making hyphal interference less likely and/or weaker and, thus, also allowing less competitive fungi or fungi better adapted to cold temperatures to establish endophytic thalli [Table 7.1(vii)] (Riesen and Sieber 1985; Sieber et al. 1988). This situation changed in spring and summer, when the growth rate of endophytic thalli increased, making intra- and inter-species hyphal interactions more probable, leading to the dominance of the few most competitive species.

Similar to mycorrhiza, strict host specificity is the exception rather than the rule for fungal root endophytes (Bruns et al. 2002; Jumpponen et al. 2004). However, the likelihood of occurrence of some endophyte species increases in the presence of particular host species, suggesting fungal host preference or shared habitat preferences. The diversity of the plant community in which the host species grows may, therefore, influence root-endophyte diversity similarly as it has been shown to affect diversity of soil microfungi (Christensen 1981, 1989). Ahlich and Sieber (1996) presented an example of the importance of the plant community in determining the spectra of fungi associated with the host. The dominant root endophytes of European beech (*Fagus sylvatica*), *Cryptosporiopsis radicicola* and *Cylindrocarpon didymum*, were rare or absent in roots of Scots pine (*Pinus sylvestris*) growing in monoculture. Likewise, *P. fortinii*, the dominant root endophyte of Scots pine, was rare or absent in monocultures of beech. However, when the roots originated from mixed stands of Scots pine and beech, Scots pine roots showed a comparatively high rate of colonisation by *C. radicicola* and *C. didymum*. Correspondingly, the roots of beech were frequently colonised by *P. fortinii* in mixed stands. In contrast, frequency of colonisation of *Betula papyrifera* and *Pseudotsuga menziesii* seedlings by DSE was not affected by whether or not the plants were grown in mixed culture or in monoculture (Jones et al. 1997).

In agriculture, the preceding crop may significantly affect endophyte diversity of the current crop. For example, species richness and the number of dominant species were significantly higher when wheat (*Triticum aestivum*) followed red clover than when it followed potatoes [Table 7.1(viii)] (Sieber et al. 1988). On average, only 59% of the endophyte species were indifferent to whether the preceding crop was clover or tomatoes. The range of indifferent endophyte species lay between 57% and 67% for other pairs of preceding crops [Table 7.1(viii)]. This observation may be related to differences in the spectra of endophytes that had colonised the preceding crop. Specific secondary metabolites and debris produced by the preceding crop, as well as the type and amount of agrochemicals (fertilisers, biocides, leafage killers) applied to the preceding crops may be other factors influencing both diversity and stimulation/inhibition of endophytes.

When different vegetables are grown in the same soil, some endophyte-host associations occur more frequently than others, suggesting host preference or adaptation. The similarity of the spectra of endophyte species among host species was as low as 44% in an experiment performed by Narisawa et al. (2002) [Table 7.1(ix)]. It is not known whether plants are able to actively recruit endophytes and vice-versa. Plant defence compounds probably select for certain rhizosphere microorganisms. Some evidence for such mechanisms comes from nematode and mycorrhiza research. Secondary metabolites released by roots of *Thuja occidentalis* upon attack by weevil larvae attracted entomopathogenic nematodes (Van Tol et al. 2001). Dormant propagules of mycorrhizal fungi were stimulated to germinate by chemical messengers from the host (Bruns et al. 2002). Correspondingly, mycelia of AMF were inhibited by non-host metabolites (Oba et al. 2002). Nothing is known about whether certain root endophytes release "pheromones" to attract roots of host plants.

7.2.4
Natural and Anthropogenic Disturbances

Anthropogenic and natural disturbances affect the species spectrum of plant communities and consequently also the communities of cohabiting microorganisms. Forest-management practices such as planting of trees, selective cutting or clearing of windthrows had a distinct effect on the endophytic mycobiota in the roots of forest trees (Görke 1998). Maximally

42% of the endophyte species were common to both planted and naturally regenerated trees [Table 7.1(x)]. Considering naturally regenerated trees only, species richness and the number of dominant species was highest in the cleared windthrow. Probably, endophyte diversity and community composition would also change as a consequence of gap formation by man and/or wind storm, which eliminates some hosts but creates habitats for many other hosts, i.e. ruderal plant species.

Mycorrhization and root-endophyte colonisation of naturally regenerated seedlings of *Betula platyphylla* var. *japonica* in soils of machine-graded ski slopes depended on the time elapsed since disturbance (Hashimoto and Hyakumachi 2000). Seedlings thrived well only in soil samples from soils disturbed more than 3 years previously and mycorrhization was significantly higher in these samples. In contrast, colonisation of roots by DSE was distinctly higher in seedlings sampled from soils disturbed only 1–3 years before sampling. In another study, the majority of naturally established seedlings of bishop pine (*Pinus muricata*) were colonised by DSE shortly after wildfire, indicating that a resident inoculum (chlamydospores, microsclerotia) survived the fire (Horton et al. 1998). Species richness of endophytes in roots of *Erica carnea* was highest at sites where soil pollution by heavy metals was high, but DSE occurred less frequently in the heavily polluted soils (Cevnik et al. 2000) [Table 7.1(xi)]. Endophytic fungi are either more competitive in disturbed or moderately polluted soils or better equipped to survive periods of adverse environmental conditions than mycorrhizal fungi.

The use of fungicides for crop protection can alter species diversity. Seed treatment with the systemic fungicide benomyl had no significant influence on endophyte species richness in wheat roots, but the frequency of roots colonised by seed borne *Septoria nodorum* was significantly reduced (Riesen and Sieber 1985). None of the fungicides applied to *Lolium perenne* fields at 18 sites in New Zealand had a significant effect on the root-endophyte communities (Skipp and Christensen 1989).

Fertilisation can affect fungal assemblages in roots. The frequency of *P. fortinii* in seedlings of potted *Picea glauca* was negatively correlated with the amount of nitrogen (N) applied (Kernaghan et al. 2003). Wilberforce et al.(2003) suspected N fertilisers to be one of the mechanisms by which management affects root endophyte communities in temperate grasslands. Emissions of air pollutants such as SO_2 and especially NO_x are thought to have a similar fertilising effect as fertilisers applied in agriculture. Adverse effects of these air pollutants on mycorrhizal fungi have been demonstrated in several studies (Cairney and Meharg 1999; Jansen and van Dobben 1987; Taylor and Read 1996).

7.3 Dark Septate Endophytes

Fungi with regularly septate and melanised hyphae probably constitute the most abundant and most widespread group of non-mycorrhizal root endophytes. In this section, we will briefly present the history of the term "DSE", outline the diversity of DSE and give an overview of current knowledge of the diversity and population genetics of the most prominent species complex of DSE: *Phialocephala fortinii* s. l.

7.3.1 History

Melin (1922, 1923) introduced the form taxon *Mycelium radicis atrovirens* (MRA) for sterile, melanised, septate mycelia that emerged from mycorrhizae and roots of *Picea abies* and *Pinus sylvestris*. The tree-fungus symbiosis was characterised by dematiaceous intra- and intercellular hyphae in the epidermal and cortical cells, but neither a Hartig net nor a mantle were formed. Melin (1923) coined the term "pseudomycorrhiza" for this relationship and considered it to form an antagonistic symbiosis. MRA-like fungi have been detected during numerous studies since Melin's pioneering work (Ahlich and Sieber 1996; Chan 1923; Freisleben 1934; Harley and Waid 1955; Jumpponen et al. 1998; Richard and Fortin 1973; Robertson 1954; Stoyke and Currah 1991). Since trinomials are not valid species names according to the International Code of Botanical Nomenclature, less stringent and more informal names are preferable. Read and Haselwandter (1981) introduced the term "DS hyphae" (DS = dark septate) for sterile, dark, septate hyphae and microsclerotia that occurred in roots of various alpine plants. Stoyke and Currah (1991) implemented the form taxon "dark septate endophyte" (DSE) and used it for fungi that form partly or entirely melanised, septate thalli within healthy root tissues. The taxon "DSE" serves primarily to differentiate these fungi from endophytes with septate, hyaline hyphae, and from fungi with sparsely septate, hyaline hyphae that are characteristic of AMF.

7.3.2 Biodiversity

The roots of more than 600 plant species representing about 320 genera in more than 110 families have been reported to be colonised by DSE (Ahlich and Sieber 1996; Barrow and Osuna 2002; Jumpponen and Trappe 1998b;

Kovacs and Szigetvari 2002; Ruotsalainen et al. 2002; Schadt et al. 2001). Dematiaceous mycelia are regularly received in culture during censuses of root endophytes, but it is often not known whether the endophytic thalli of these fungi are hyaline or melanised. This being the case, we must assume that DSE are much more widespread than previously assumed.

Species identity of some DSE is known because they readily sporulate in culture, e.g. *Microdochium bolleyi* and several *Phialophora* species in grasses and sedges. Many non-pathogenic *Phialophora* endophytes are related to the take-all fungi (*Gaeumannomyces graminis* var. *tritici* and var. *avenae*) of cereals and grasses in temperate areas and to *G. graminis* var. *graminis*, which causes crown sheath rot of rice in the tropics. *Phialophora radicicola* forms melanised sclerotia in cortical cells of maize roots without causing any apparent harm (Cain 1952). *P. radicicola* was also observed in the roots of three alpine grasses growing at the timberline in Bavaria (Blaschke 1986) or in roots of *Lolium perenne* in New Zealand (Skipp and Christensen 1989). The DSE abundantly observed in many alpine sedges in the Tyrolean Alps may also belong to *P. radicicola* (Haselwandter and Read 1980; Read and Haselwandter 1981). *P. radicicola* and *P. zeicola*, the maize take-all fungi from China, were recently shown to be the same species (Ward and Bateman 1999). *P. graminicola*, another non-pathogenic DSE of cereal and grass roots (Newsham 1999), provided significant control of the take-all disease by competition for senescing root tissues (Deacon 1981).

Taxonomic assignment of many DSE is problematic because sexual and asexual reproductive structures are either absent, rare, or are produced only under specific conditions. Cold treatment for up to 1 year was shown to induce sporulation in some DSE isolates, e.g. in isolates of *Chloridium paucisporum*, *Phialophora finlandica*, and *Phialocephala fortinii* (Wang and Wilcox 1985). Unfortunately, even then many DSE strains remain sterile and classification is complicated. Many mycologists have tried to bring some order into this difficult group of DSE (Harney et al. 1997; Melin 1923; Richard and Fortin 1973). Culture morphology is often used for an initial classification (Ahlich and Sieber 1996; Girlanda et al. 2002; Steinke et al. 1996; Stoyke et al. 1992). However, modern molecular biology offers a multitude of additional and potentially more reliable methods for the identification and typing of species, varieties and individuals (Carter et al. 1997; Geiser et al. 1994; White et al. 1990; Zietkiewicz et al. 1994). Some of these methods have been used to type DSE. Restriction patterns of a region on the ribosomal RNA (rRNA) genes indicated that two-thirds of the DSE from roots of subalpine plants were closely related to or conspecific with *P. fortinii* (Stoyke et al. 1992). Similarly, in a study by Harney et al. (1997), restriction site mapping of the nuclear rDNA internal transcribed spacer (ITS) regions showed that the majority of the isolates was *P. fortinii*-like and only two isolates were *Phialophora finlandica*.

According to isozyme analysis, DSE from various woody plant species belonged to two distinct groups (Ahlich-Schlegel 1997; Grünig et al. 2001; Sieber 2002). Members of the larger group were conspecific with *P. fortinii*, whereas those of the other group represented the sterile Type 1, which has been recently described as *Acephala applanata* (Ahlich and Sieber 1996; Grünig and Sieber 2005). Phylogenetic analysis of the ITS regions showed that *P. fortinii* and *A. applanata* are closely related and have *Phialocephala compacta*, *P. dimorphospora* and *P. scopiformis* as closest relatives (Grünig et al. 2002b). These five species are more closely related to members of the Leotiales such as *Gremmeniella abietina*, the causal agent of sclerroderis canker on pines, than to other *Phialocephala* species. The "*P. fortinii*-group" was also positioned within the Leotiales by phylogenetic analyses of the sequence data of the 18S and 28S subunits of the nuclear rRNA genes (Jacobs et al. 2003).

7.3.3
Diversity of *Phialocephala fortinii*

Phialocephala fortinii was shown to be the dominant DSE in coniferous and ericaceous roots in heathlands, forests and alpine ecosystems of the Northern temperate zones (Ahlich and Sieber 1996; Stoyke and Currah 1991). There is strong evidence that the roots of every Norway spruce (*Picea abies*) tree in natural forest habitats of Central Europe are colonised by this fungus (Ahlich and Sieber 1996; Grünig et al. 2004). The nature of root–*P. fortinii* symbioses and their ecological significance are largely unknown.

P. fortinii may function as a mycorrhizal fungus and mediate nutrient uptake, synthesise secondary metabolites, stimulate plant growth and/or play an important role in plant defence against root pathogens (Fernando and Currah 1996; Jumpponen and Trappe 1998a; O'Dell et al. 1993; Yu et al. 2001). Alternatively, it may behave as an opportunistic pathogen (Wilcox and Wang 1987). However, considering its widespread distribution and abundance it is very unlikely that *P. fortinii* is a primary pathogen.

We will provide a compilation of the newest findings on the genetic diversity within and among populations of *P. fortinii* and will conclude this section by forwarding some ideas and thoughts that could explain the observed diversity of this ecologically very successful species.

Genetic diversity of *P. fortinii* strains was examined on different spatial scales using isozymes, PCR-fingerprinting and analysis of the rDNA ITS regions either by polymerase chain reaction -restriction fragment length polymorphism (PCR-RFLP) analysis or sequencing. Ahlich-Schlegel (1997) studied the allelic diversity at seven isozyme loci and detected 108 different allozyme phenotypes among 194 European and North-American DSE strains. Allozyme patterns were neither host- nor site-specific. Harney et al. (1997) found many polymorphisms in the rDNA ITS regions of *P. fortinii* strains from Europe and North America by restriction mapping. Similarly, variability among rDNA ITS sequences was high (up to 12 substitutions) among 18 strains of *P. fortinii* from Central and Northern Europe (Grünig et al. 2002b). In contrast, Addy et al. (2000) detected a high degree of homogeneity among the rDNA ITS sequences of six strains of *P. fortinii* from Canada and Japan.

Strain-specific markers are necessary to study the genetic diversity at small spatial scales. In contrast to allozyme markers, ISSR-PCR markers were strain specific and allowed discrimination among isolates with identical allozyme phenotypes (Grünig et al. 2001). These markers were used to detect the population structure of DSE isolated from Norway spruce (*Picea abies*) roots collected within a 3 × 3 m plot of a 40-year-old plantation (Grünig et al. 2002a). Twenty-one unique ISSR-PCR genets were present among 144 strains. Identity of the isolated DSE as *P. fortinii* was confirmed by the morphology of the conidiogenous apparatus and by sequence comparisons of the rDNA ITS regions. Two genets dominated and were isolated from all sampling points within contiguous areas of at least 6.8 m^2 and 5.3 m^2 that overlapped by 3.6 m^2. Other genets were rare and were isolated only once or twice.

Jumpponen (1999) employed the random amplified polymorphic DNA (RAPD) technique to determine the population structure of *P. fortinii* at a primary succession site on a glacier forefront. In one year, 23 genets of *P. fortinii* were detected in 34 strains, in the next year 10 genets were found in 40 strains, but none of the genets was isolated in both years. Diversity of *P. fortinii* can be high even within single root pieces. For example, 8- to 10-cm-long pieces of fine root of *Picea abies* were colonised by up to six different inter-simple sequence repeat (ISSR) phenotypes (N. Nüssli and C.R. Grünig, unpublished) (Fig. 7.1). In summary, genetic diversity of *P. fortinii* seems to be high at every level. This is surprising for a supposedly asexual fungus. Therefore, studies on population genetics were initiated to find the sources of this high diversity.

ISSR-PCR and RAPD markers have many analytical drawbacks, such as dominance, and they cannot be used to infer population differentiation and recombination. In contrast, single-locus RFLP markers are codominant and supply robust data for precise population genetic analyses. In addition, data are comparable among studies and thus may be used for global analyses (Sunnucks 2000). Therefore, single-locus RFLP probes were developed for population genetic analysis of *P. fortinii* and used to find evidence for recombination, gene and genotype flow in *P. fortinii* (Grünig et al. 2003, 2004). Strains collected from three Norway-spruce plots up to 10 km apart

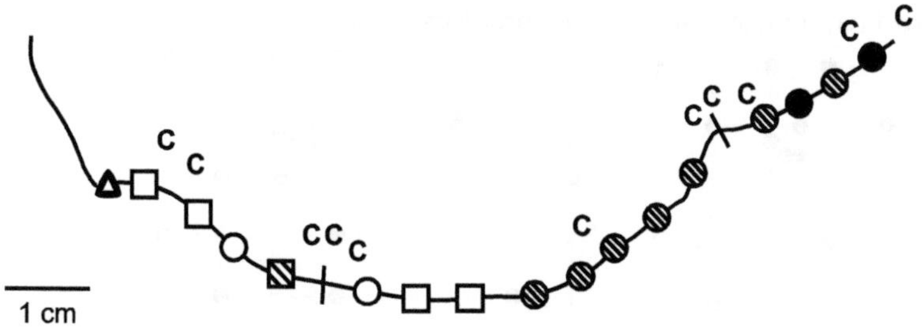

Fig. 7.1. Distribution of six inter-simple sequence repeat-polymerase chain reaction (ISSR-PCR) phenotypes belonging to three cryptic species of the root endophyte *Phialocephala fortinii* s. l. along a healthy fine root of Norway spruce (*Picea abies*). Identical symbols indicate positions on the root where the same phenotype was isolated. Symbols with identical shape represent the same cryptic species. *C* Positions on the root where *Cylindrocarpon didymum* was isolated as an endophyte

from each other were studied using 11 single-locus RFLP probes. The average gene diversity was high and up to 96 multilocus haplotypes (MLH) were observed per study plot. Significant population subdivision was detected among groups of MLH within plots, suggesting that groups were reproductively isolated and should be considered cryptic species. The RFLP data of more than 1,000 European strains indicate that *P. fortinii* s. l. is a species complex of at least eight cryptic species (C.R. Grünig, unpublished). The index of association (I_A) did not deviate significantly from zero within any cryptic species, suggesting that recombination occurs, or has occurred, within these species. Although evidence for recombination is strong for all cryptic species, it remains unclear whether sexual or parasexual processes are involved, and how often and where recombination occurs or when it last occurred (Taylor et al. 1999). Even a little sex is, however, already enough to give an organism the appearance of a recombining population (Brown 1999).

The sympatric occurrence of up to four reproductively isolated, cryptic species within a few square metres of forest floor, and sometimes even in the same root segment, is a highly interesting phenomenon and deserves a brief discussion (Grünig et al. 2004) (Figs. 7.1, 7.2). Reproductive isolation is essential for speciation. Geographically isolated populations are often reproductively isolated, and may experience allopatric speciation through genetic drift (Carter et al. 2001). On the other hand, niche or habitat specialisation may lead to sympatric speciation when local populations are confronted with heterogeneous habitats or several niches within habitats (Futuyma and Moreno 1988; Maynard Smith 1966). The patterns observed by Grünig et al.(2004) are clearly indicative of speciation. Possibly, the cryptic species were the products of allopatric speciation in the

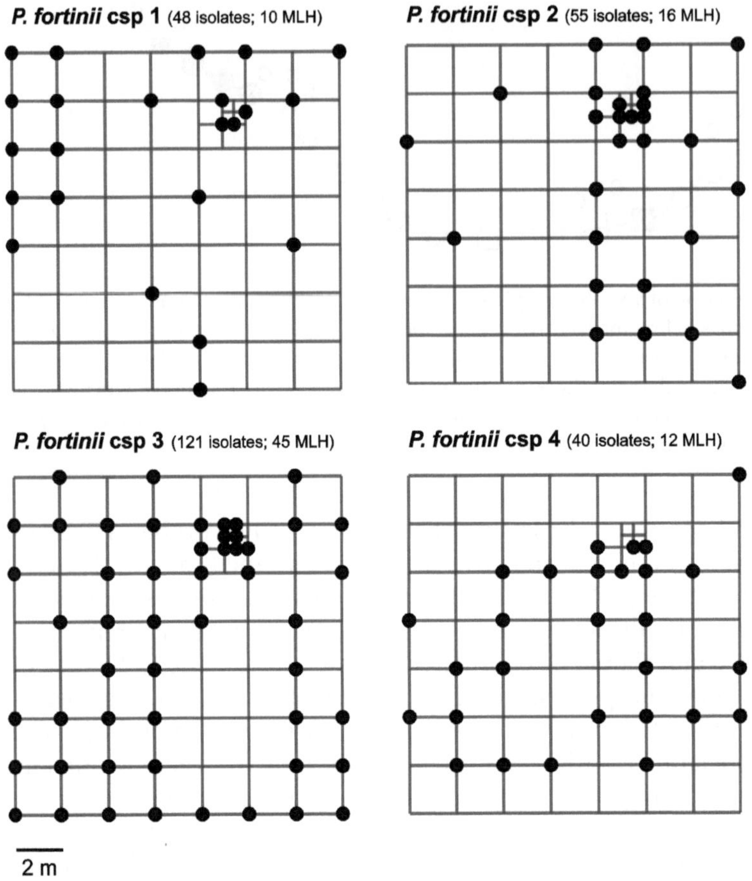

Fig. 7.2. Distribution of the four most frequently observed cryptic species (*csp*) of *Phialocephala fortinii* s. l. within healthy fine roots of Norway spruce (*Picea abies*) collected at the intersections of a 2 × 2 m grid superimposed on a forest plot (14 × 14 m) at Zürichberg, Switzerland. The four graphs represent the same study plot; the distributions of the four cryptic species are presented in separate graphs to maintain clarity. The number of isolates and the multilocus haplotypes (MLH) of each cryptic species are given in brackets

past due to geographical isolation. The ranges of these species may have subsequently overlapped (Brasier 1987). In this respect it is interesting to study the role of Quaternary climatic changes (Hewitt 2000). The succession of several glaciations and warmer inter-glacial periods had profound effects on animals, plants, and, consequently, on fungi. During the Quaternary, each species experienced many contractions/expansions of range, leading to extinctions and foundations of populations, decreases and increases in diversity and, thus, also to speciation (Taberlet et al. 1998). Refugia of relevant hosts of *P. fortinii* were often geographically isolated, making allopatric speciation of *P. fortinii* possible. Alternatively, habitat heterogeneities are certainly present even within very small compartments of root tissues, the rhizosphere, and the surrounding soil. These heterogeneities may be pronounced enough for ecological isolation and for the development of cryptic species. Some cryptic species may be interspecific hybrids. For example, most asexual *Epichloë*-related grass endophytes appear to be such hybrids (Scott 2001). Interspecific hybrids may be better adapted to new niches such as new hosts and can provide greater or more diverse benefits to host plants (Schardl and Craven 2003). However, such hybrids were never observed for *P. fortinii* using codominantly inherited single-copy RFLP markers.

MLH with identical ISSR fingerprinting patterns were common to at least two of the sites in the study of Grünig et al. (2004). These results indicate that not only gene flow but also genotype flow most likely occurs in cryptic species of *P. fortinii*. Gene and genotype flow occur either naturally via conidia or microsclerotia transported by wind or micro- and macrofauna, or by silvicultural practices. Genotypes may be introduced by planting plants from nurseries located up to several hundreds of kilometers away (Bürgi and Schuler 2003), since nursery plants are frequently colonised by DSE including *P. fortinii* (Danielson and Visser 1990). Alternatively, machinery used during thinning and harvesting could be responsible for the import of genotypes.

Nothing is known about the significance of mutations, the ultimate source of genetic variation, for speciation within *P. fortinii* s. l. If a population is large and the mutation rate high, it is likely that mutants with higher fitness, e.g. better mutualists, will emerge (McDonald and Linde 2002). Non-lethal somatic mutations in the mitotic phase may affect the genetic diversity of a population since each nucleus has the capacity to be the founder genome of another, new mycelium (Burnett 2003). The diversity thus generated may supplement diversity generated by recombination.

7.4 Conclusions

Colonisation of roots by fungal endophytes is a common feature in the plant kingdom. In contrast to classical mycorrhizae, endophytes are regularly present in roots undergoing secondary growth. Root-endophyte species diversity is affected by climatic, physical, chemical, biological and anthropogenic factors. DSEs are among the most abundant root endophytes. They constitute a taxonomically very heterogeneous group of fungi, mostly ascomycetes, that form melanised, septate hyphae, chlamydospores or microsclerotia within the roots of the host.

Phialocephala fortinii is the most prominent DSE, especially in woody plant species. *P. fortinii* s. l. is genotypically very diverse and forms a complex of several cryptic species that can occur sympatrically. Cryptic species and selected genotypes of *P. fortinii* s. l. can now be used to test the ecological significance of these extremely abundant and successful organisms and to explain some of the contradictory results on fungus-host interactions reported in earlier studies. The elucidation of the mating mechanism(s) and the evolutionary forces that govern speciation in *P. fortinii* s. l. are other fascinating topics for future research.

We have reviewed patterns of species diversity and within-species genotypic diversity and presented several plausible explanations for these patterns, although conclusive evidence for cause and effect are still virtually lacking. Nevertheless, we would like to conclude with a motivating citation by Begon et al. (1990): "This is not so much a disappointment as a challenge to ecologists and biologists of the future. Much of the fascination of ecology and biology lies in the fact that many problems are blatant and obvious for everybody to see, while the solutions have as yet eluded us".

References

Addy HD, Hambleton S, Currah RS (2000) Distribution and molecular characterization of the root endophyte *Phialocephala fortinii* along an environmental gradient in the boreal forest of Alberta. Mycol Res 104:1213–1221

Ahlich K, Sieber TN (1996) The profusion of dark septate endophytic fungi in non-ectomycorrhizal fine roots of forest trees and shrubs. New Phytol 132:259–270

Ahlich K, Rigling D, Holdenrieder O, Sieber TN (1998) Dark septate hyphomycetes in Swiss conifer forest soils surveyed using Norway-spruce seedlings as bait. Soil Biol Biochem 30:1069–1075

Ahlich-Schlegel K (1997) Vorkommen und Charakterisierung von dunklen, septierten Hyphomyceten (DSH) in Gehölzwurzeln. PhD dissertation, Swiss Federal Institute of Technology, Department of Forest Sciences, Zürich, Switzerland

Ananda K, Sridhar KR (2002) Diversity of endophytic fungi in the roots of mangrove species on the west coast of India. Can J Microbiol 48:871–878

Bååth E, Söderström B (1979) Fungal biomass and fungal immobilization of plant nutrients in Swedish coniferous forest soils. Rev Ecol Biol Sol 16:477–489

Barrow JR, Osuna P (2002) Phosphorus solubilization and uptake by dark septate fungi in fourwing saltbush, *Atriplex canescens* (Pursh) Nutt. J Arid Environ 51:449–459

Bazin MJ, Markham P, Scott EM, Lynch JM (1990) Population dynamics and rhizosphere interactions. In: Lynch JM (ed) The rhizosphere. Wiley, Chichester, UK, pp 99–127

Begon M, Harper JL, Townsend CR (1990) Ecology – individuals, populations, communities, 2nd edn. Blackwell, Oxford, UK

Bever JD, Westover KM, Antonovics J (1997) Incorporating the soil community into plant population dynamics: the utility of the feedback approach. J Ecol 85:561–573

Blaschke H (1986) Vergleichende Untersuchungen über die Entwicklung mykorrhizierter Feinwurzeln von Fichten in Waldschadensgebieten. Forstw Cbl 105:477–487

Bledsoe C, Klein P, Bliss LC (1990) A survey of mycorrhizal plants on Truelove Lowland, Devon Island, N. W. T., Canada. Can J Bot 68:1848–1856

Brasier CM (1987) The dynamics in fungal speciation. In: Rayner ADM, Brasier CM, Moore D (eds) Evolutionary biology of the fungi. Cambridge University Press, Cambridge, UK, pp 232–260
Bridge P, Spooner B (2001) Soil fungi: diversity and detection. Plant Soil 232:147–154
Brown JKM (1999) The evolution of sex and recombination in fungi. In: Worrall JJ (ed) Structure and dynamics of fungal populations. Kluwer, Dordrecht, pp 73–95
Bruns TD, Bidartondo MI, Taylor DL (2002) Host specificity in ectomycorrhizal communities: What do the exceptions tell us? Integr Comp Biol 42:352–359
Bürgi M, Schuler A (2003) Driving forces of forest management – an analysis of regeneration practices in the forests of the Swiss Central Plateau during the 19th and 20th century. For Ecol Manage 176:173–183
Burnett J (2003) Fungal populations and species. Oxford University Press, Oxford, UK
Cain RF (1952) Studies of fungi imperfecti. I. *Phialophora*. Can J Bot 30:338–343
Cairney JWG, Meharg AA (1999) Influences of anthropogenic pollution on mycorrhizal fungal communities. Environ Pollut 106:169–182
Cannon PF (1997) Strategies for rapid assessment of fungal diversity. Biodivers Conserv 6:669–680
Cannon PF, Hawksworth DL (1995) The diversity of fungi associated with vascular plants: the known, the unknown and the need to bridge the knowledge gap. Adv Plant Pathol 11:277–302
Carroll GC (1988) Fungal endophytes in stems and leaves – from latent pathogen to mutualistic symbiont. Ecology 69:2–9
Carter DA, Burt A, Taylor JW, Koenig GL, Dechairo B, White TJ (1997) A set of electrophoretic molecular markers for strain typing and population genetic studies of *Histoplasma capsulatum*. Electrophoresis 18:1047–1053
Carter DA, Taylor JW, Dechairo B, Burt S, Koenig GL, White TJ (2001) Amplified single-nucleotide polymorphisms and a (GA)(n) microsatellite marker reveal genetic differentiation between populations of *Histoplasma capsulatum* from the Americas. Fungal Genet Biol 34:37–48
Cevnik M, Jurc M, Vodnik D (2000) Filamentous fungi associated with the fine roots of *Erica herbacea* L. from the area influenced by the Zerjav lead smelter (Slovenia). Phyton Ann Rei Bot 40:61–64
Chan TAB (1923) Über die Mykorrhiza der Buche. Allg Forst J Ztg 99:25–52
Christensen M (1981) Species diversity and dominance in fungal communities. In: Wicklow DT, Carroll GC (eds) The fungal community, its organization and role in the ecosystem. Dekker, New York, pp 201–232
Christensen M (1989) A view of fungal ecology. Mycologia 81:1–19
Christie P, Nicolson TH (1983) Are mycorrhizas absent from the Antarctic? Trans Br Mycol Soc 80:557–560
Coûteaux M-M, Kurz C, Bottner P, Raschi A (1999) Influence of increased atmospheric CO_2 concentration on quality of plant material and litter decomposition. Tree Physiol 19:301–311
Danielson RM, Visser S (1990) The mycorrhizal and nodulation status of container-grown trees and shrubs reared in commercial nurseries. Can J For Res 20:609–614
Deacon JW (1981) Ecological relationships with other fungi: competitors and hyperparasites. In: Asher MJC, Shipton PJ (eds) Biology and control of Take-all. Academic, London, UK, pp 75–101
Fernando AA, Currah RS (1996) A comparative study of the effects of the root endophytes *Leptodontidium orchidicola* and *Phialocephala fortinii* (fungi imperfecti) on the growth of some subalpine plants in culture. Can J Bot 74:1071–1078

Fisher PJ, Petrini O, Webster J (1991) Aquatic hyphomycetes and other fungi in living aquatic and terrestrial roots of *Alnus glutinosa*. Mycol Res 95:543–547
Fisher PJ, Graf F, Petrini LE, Sutton BC, Wookey PA (1995) Fungal endophytes of *Dryas octopetala* from a high arctic polar semidesert and from the Swiss Alps. Mycologia 87:319–323
Fitter AH, Atkinson D, Read DJ, Usher MB (1985) Ecological interactions in soil. Blackwell, Oxford
Foster RC, Rovira AD, Cock TW (1983) Ultrastructure of the root-soil interface. APS Press, St. Paul, MN
Freisleben R (1934) Zur Frage der Mykotrophie in der Gattung *Vaccinium* L. Jahrb wissenschaftl Bot 80:421–456
Futuyma DJ, Moreno G (1988) The evolution of ecological specialization. Annu Rev Ecol Syst 19:207–233
Geiser DM, Arnold ML, Timberlake WE (1994) Sexual origins of British *Aspergillus nidulans* isolates. Proc Natl Acad Sci USA 91:2349–2352
Girlanda M, Ghignone S, Luppi AM (2002) Diversity of sterile root-associated fungi of two Mediterranean plants. New Phytol 155:481–498
Gochenaur SE (1984) Fungi of a Long Island oak-birch forest. II. Population dynamics and hydrolase patterns for the soil Penicillia. Mycologia 76:218–231
Görke C (1998) Mykozönosen von Wurzeln und Stamm von Jungbäumen unterschiedlicher Bestandsbegründungen. Bibl Mycol 173:1–462
Grünig CR, Sieber TN (2005) Molecular and phenotypic description of the widespread root symbiont *Acephala applanata* gen. et sp. nov., formerly known as "Dark Septate Endophyte Type 1". Mycologia 97:628–640
Grünig CR, Sieber TN, Holdenrieder O (2001) Characterisation of dark septate endophytic fungi (DSE) using inter-simple-sequence-repeat-anchored polymerase chain reaction (ISSR-PCR) amplification. Mycol Res 105:24–32
Grünig CR, Sieber TN, Rogers SO, Holdenrieder O (2002a) Spatial distribution of dark septate endophytes in a confined forest plot. Mycol Res 106:832–840
Grünig CR, Sieber TN, Rogers SO, Holdenrieder O (2002b) Genetic variability among strains of *Phialocephala fortinii* and phylogenetic analysis of the genus *Phialocephala* based on rDNA ITS sequence comparisons. Can J Bot 80:1239–1249
Grünig CR, Linde CC, Sieber TN, Rogers SO (2003) Development of single-copy RFLP markers for population genetic studies of *Phialocephala fortinii* and closely related taxa. Mycol Res 107:1332–1341
Grünig CR, McDonald BA, Sieber TN, Rogers SO, Holdenrieder O (2004) Evidence for subdivision of the root-endophyte *Phialocephala fortinii* into cryptic species and recombination within species. Fungal Genet Biol 41:676–687
Hadacek F, Kraus GF (2002) Plant root carbohydrates affect growth behaviour of endophytic microfungi. FEMS Microbiol Ecol 41:161–170
Harley JL, Waid JS (1955) A method of studying active mycelia on living roots and other surfaces in soil. Trans Br Mycol Soc 38:104–118
Harney SK, Rogers SO, Wang CJK (1997) Molecular characterization of dematiaceous root endophytes. Mycol Res 101:1397–1404
Haselwandter K, Read DJ (1980) Fungal associations of roots of dominant and sub-dominant plants in high-alpine vegetation systems with special reference to mycorrhiza. Oecologia 45:57–62
Haselwandter K, Read DJ (1982) The significance of a root-fungus association in two *Carex* species of high-alpine plant communities. Oecologia 53:352–354

Hashimoto Y, Hyakumachi M (2000) Quantities and types of ectomycorrhizal and endophytic fungi associated with *Betula platyphylla* var. *japonica* seedlings during the initial stage of establishment of vegetation after disturbance. Ecol Res 15:21–31

Hawksworth DL (1991) The fungal dimension of biodiversity: magnitude, significance, and conservation. Mycol Res 95:641–655

Hewitt GM (2000) The genetic legacy of the Quaternary ice ages. Nature 405:907–913

Hill MO (1973) Diversity and evenness: a unifying notation and its consequences. Ecology 54:427–432

Horton TR, Cázares E, Bruns TD (1998) Ectomycorrhizal, vesicular-arbuscular and dark septate fungal colonisation of bishop pine (*Pinus muricata*) seedlings in the first 5 months of growth after wildfire. Mycorrhiza 8:11–18

Hurlbert SH (1971) The non-concept of species diversity: a critique and alternative parameters. Ecology 52:577–586

Jacobs A, Coetzee MPA, Wingfield BD, Jacobs K, Wingfield MJ (2003) Phylogenetic relationships among *Phialocephala* species and other ascomycetes. Mycologia 95:637–645

Jansen E, van Dobben HF (1987) Is decline of *Cantharellus cibarius* in the Netherlands due to air pollution. Ambio 16:27–29

Jones MD, Durall DM, Harniman SMK, Classen DC, Simard SW (1997) Ectomycorrhizal diversity on *Betula papyrifera* and *Pseudotsuga menziesii* seedlings grown in the greenhouse or outplanted in single-species and mixed plots in southern British Columbia. Can J For Res 27:1872–1889

Jumpponen A (1999) Spatial distribution of discrete RAPD phenotypes of a root endophytic fungus, *Phialocephala fortinii*, at a primary successional site on a glacier forefront. New Phytol 141:333–344

Jumpponen A, Trappe JM (1998a) Performance of *Pinus contorta* inoculated with two strains of root endophytic fungus, *Phialocephala fortinii*: effects of synthesis system and glucose concentration. Can J Bot 76:1205–1213

Jumpponen A, Trappe JM (1998b) Dark septate endophytes: A review of facultative biotrophic root-colonising fungi. New Phytol 140:295–310

Jumpponen A, Mattson KG, Trappe JM (1998) Mycorrhizal functioning of *Phialocephala fortinii* with *Pinus contorta* on glacier forefront soil: interactions with soil nitrogen and organic matter. Mycorrhiza 7:261–265

Jumpponen A, Claridge AW, Trappe JM, Lebel T, Claridge DL (2004) Ecological relationships among hypogeous fungi and trees: inferences from association analysis integrated with habitat modeling. Mycologia 96:510–525

Kattner D, Schönhar S (1990) Untersuchungen über das Vorkommen mikroskopischer Pilze in Feinwurzeln optisch gesunder Fichten (*Picea abies* Karst.) auf verschiedenen Standorten. Mitt Ver Forstl Standortskde Forstpflanzenzücht 35:39–43

Kernaghan G, Sigler L, Khasa D (2003) Mycorrhizal and root endophytic fungi of containerized *Picea glauca* seedlings assessed by rDNA sequence analysis. Microb Ecol 45:128–136

Kirk JJ, Deacon JW (1987) Control of the take-all fungus by *Microdochium bolleyi*, and interactions involving *M. bolleyi*, *Phialophora graminicola* and *Periconia macrospinosa* on cereal roots. Plant Soil 98:231–237

Körner C (2000) Biosphere responses to CO_2 enrichment. Ecol Appl 10:1590–1619

Kovacs GM, Szigetvari C (2002) Mycorrhizae and other root-associated fungal structures of the plants of a sandy grassland on the Great Hungarian Plain. Phyton Ann Rei Bot 42:211–223

Maynard Smith J (1966) Sympatric speciation. Am Nat 100:637–650

McDonald BA, Linde C (2002) Pathogen population genetics, evolutionary potential, and durable resistance. Annu Rev Phytopathol 40:349–379

Melin E (1922) On the mycorrhizas of *Pinus silvestris* L. and *Picea abies* (L.) Karst. J Ecol 9:254–257

Melin E (1923) Experimentelle Untersuchungen über die Konstitution und Ökologie der Mycorrhizen von *Pinus silvestris* L. und *Picea abies* (L.) Karst. Falk Mykol Unters 2:73–331

Narisawa K, Kawamata H, Currah RS, Hashiba T (2002) Suppression of *Verticillium* wilt in eggplant by some fungal root endophytes. Eur J Plant Pathol 108:103–109

Newsham KK (1999) *Phialophora graminicola*, a dark septate fungus, is a beneficial associate of the grass *Vulpia ciliata* ssp. *ambigua*. New Phytol 144:517–524

Oba H, Tawaraya K, Wagatsuma T (2002) Inhibition of pre-symbiotic hyphal growth of arbuscular mycorrhizal fungus *Gigaspora margarita* by root exudates of *Lupinus* spp. Soil Sci Plant Nutr 48:117–120

Oberholzer-Tschütscher B (1982) Untersuchungen über endophytische Pilze von *Erica carnea* L. PhD dissertation, Swiss Federal Institute of Technology, Institute of Microbiology, Zürich, Switzerland

O'Dell TE, Massicotte HB, Trappe JM (1993) Root colonisation of *Lupinus latifolius* Agardh. and *Pinus contorta* Dougl. by *Phialocephala fortinii* Wang & Wilcox. New Phytol 124:93–100

Papritz A, Flühler H (1991) Räumliche Verteilung von bodenchemischen Grössen auf Transsekten zwischen Bäumen (Beobachtungsfläche Lägern). In: Pankow W (ed) Lufthaushalt, Luftverschmutzung und Waldschäden in der Schweiz, Band 6, Belastung von Waldböden. Verlag der Fachvereine, Zürich, pp 125–136

Petrini O, Fisher PJ, Petrini LE (1992) Fungal endophytes of bracken (*Pteridium aquilinum*), with some reflections on their use in biological control. Sydowia 44:282–293

Read DJ, Haselwandter K (1981) Observation on the mycorrhizal status of some alpine plant communities. New Phytol 88:341–352

Reinecke P (1978) *Microdochium bolleyi* at the stem base of cereals. Z Pflanzenkr Pflanzenschutz 85:679–685

Richard C, Fortin J-A (1973) The identification of *Mycelium radicis atrovirens* (*Phialocephala dimorphospora*). Can J Bot 51:2247–2248

Riesen T, Sieber TN (1985) Endophytic fungi in winter wheat (*Triticum aestivum* L.). Institute of Microbiology, Swiss Federal Institute of Technology, Zürich, Switzerland

Rillig MC, Wright SF, Allen MF, Field CB (1999) Rise in carbon dioxide changes in soil structure. Nature 400:628–628

Robertson NF (1954) Studies on the mycorrhiza of *Pinus silvestris*. I. Pattern of development of mycorrhizal root and its significance for experimental studies. New Phytol 53:253–283

Ruotsalainen AL, Väre H, Vestberg M (2002) Seasonality of root fungal colonisation in low-alpine herbs. Mycorrhiza 12:29–36

Schadt CW, Mullen RB, Schmidt SK (2001) Isolation and phylogenetic identification of a dark-septate fungus associated with the alpine plant *Ranunculus adoneus*. New Phytol 150:747–755

Schardl CL, Craven KD (2003) Interspecific hybridization in plant-associated fungi and oomycetes: a review. Mol Ecol 12:2861–2873

Schulz B, Römmert AK, Dammann U, Aust H-J, Strack D (1999) The endophyte-host interaction: a balanced antagonism? Mycol Res 103:1275–1283

Scott B (2001) *Epichloë* endophytes: fungal symbionts of grasses. Curr Opin Microbiol 4:393–398

Sieber TN (2002) Fungal root endophytes. In: Waisel Y, Eshel A, Kafkafi U (eds) Plant roots: The hidden half, 3rd edn. Dekker, New York, pp 887–917

Sieber TN, Riesen TK, Müller E, Fried PM (1988) Endophytic fungi in four winter wheat cultivars (*Triticum aestivum* L.) differing in resistance against *Stagonospora nodorum* (Berk.) Cast. & Germ. = *Septoria nodorum* (Berk.) Berk. J Phytopathol 122:289–306
Skipp RA, Christensen MJ (1989) Fungi invading roots of perennial ryegrass (*Lolium perenne* L.) in pasture. N Z J Agric Res 32:423–431
Soerensen T (1948) A method of establishing groups of equal amplitude in plant sociology. Vid Selsk Biol Skr 5:4–4
Steinke E, Williams PG, Ashford AE (1996) The structure and fungal associates of mycorrhizas in *Leucopogon parviflorus* (Andr.) Lindl. Ann Bot 77:413–419
Stoyke G, Currah RS (1991) Endophytic fungi from the mycorrhizae of alpine ericoid plants. Can J Bot 69:347–352
Stoyke G, Egger KN, Currah RS (1992) Characterization of sterile endophytic fungi from the mycorrhizae of subalpine plants. Can J Bot 70:2009–2016
Sunnucks P (2000) Efficient genetic markers for population biology. Trends Ecol Evol 15:199–203
Swift MJ (1976) Species diversity and the structure of microbial communities in terrestrial habitats. In: Anderson JM, Macfayden A (eds) The role of terrestrial and aquatic organisms in decomposition processes. Blackwell, Oxford, UK, pp 185–222
Taberlet P, Fumagalli L, Wust-Saucy AG, Cosson JF (1998) Comparative phylogeography and postglacial colonisation routes in Europe. Mol Ecol 7:453–464
Taylor AFS, Read DJ (1996) A European north-south survey of ectomycorrhizal populations on spruce. In: Azcon-Aguilar C, Barea JM (eds) Mycorrhizas in integrated systems from genes to plant development, Proc 4th European Symposium on Mycorrhizas. Office for official publications of the European Community, Luxembourg, pp 144–147
Taylor JW, Jacobson DJ, Fisher MC (1999) The evolution of asexual fungi: reproduction, speciation and classification. Annu Rev Phytopathol 37:197–246
Torsvik V, Salte K, Sorheim R, Goksoyr J (1990) Comparison of phenotypic diversity and DNA heterogeneity in a population of soil bacteria. Appl Environ Microbiol 56:776–781
Van Der Putten WH (2003) Plant defense belowground and spatiotemporal processes in natural vegetation. Ecology 84:2269–2280
Van Tol RWHM, Van der Sommen ATC, Boff MIC, Van Bezooijen J, Sabelis MW, Smits PH (2001) Plants protect their roots by alerting the enemies of grubs. Ecol Lett 4:292–294
Väre H, Vestberg M, Eurola S (1992) Mycorrhiza and root associated fungi in Spitsbergen. Mycorrhiza 1:93–104
Wang CJK, Wilcox HE (1985) New species of ectendomycorrhizal and pseudomycorrhizal fungi: *Phialophora finlandia*, *Chloridium paucisporum*, and *Phialocephala fortinii*. Mycologia 77:951–958
Ward E, Bateman GL (1999) Comparison of *Gaeumannomyces*- and *Phialophora*-like fungal pathogens from maize and other plants using DNA methods. New Phytol 141:323–331
Watanabe T (1994) Pictorial atlas of soil and seed fungi. Lewis, Boca Raton, FL
White TJ, Bruns TD, Lee S, Taylor J (1990) Amplification and direct sequencing of fungal ribosomal RNA genes for phylogenetics. In: Innis MA, Gelfand DH, Sninsky JJ, White TJ (eds) PCR Protocols: a guide to methods and applications. Academic, New York, pp 315–322
Wilberforce EM, Boddy L, Griffiths R, Griffith GW (2003) Agricultural management affects communities of culturable root-endophytic fungi in temperate grasslands. Soil Biol Biochem 35:1143–1154
Wilcox HE, Wang CJK (1987) Mycorrhizal and pathological associations of dematiaceous fungi in roots of 7-month-old tree seedlings. Can J For Res 17:884–899
Wirsel SGR, Leibinger W, Ernst M, Mendgen K (2001) Genetic diversity of fungi closely associated with common reed. New Phytol 149:589–598

Yu T, Nassuth A, Peterson RL (2001) Characterization of the interaction between the dark septate fungus *Phialocephala fortinii* and *Asparagus officinalis* roots. Can J Microbiol 47:741–753

Zietkiewicz E, Rafalski A, Labuda D (1994) Genome fingerprinting by simple sequence repeat (SSR)-anchored polymerase chain reaction amplification. Genomics 20:176–183

8 Endophytic Root Colonization by *Fusarium* Species: Histology, Plant Interactions, and Toxicity

Charles W. Bacon, Ida E. Yates

8.1 Introduction

Fusarium species have adapted to a wide range of geographical sites, climatic conditions, ecological habitats, and host plants, and species of this polyphyletic genus have been documented to occur worldwide (Backhouse et al. 2001). In spite of the information available on the extremes in geographic distribution and climatic conditions, appropriate data to predict the center of origin(s) or the mode(s) of dispersion of this genus have not been obtained. Much of the information on distribution patterns has been determined from analyses of soil samples, a common habitat, in addition to colonization of many plant species. The diversity of plant species colonized by members of the genus *Fusarium* is amazing. A recent literature survey determined that *Fusarium* species have been isolated from plants belonging to the gymnosperms and the monocotyledonous and dicotyledonous angiosperms (Kuldau and Yates 2000). They are the primary incitants of root, stem, and ear rots in many agriculturally important crops. For example, *F. verticillioides* (= *F. moniliforme*) is capable of colonizing well over 1,000 plant species, including maize (*Zea mays* L.), one of the world's most important food crops. Another species, *F. oxysporum*, is cosmopolitan; certain strains are usually host specific and pose a severe threat to most of the world's supply of food crops. Furthermore, species such as *F. graminearum*, along with *F. verticillioides* and related species within the *Liseola* section, are notorious for the production of mycotoxins on wheat, maize, barley, rice and other cereal grains and foodstuffs (Marasas et al. 1984). Consequently, studies on the association of *Fusarium* species with plants are critical in order to develop control measures for this group of fungi that affects the quality and quantity of the world's food supply.

Charles W. Bacon: Richard B. Russell Research Center, ARS, United States Department of Agriculture, Toxicology and Mycotoxin Research Unit, SAA, P.O. Box 5677, Athens, GA 30604, USA, E-mail: cbacon@saa.ars.usda.gov

Ida E. Yates: Richard B. Russell Research Center, ARS, United States Department of Agriculture, Toxicology and Mycotoxin Research Unit, Athens, GA 30604, USA

Soil Biology, Volume 9
Microbial Root Endophytes
B. Schulz, C. Boyle, T. N. Sieber (Eds.)

The discussion of *Fusarium* root endophytes in this chapter is based on, and will be discussed relative to, our knowledge of fungal endophytes of grasses. Noted examples of fungal endophytes include the species of the Balansieae that show various degrees of tissue specificity and often display evidence of infection by the production of sporulation structures on the adaxial or abaxial leaf surface of grasses (Diehl 1950), as well as the production of characteristic toxic secondary metabolites (Bacon et al. 1986). For example, fungi of the genus *Neotyphodium* (teleomorph = *Epichloë*) are found only in the stems, leaves and seed of grasses but are not found in roots, while species of *Myriogenospora* are restricted to the leaves, and species of *Balansia* are restricted to stems or leaves, but all produce ergot alkaloids.

Some research suggests that there are similar positive interactions of endophytic *Fusarium* species with plants (Damicone and Manning 1982; Hallmann and Sikora 1994a, 1994b; Blok and Bollen 1995). We use the definition of fungal endophytes as indicated by Stone et al. (2000) to include those *Fusarium* species that are associated with roots as intercellular, symptomless fungi (Fig. 8.1a–e), although the endophytic association may extend to above ground plant parts and there may be a differential expression of infection with different host tissue types (Bergman and Bakker-Van der Voort 1979; Fisher et al. 1992; Foley 1962; Yates and Jaworski 2000). In addition to grass endophytes, specific attention will be directed to our past and present toxicological, physiological, and morphological studies on *Fusarium verticillioides* [synonym *F. moniliforme*, teleomorph *Gibberella fujikuroi* (Sawada) Ito in Ito & K. Kimura] and its association with maize. Thus, species of *Fusarium* endophytes include those fungi that occupy the intercellular spaces of plants, and the intercellular infections may be localized to roots. However, localization to roots is not mutually exclusive as some *Fusarium* species, while living as root endophytes, may also infect above ground plant organs, although the foliage origin of such hyphae from the endophytic infections in the roots has not been established for all species. Indeed, secondary infections from aerial spores are suspected of contributing to most foliage infection (Adams 1921; Boshoff et al. 1996; Kang and Buchenauer 2000) and, as discussed below, there is the possibility of different specific fungal strains that infect specific tissue types, especially the flower infections.

8.2 Plant and Fungus Interactions

Contrary to the association found in *Neotyphodium* grass species, *Fusarium* species are not necessarily obligate endophytes. Indeed, as presented

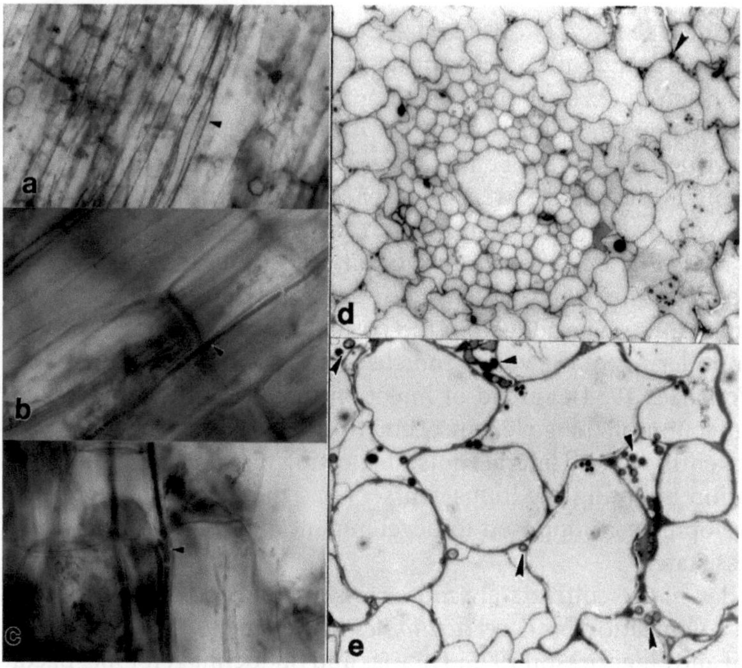

Fig. 8.1. a–e Light micrographs of the endophytic habit of a non-virulent isolate of *Fusarium verticillioides*, RRC 826, in roots of 2-week-old maize seedlings. **a** Hyphae (*arrowhead*) running parallel within intercellular spaces of the first internodes and junction of primary root of maize (54.4X). **b** Higher magnification of maize root showing a branching septate hypha (*arrowhead*) between two cell walls (272X, phase contrast). **c** Hyphae running parallel within intercellular spaces with a branching hypha at arrowhead (272X, phase contrast). **d** Cross section of a secondary root with hyphae (*arrowhead*) in the intercellular spaces (54.5X) (very dark spherical bodies within cells are artifacts of staining). **e** Cross section through the cortex of a primary root with groups of hyphae (*arrowheads*) in the intercellular spaces (109X, phase contrast) (from Bacon and Hinton 1996, with permission)

below, their nutritional physiology, i.e., parasitic and saprophytic, predicts a transitory nature of the endophytic phase of *Fusarium* associations. Understanding to what extent species or isolates of this genus are endophytic is hampered by studies, histological or otherwise, that inadequately describe the qualitative and quantitative distribution of *Fusarium* within hosts. Nor are there studies to indicate any host requirements or benefits derived from such association that are indicated and characteristic of those derived from the *Neotyphodium* grass endophytes (for review, see Bacon and White 2000). However, some studies are highly suggestive of benefits, at least to the fungus (Lee et al. 1995; Yates et al. 1997; Munkvold and Carlton 1997; Kuldau and Yates 2000; Pinto et al. 2000; Bacon et al. 2004).

Reports about the interactions of *F. verticillioides* with maize have been contradictory since the initial description of this fungus as both a pathogenic and a symptomless infection (Sheldon 1902; Voorhees 1934). Anatomical features of pathogenic infections by *Fusarium* species, including *F. verticillioides*, have been reviewed (Pennypacker 1981; Bacon and Hinton 1996), although some of the reports were concerned mainly with above ground plant parts. Voorhees (1934) described an initial infection of *F. verticillioides* into roots that occurred from the soil. Infection took place via the primary radicle by the fungus entering the epidermis, although it can also enter through ruptures produced in the cortex by emerging lateral roots. Since the endodermis of the young radicle acts as the barrier against penetration (Voorhees 1934), spread of infection into the stele is prevented. However, infection of the stele can occur from soil via the wounds produced by adventitious and lateral roots, suggesting that the degree and rate of suberization in young seedlings can serve as the key to the nature of disease development as opposed to development and the duration of the symptomless state.

Modern day maize cultivars are more resistant and, thus, symptomless endophytic colonization by *F. verticillioides* is the rule today (Foley 1962; Kommedahl and Siggerirsson 1975; Thomas and Buddenhagen 1980; Bacon and Hinton 1988; Ayers et al. 1989; Leslie et al. 1990; Corell et al. 1992; Bentley et al. 1995; Ahmed et al. 1996; Anaya and Roncero 1996; Bacon and Hinton 1996; Bai 1996; Bakan et al. 2002; Anjaiah et al. 2003). Maize kernels are universally infected by *F. verticillioides* and related species, but disease symptoms are rarely exhibited. Possibly, plant breeding and selection may also form the basis for symptomless root infection in other agricultural species and cultivars. Further, species and strain genetics are important within the overall population of each species, and are important in the overall impact of a species on a host. Indeed, there is now information to indicate that a genetic change that will convert a *Fusarium* pathogen to a nonpathogenic endophytic mutualist can occur (Freeman and Rodriguez 1993). The following are brief descriptions of both the symptomless and pathogenic infections.

Molecular tools have also been utilized to study the association of *F. verticillioides* with maize and other plants. In situ studies of this species were made possible with avirulent isolates of *F. verticillioides* transformed with a plasmid containing the *gus*A gene coding for β-glucoronidase (GUS) and the *hygr* gene coding for hygromycin resistance (Yates et al. 1999). Subsequent GUS activity is detectable by histochemical and fluorometric enzymatic assays during the colonization period. The results indicated that *F. verticillioides* could be traced from the initial seed, to recovery from roots of plants produced from these seed, up through the stem, and finally isolated internally from seed (Bacon et al. 2001). Further, it was

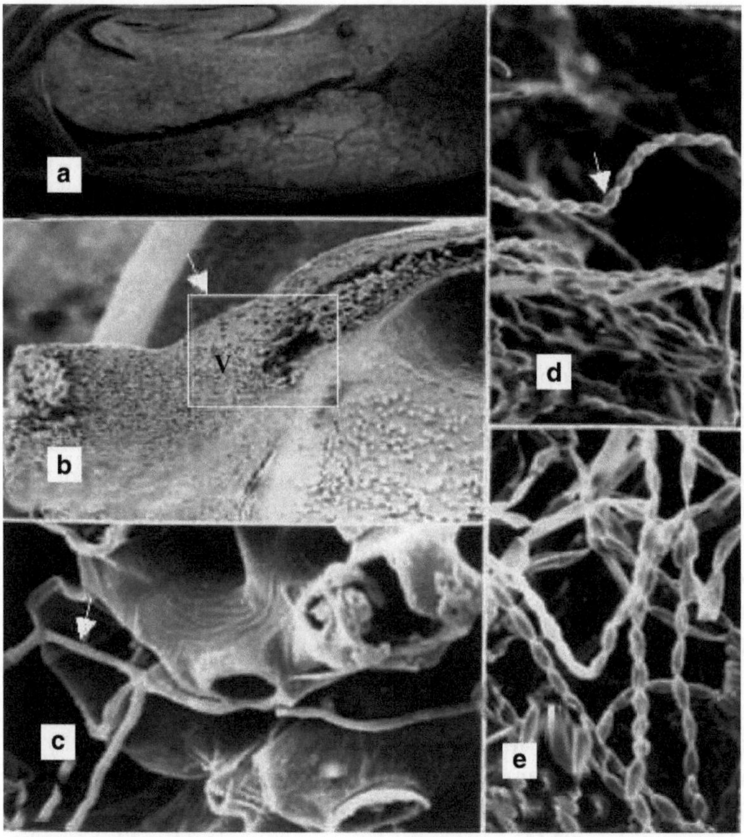

Fig. 8.2. **a–e** Scanning electron micrographs of a symptomless *F. verticillioides*-infected maize kernel. **a** A longitudinal section showing the location of the fungus at the tip cap (x13). **b** Magnified version of **a**, showing the location of the fungus in box (*arrow*) below the vascular tissues (*v*). **c** A hypha of *F. verticillioides* (*arrow*) within the boxed area of **b**. **d** Growth of the fungus on culture medium showing chains of microconidia (*arrow*) characteristic of this species. **e** Growth of the fungus on the other half of the sectioned maize kernel in **a**, above, incubated aseptically for 2 days on damp filter paper, and showing the chains of microconidia, which established that the hypha in **a** was that of *F. verticillioides* (from Bacon et al. 1992, with permission)

demonstrated that seed produced from these plants are sound, the fungus is internally seed borne (Fig. 8.2a–f), and produce seedlings that are infected with the transformed fungus (Bacon et al. 2001). Thus, this species is disseminated vertically, but horizontal dissemination is expected to occur through wounds due to the activity of insects (Munkvold and Carlton 1997), from soil to roots and other injured plant parts (Foley 1962), as well as via aerial borne spores. Most *Fusarium* species are also disseminated horizontally (Adams 1921), although vertical transmission is equally possible since several species have been recovered from seed as endophytes (Gordon 1952; Fisher and Petrini 1992).

8.2.1
Hemibiotrophic Characteristics

Hemibiotrophic fungi include those that infect living tissue, similar to biotrophs, but after an extended incubation period of days or weeks the infected tissue dies, within which the fungus now continues to develop as a saprotroph, usually resulting in sporulation (Luttrell 1974). The distinction between a hemibiotrophic parasite and a necrotrophic parasite is one of tissue infections. Necrotrophs kill host tissue in advance of penetration; while biotrophs penetrate, infect and obtain their food from living tissue, and this includes sporulation on living tissue. Having made the distinction between the basic terminologies used to distinguish the nutrition association of parasitic fungi with vascular plants, we now can apply this to endophytic *Fusarium* species.

Following the definition above, *Fusarium* species should be considered hemibiotrophs. That is, they are fungi that infect living tissue as biotrophs but after a latency period, which may last for a period of days to weeks, can cause host tissue to die, at which point the fungus becomes a saprotroph. Thus, endophytic *Fusarium* species, along with other fungal endophytes, are facultative biotrophic parasites, although this mode of nutrition can be a transient feature. Fungi belonging to *Claviceps* and *Colletotrichum*, as well as *Ustilago maydis* and *Magnaporthe grisea*, are typical hemibiotrophs. However, as we shall see below, not all isolates of the species *F. verticillioides* are hemibiotrophic and this might be due to either host and fungus genetics or environmental factors, or both. Certainly, in those situations where the saprotrophic stage is prevented, mycotoxin accumulation might be reduced, providing one point of reducing the concentration of these toxins.

The parasitic association of *Fusarium* species with roots as endophytes may be viewed as a long-term event, perhaps confounded by plant breeding, and selection not for endophytic infection but for disease expression. Plant breeding might make it difficult to clearly define the degree of biotrophy characteristic for each *Fusarium* species relative to specific modern agricultural host cultivars that differ in disease resistance. In addition, the degree of biotrophy depends on the genetics of fungal strains and hosts, and the site of inoculation (Corell et al. 1992; Munkvold and Carlton 1997; Carter et al. 2000; Mesterhaszy et al. 2003). Thus, the association of *Fusarium* species with plants as symptomless endophytes cannot be explained entirely on the basis of the nutritional parasitic relationships described above. Complete understanding of *Fusarium* species as symptomless root endophytes is hampered due to the complex of interactions, compounded by the fact that symptomless associations are altered by damaged portions of roots, due to predation by insects and wounds from emerging lateral roots. This results in a transformation of the symptomless state of the root-infecting species to the biotrophic and saprophytic state that infects dying and dead tissue. Most *Fusarium* rot diseases, i.e., root rots, are characterized during this phase. However, this appearance of biotrophic and saprophytic states is not necessarily obligatory and, indeed, a plant may have the symptomless state in roots while other parts of the same plant may be diseased. It is this aspect of the association that is of major concern in this review.

8.2.2 Histology

Symptomless root infections are characteristic of many *Fusarium* species (Foley 1962; Pennypacker 1981; Wong et al. 1992; Leslie 1994; Bacon and Hinton 1996; Bowden and Leslie 1994; Gang et al. 1998; Nicholson et al. 1998; Kedera et al. 1999; Kuldau and Yates 2000; Bai et al. 2002), and there are very aggressive or virulent strains of all species that serve as incitants of several plant diseases (Foley 1962; Malalasekera et al. 1973; Pennypacker 1981; Manaka and Chelkowski 1985; Liddell and Burgess 1985; Wilcoxon et al. 1988; Fisher and Petrini 1992; Bacon and Hinton 1996; Bai 1996; Boshoff et al. 1996; Parry and Nicholson 1996; Kosiak et al. 1997; Nicholson et al. 1998; Wildermuth et al. 1999; Hysek et al. 2000; Ribichich et al. 2000). Disease development is considered to be a consequence of fungal and host genetics, while environmental biotic and abiotic factors negatively affect overall survival of the host (Schroeder and Christensen 1963; Bacon et al. 1996; Ribichich et al. 2000).

Detailed histological studies of specific *Fusarium* species are lacking since it is the disease state that is most often studied. Further, molecular analyses now suggest that these studies involved either different species or more than one species. For example, we now know that studies of scab or head blight of wheat consisted of several cryptic species. The flower-foliage disease or scab is caused primarily by *F. graminearum* and, to a lesser extent, by a new species *F. pseudograminearum*, while crown rot of wheat is caused by yet another new species, *Gibberella coronicola* (Aoki and O'Donnell 1999; O'Donnell et al. 2000). Nevertheless, there are similarities and the following discussion takes these into account.

Bacon and Hinton (1996) compared root and shoot infections by both virulent and non-virulent strains of *F. verticillioides* on maize using light- and electron-transmission-microscopy and concluded that this strain, and perhaps most others, should be considered as symptomless endophyte(s)

(Fig. 8.1a–e), especially since most isolates produce symptomless infections with most modern day cultivars of maize. It was also concluded that this species was not a vascular rot fungus, agreeing with the earlier observations of Pennypacker (1981) that *F. verticillioides* has the potential to be a cortical rot fungus (Fig. 8.1d–e). The hyphae of *F. verticillioides* run parallel within intercellular spaces, although branching hyphae are observed, especially in areas of branching roots (Fig. 8.1c). A study of the symptomless state in seedling roots suggests that the symptomless state persists beyond the seedling stage and contributes, without visual signs, to mycotoxin contamination of maize both before and during kernel development (Bacon and Hinton 1996). Symptomless root infections have been reported in plants infected by several species of *Fusarium* including *F. graminearum* (Gordon 1952; Sieber et al. 1988), *F. oxysporum* f. sp. *melonis* (Katan 1971), *F. oxysporum* (Lemanceau et al. 1993), *F. nivale*, *F. culmorum* (Sieber et al. 1988), *F. crookwellense* (Boshoff et al. 1996), *F. culmorum* (Kang and Buchenauer 1999); see additional species in the review of Kuldau and Yates (2000), and in Warren and Kommedahl (1973).

Symptomless infection by *Fusarium* species varies and is related to the infection age of the hyphae (Baayen and Rykenbuerg 1999), and compartmentalization of the disease within resistant host tissues (Baayen et al. 1996) while others remain symptomless and biotrophic and are characterized as having interfacial membranes that separate fungal and plant plasma membranes (Fig. 8.3a–c) (Marchant 1966; Malalasekera et al. 1973; Baayen and Elgersma 1983; Bacon and Hinton 1996; Boshoff et al. 1996) The association is compatible over an extended time period, and non-specialized hyphae, which appear not to differ morphologically from the intercellular hyphae of symptomless infection (Bacon and Hinton 1996; Figs. 8.3a–c, Figs. 8.4a–f), are used for nutrient absorption. Endophytic hyphae of *Fusarium* spp. are not dormant (or quiescent), but are metabolically active throughout the association and within the intercellular spaces (Miller and Young 1985; Evans et al. 2000; Kang and Buchenauer 2000; Bacon et al. 2001). In such hyphae, there is a consistent lack of distinct nutrient absorbing structures characteristic of the usual pathogenic infections in rust and powdery mildews such as haustoria. Virulent strains of *F. verticillioides* do have intracellular hyphae that are not distinct haustoria (Fig. 8.4b–f) (Bacon and Hinton 1996). However, haustoria represent only one type of nutrient absorption structure. Fungal cell-wall-to-plant-wall appositions, as observed in most strains of *F. verticillioides* and other *Fusarium* species as well as other fungal endophytes, are one of three fungus-plant cell nutrient absorbing structures (Honneger 1986). Thus, nutrient absorption by wall-to-wall contact and nutrients from the apoplasm are apparently just as efficient as intracellular hyphae since growth and colonization of the host is still accomplished, along with the production of secondary metabolites (Bacon et

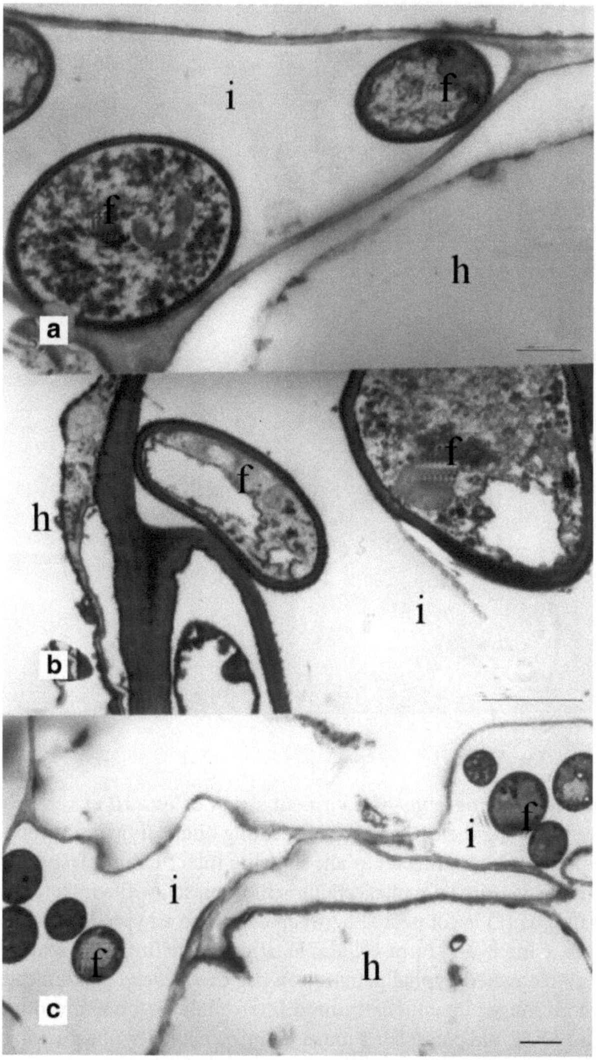

Fig. 8.3. a–c Transmission electron micrographs of the symptomless endophytic hyphae of *F. verticillioides*, RRC 826, as they appear in cross-section of 1- to 8-week-old plants of maize. **a** Two hyphae (*f*) in an intercellular space (*i*) of maize cells (*h*); *bar* 1 μm. **b** Another view of the intercellular nature of hyphae in roots of maize plants approximately 8 weeks old; *bar* 1 μm. **c** Two groups of hyphae in intercellular spaces of the root cortex; *bar* 1 μm (from Bacon and Hinton 1996, with permission)

al. 2001). The accumulation of secondary compounds must require, a priori, a tremendous amount of energy. Certainly, the greater distribution of *Fusarium* hyphae in roots (Foley 1962; Warren and Kommedahl 1973; Kommedahl and Siggerirsson 1975; Blok and Bollen 1995; Bacon and Hin-

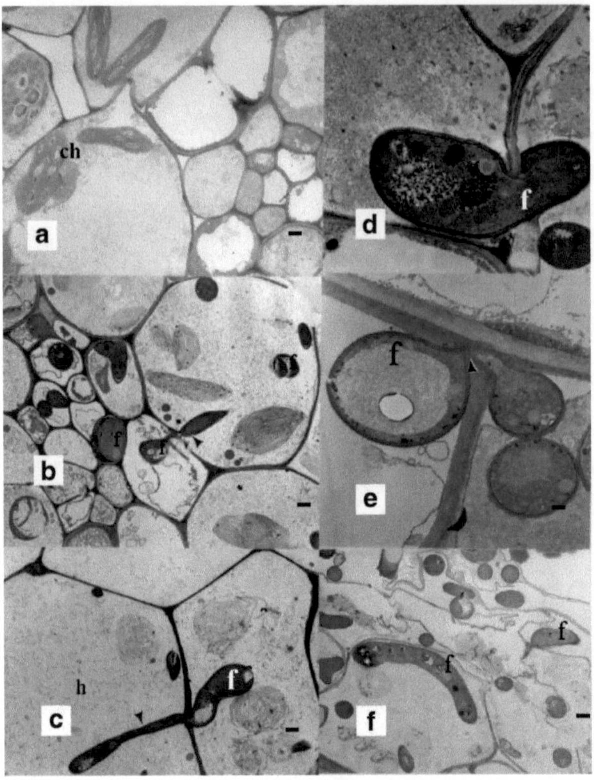

Fig. 8.4. a–f Transmission electron micrographs of a virulent strain of *Fusarium verticillioides*, RRC pat, in maize plants during the early stages of seedling blight showing several variations of intracellular hyphae. **a** A noninfected plant showing intact chloroplast (*ch*); *bar* 1 μm. **b** At 3 weeks, the fungus is primarily intracellular; chloroplasts, while intact, are becoming disorganized; one fungus (*f*) is connected between cells with a hyphal penetration peg (*arrowhead*); *bar* 1 μm. **c** Inter- and intra-cellular location of the fungus in leaves; intracellular fungus (*f*) with an elongated hyphal penetration peg (*arrowhead*) within cell of maize (*h*); chloroplasts are no longer intact; *bar* 1 μm. **d** Intercellular hyphae invading another cell in stem tissue of a 3-week-old plant; *bar* 1 μm. **e** An intracellular hypha growing between two cells along and between the middle lamellae; *bar* 1 μm. **f** Maize tissue showing extensive inter- and intra-cellular colonization; *bar* 1 μm (from Bacon and Hinton 1996, with permission)

ton 1996) reflects the larger amounts of nutrients within the root apoplasm that accumulate as a sink from photosynthesis.

Reports on symptomatic infections are numerous, although detailed histological studies of these types of infections tend to be restricted to the specific plant organs showing the effects (Kang and Buchenauer 2000; Boshoff et al. 1996; Ribichich et al. 2000). However, there are several histological similarities during the change from biotrophic phase to necrotrophic phase

(Adams 1921; Bennet 1931; Malalasekera et al. 1973; Wong et al. 1992; Bacon and Hinton 1996; Baayen and Rykenbuerg 1999; Kang and Buchenauer 2000), reinforcing our concept that most *Fusarium* species are hemibiotrophic.

Specialized intracellular infection and nutrient absorbing hyphae are absent during the biotrophic state of *F. verticillioides* infecting maize (Figs. 8.1a–e, 8.3a–c) and in other hosts as well (Malalasekera et al. 1973; Manaka and Chelkowski 1985; Honneger 1986; Boshoff et al. 1996; Kang and Buchenauer 1999, 2000), and in general these fall within the type 1 category that ranges from the simple to the appressorium configuration (Honegger 1986). This type of fungus-plant cell interaction is characteristic of interactions that occur in different symbiotic systems (Honegger 1986). We do not know if the lack of specialized nutrient absorption structures during this phase is characteristic of all *Fusarium* species. However, infection of host cells by specialized hyphae is described for *F. culmorum* and other species during the change to the intracellular stage of infection (Malalasekeraet al. 1973; Kang and Buchenauer 2000) and these would fall within the definition of one of the type 2 intracellular haustoria without sheath and papilla described by Honegger (1986) as indicative of mutualistic symbioses commonly found in lichens. However, the nature of the association with a particular host may vary, as some endophytic associations appear to be more epicuticular (on glumes) than endophytic and systemic (Kang and Buchenauer 1999). A lack of specialized intracellular absorbing structures assures less injury to the host, thus insuring compatibility, and presumably nutrients are obtained from the apoplasm of the intercellular spaces, although there may be a fungus-directed source and sink relationship.

8.2.3 Mycotoxins

Fusarium species are a highly successful group of fungi that produce a variety of secondary metabolites, some of which might be important in both the long- and short-term strategies of the species. Most biochemical studies have concentrated on the production of, and factors leading to the accumulation of, mycotoxins and related compounds. However, there are sporadic and often observational reports on the positive and negative effects of symptomless infections by *Fusarium* species on hosts, and on competing organisms, which may be activities of secondary metabolites. *Fusarium* species produce a wide diversity of mycotoxins, resulting in a variety of effects on animals (Marasas et al. 1984, 1988; Voss et al. 1990; Norred et al. 1992; Riley et al. 1993). Mycotoxins have been speculated to play an important role in the long-term survival strategy of the producing fungus. However, in some instances, some mycotoxins might play an even greater role in the day-to-day competitive fitness strategies of *Fusarium* species and these are presented briefly below.

8.2.4
Mycotoxins and Host Relationships

Fumonisins are mycotoxins and are the subject of current concern due to their widespread occurrence in maize and maize products. *F. verticillioides* and other fungi of the Liseola section produce fumonisins in maize and other commodities. Fumonisins were shown to accumulate in colonized maize roots early during maize seedling development, and more fumonisins are isolated from roots than from shoots at this early stage of growth (Bacon et al. 2001). Fumonisin mycotoxins probably also occur in the roots of other plant species, and a rooting response has been shown for one cultivar of tomato (Bacon and Williamson 1992). A genetic and morphological study of conidiation mutants of *F. verticillioides* yielded interesting results concerning the role of fumonisins in plants (Glenn et al. 2004). Wild type isolates produced enteroblastic phialidic conidia, while mutants incapable of enteroblastic conidiogenesis produced undulating germ tube-like outgrowths. These mutants were not capable of infecting maize roots, and varied in their fumonisin production. Although they could not infect the plant, only those mutants that could produce the fumonisin were able to cause death of seedlings, but only in a small sampling of maize cultivars. This suggests that fumonisins play a role in the pathogenic processes of some *Fusarium* species but expression of toxicity on the host depends on the host genotype.

Other *Fusarium* mycotoxins that might have a function in the physiology of the association include those produced by *F. graminearum* and related species. These include deoxynivalenol (DON or vomitoxin, a type B trichothecene), other related trichothecenes, and zearalenone. The mycotoxin DON is by far the most dominant of the trichothecenes, occurring on oats, rye, and maize, and occasionally on rice, sorghum, and triticale. In addition to the economic impact of this toxin in reducing animal performance, DON has adverse effects on plant performance.

DON has been implicated as a virulence factor for some hosts (Bai et al. 2002; Desjardins and Hahn 1997; Harris et al. 1999), although the concentrations produced may or may not directly correlate with the degree of fungal virulence (Desjardins et al. 1996; Carter et al. 2000; Bai et al. 2001; Mesterházy et al. 2003). DON was isolated from florets and grains of rye, wheat, and maize (Desjardins et al. 1996), but nothing is known about its distribution and early toxin production within root tissue, or about how much of this toxin is produced during any endophytic stage of wheat or maize root colonization by *F. graminearum*, the main cause of blight disease of cereals. Nevertheless, this species also colonizes maize stalks and all tissues of wheat asymptomatically (Sieber et al. 1988; Dodd 1992; Fisher et al. 1992). McLean (1996) has reviewed additional information on the phytotoxicity of various *Fusarium* metabolites.

8.2.5
Physiological Interactions and Defense Metabolites

Positive physiological interactions with maize were recorded for several strains of *F. verticillioides* These include increased rooting (Bacon and Williamson 1992), and earlier lignification of roots in seedling plants (Yates et al. 1997). Biocontrol uses of several *Fusarium* species not only indicate the utility of the genus but also suggest possible mutualistic interactions derived from the associations, including insecticidal (Abado-Becognee et al. 1998), nematocidal (Hallmann and Sikora 1994a, 1994b), and fungicidal (Damicone and Manning 1982; Lemanceau et al. 1993) activities.

The endophytic association of *F. verticillioides* with maize has evolutionary and physiological implications since it has been established that all maize isolates of this fungus can detoxify the host's native antimicrobial compounds, the benzoxazinoids (Glenn and Bacon 1998; Glenn et al. 2001, 2002). The ability to detoxify these compounds, the concentrations of which may be high in roots (Xie et al. 1991), has been interpreted as one mechanism by which endophytic colonization is not prevented, since the benzoxazinoids are especially toxic to fungi (Xie et al. 1991; Schulz and Wieland 1999; Sicker et al. 2000; Glenn et al. 2001, 2002). Further, strains of *F. verticillioides* isolated from banana cannot detoxify the benzoxazinoids (Glenn et al. 2001, 2002). Several other species of *Fusarium* can detoxify the benzoxazinoids, suggesting the importance of this mechanism to this group, as well as to other maize pathogens (Schulz and Wieland 1999), for the endophytic association with grasses (Glenn 2000; Sicker et al. 2000).

It is well known that several species of bacteria are endophytic in plants, also occupying intercellular spaces [for review, see Chanway 1998; see also Chaps. 2 (Hallmann and Berg), 3 (Kloepper and Ryu), and 6 (Anand et al.)]. Endophytic bacteria are ecological homologues for most species of endophytic *Fusarium* species, and compete for nutrients within the apoplasm. Several biocontrol strategies are based on the use of bacterial endophytes [Chanway 1998; Kobayashi and Palumbo 2000; see Chaps. 3 (Kloepper and Ryu), and 4 (Berg and Hallmann)]. However, all *Fusarium* species examined produce fusaric acid (Bacon et al. 1996), and possibly other unknown antibiotics that serve to control bacteria in planta. Fusaric acid is inhibitory to most strains of endophytic and rhizosphere bacteria (Notz et al. 2002; Bacon et al. 2004). Fusaric acid is moderately toxic to mammals (Porter et al. 1995), but is produced by most isolates of *Fusarium* (Bacon et al. 1996). However, its activity is broad and it has pronounced effects on microorganisms. Fusaric acid is an antibiotic, showing activity against both Gram-negative and -positive bacteria, especially those used for biocontrol. Fusaric acid was shown to interact with genes specifically used by the biocontrol bacterium *Pseudomonas fluorescens* by preventing expression of genes encoding specific inhibitors of fungi (Notz et al 2002). A similar mode of action has been established for DON, which prevents the production of the chitinase gene expressed by the biocontrol agent *Trichoderma atroviride* (Lutz et al. 2003).

Knowledge of the complete spectra of activity for most *Fusarium* secondary metabolites is incomplete. Moniliformin is a weak mycotoxin and an even better phytotoxin for specific hosts (Cole et al. 1973), but any overall effects pertaining to *Fusarium* species are unknown. The beauvericins show strong selective insecticidal properties (Vesonder and Hesseltine 1981; Plattner and Nelson 1994; Logrieco et al. 1998). The beauvericins are produced by at least 12 *Fusarium* species (Logrieco et al. 1998), and can protect *Fusarium*-infected plants from insect predation. The chemical isolation and some biological activities of numerous other metabolites have been reviewed by Marasas et al. (1984) and Thrane (1989) and include the enniatins, butenolide, wortmannin, diacetoxyscripenol, nivalenol, visoltricin, chrysogine, culmorin, aurofusarin, equisetin, fusoproliferatum, fusarochromanone, acuminatopyrone, fusamarin, chlamydosporol, additional derivatives of T-2 toxins, and several unidentified antibiotics. The identities of the *Fusarium* species producing these metabolites and a more comprehensive listing of the activities of these and other mycotoxins can be obtained from Thrane (1989, 2001) and Summerell et al. (2001).

8.3 Summary

Fusarium is a very important genus from the point of view of food production and food safety. *Fusarium* species exist as intercellular root endophytes in both cultivated and wild plants and their role during the symptomless state of infection is ambiguously defined. However, many species are pathogenic, causing diseases such as root, stem, and ear rot on crop plants, thereby reducing plant productivity. *F. verticillioides* and other *Fusarium* species are unique endophytes, with similarities to other endophyte species such as the foliage endophytes of forage grasses, but are more versatile since they are also hemibiotrophic in their associations with plants. The problems associated with this genus are global as their distribution is worldwide, and most host plants are susceptible to infection by one or more *Fusarium* species. Certain biotic and abiotic factors may alter *Fusarium* relationships with plants from a symptomless endophytic association to a hemibiotrophic and finally a saprotrophic association where mycotoxins might accumulate, leading to animal and human health concerns. To summarize, a major concern is that many *Fusarium* species produce mycotoxins that are harmful to humans and animals ingesting food or feed products colonized by the fungus. The mycotoxins are produced during the pathogenic and saprophytic states, but the infections caused by these two states can be initiated from the symptomless root endophytic biotrophic state.

The dual characterization of *F. verticillioides* as both a pathogen and a symptomless endophyte indicates both the complex relationship of this species with plants as well as suggesting similar complexities for other *Fusarium*-plant interactions. Consequently, the development of appropriate control measures for virulent *Fusarium* isolates are expected to be difficult. For example, on the one hand, diseases and mycotoxins produced by *F. verticillioides* must be controlled, while on the other hand, the intimate association of the endophytic state of this species appears to confer some positive competitive fitness traits to certain plants. The extent to which this occurs due to present-day plant breeding efforts will be determined with more detailed studies. The association of this genus with roots as symptomless endophytes indicates a role for these fungi in nutrition, and suggests the importance of the root as an endophytic niche during the co-evolution of *Fusarium* species and plants.

References

Abado-Becognee K, Fleurat-Lessard F, Creppy EE, Melcion D (1998) Effects of fumonisin B_1 on growth and metabolism of larvae of the yellow mealworm, *Tenebrio molitor*. Entomol Exp Appl 86:135–143

Aoki T, O'Donnell K (1999) Morphological and molecular characterization of *Fusarium pseudograminearum* sp nov. formerly recognized as the Group 1 population of *F. graminearum*. Mycologia 91:597–609

Adams JE (1921) Observation on wheat scab in Pennsylvania and its pathological histology. Phytopathology 11:115–124

Ahmed KZ, Mesterhazy A, Bartok T, Sagi F (1996) In vitro techniques for selecting wheat (*Triticum aestivum* L) for *Fusarium*-resistance 2. Culture filtrate technique and inheritance of *Fusarium*-resistance in the somaclones. Euphytica 91:341–349

Anaya N, Roncero MIG (1996) Stress-induced rearrangement of *Fusarium* retrotransposon sequences. Mol Gen Genet 253:89–94

Anjaiah V, Cornelis P, Koedam N (2003) Effect of genotype and root colonization in biological control of fusarium wilts in pigeonpea and chickpea by *Pseudomonas aeruginosa* PNA1. Can J Microbiol 49:85–91

Ayers JE, Nelson PE, Krause RA (1989) Fungi associated with maize stalk rot in Pennsylvania in 1070 and 1971. Plant Dis Rep 56:836–839

Baayen RP, Elgersma DM (1983) Colonization and histopathology of susceptible and resistant carnation cultivars infected with *Fusarium oxysporum* f.sp. *dianthi*. Neth J Plant Pathol 91:119–135

Baayen RP, Rykenbuerg FHJ (1999) Fine structure of the early interaction of lily roots with *Fusarium oxysporum* f.sp. *lilii*. Eur J Plant Pathol 105:431–443

Baayen RP, Ouellette GB, Rioux D (1996) Compartmentalization of decay in carnations resistant to *Fusarium oxysporum* f.sp. *dianthi*. Phytopathology 86:1018–1031

Backhouse D, Burgess LW, Summerell BA (2001) Biogeography of *Fusarium*. In: Summerell BA, Leslie JF, Backhouse D, Bryden WL, Burgess LW (eds) *Fusarium* Paul E. Nelson Memorial Symposium APS Press, St. Paul, MN, pp 122–137

Bacon CW, Hinton DM (1988) Pathogenic differences in ten isolates of *Fusarium moniliforme* Sheldon. Mycol Soc Am Newslett 39:S19–S20

Bacon CW, Hinton DM (1996) Symptomless endophytic colonization of maize by *Fusarium moniliforme*. Can J Bot 74:1195–1202

Bacon CW, White JF Jr (eds) (2000) Microbial endophytes. Dekker, New York

Bacon CW, Williamson JW (1992) Interactions of *Fusarium moniliforme*, its metabolites and bacteria with maize. Mycopathologia 117:65–71

Bacon CW, Lyons PC, Porter JK, Robbins JD (1986) Ergot toxicity from endophyte-infected grasses: a review Agron J 78:106–116

Bacon CW, Porter JK, Norred WP, Leslie JF (1996) Production of fusaric acid by *Fusarium* species. Appl Environ Microbiol 62:4039–4043

Bacon CW, Yates IE, Hinton DM, Meredith R (2001) Biological control of *Fusarium moniliforme* in maize. Environ Health Perspect 109:325–332

Bacon CW, Hinton DM, Porter JK, Glenn AE, Kuldau GA (2004) Fusaric acid, a *Fusarium verticillioides* metabolite, antagonistic to the endophytic biocontrol bacterium *Bacillus mojavensis*. Can J Bot 82:878–885

Bai GH (1996) Variation in *Fusarium graminearum* and cultivar resistance to wheat scab. Plant Dis 80:975–979

Bai GH, Plattner R, Desjardins A, Kolb F (2001) Resistance to Fusarium head blight and deoxynivalenol accumulation in wheat. Plant Breeding 120:1–6

Bai GH, Desjardins AE, Plattner RD (2002) Deoxynivalenol-nonproducing *Fusarium graminearum* causes initial infection, but does not cause disease spread in wheat spikes. Mycopathologia 153:91–98

Bakan B, Melcion D, Richard-Molard D, Cahagnier B (2002) Fungal growth and *Fusarium* mycotoxin content in isogenic traditional maize and genetically modified maize grown in France and Spain. J Agric Food Chem 50:728–731

Bentley S, Pegg KG, Dale JL (1995) Genetic variation among a world-wide collection of isolates of *Fusarium oxysporum* f sp *cubense* analysed by RAPD-PCR fingerprinting. Mycol Res 99:1378–1384

Bergman BHH, Bakker-Van der Voort MAM (1979) Latent infection of tulip bulbs by *Fusarium oxysporum*. Neth J Plant Pathol 85:187–195

Bennett FT (1931) *Gibberella saubinetti* (Mont) Sacc. on British cereals. II. Physiological and pathological studies. Ann Appl Biol 18:158–177

Blok WJ, Bollen GJ (1995) Fungi on the roots and stem bases of asparagus in the Netherlands: species and pathogenicity. Eur J Plant Pathol 101:15–24

Boshoff WHP, Pretorius ZA, Swart WJ, Jacobs AS (1996) A comparison of scab development in wheat infected by *Fusarium graminearum* and *Fusarium crookswellense*. Phytopathology 86:558–563

Bowden RL, Leslie JF (1994) Diversity of *Gibberella zeae* at small spatial scales. Phytopathology 86:1140

Carter JP, Rezanoor HN, Desjardins AE, Nicholson P (2000) Variation in *Fusarium graminearum* isolates from Nepal associated with their host of origin. Plant Pathol 49:452–460

Chanway CP (1998) Bacterial endophytes: ecological and practical implications. Sydowia 50:149–170

Cole RJ, Kieksey JW, Cutler HG, Doupnik BL, Peckham JC (1973) Toxin from *Fusarium moniliforme*: effects on plants and animals. Science 179:1324–1326

Corell JC, Gordon TR, McCain AH (1992) Genetic diversity in California and Florida populations of the pitch canker fungus *Fusarium subglutinans*. f. sp. *pina* Phytopathology 82:415–420

Damicone JP, Manning WJ (1982) Avirulent strains of *Fusarium oxysporum* protect asparagus seedling from crown rot. Can J Plant Pathol 4:143–146

Desjardins AE, Hahn TM (1997) Mycotoxins in plant pathogenesis. Mol Plant Microbe Interact 10:147–152

Desjardins AE, Proctor RH, Bai GH, McCormick SP, Shaner G, Buechley G, Hahn TM (1996) Reduced virulence of trichothecenes antibiotics-nonproducing mutants of *Gibberella zeae* in wheat field tests. Mol Plant Microbe Interact 9:775–781

Diehl WW (1950) *Balansia* and the Balansiae in America. In: USDA Agric Monogr 4, US Govt Print Office, Washington, DC, pp 1–82

Dodd JL (1992) The role of plant stresses in development of maize stalk rots. Plant Dis 64:533–537

Evans CK, Xie W, Dill-Macky R, Mirocha CJ (2000) Biosynthesis of deoxynivalenol in spikelets of barley inoculated with macroconidia of *Fusarium graminearum*. Plant Dis 84:654–660

Fisher PJ, Petrini O (1992) Fungal saprobes and pathogens as endophytes of rice (*Oryza sativa*) New Phytol 120:137–143

Fisher PJ, Petrini O, Scott HML (1992) The distribution of some fungal and bacterial endophytes in maize (*Zea mays* L.). New Phytol 122:299–305

Foley DC (1962) Systemic infection of corn by *Fusarium moniliforme*. Phytopathology 52:870–872

Freeman S, Rodriguez RJ (1993) Genetic conversion of a fungal plant pathogen to a nonpathogenic, endophytic mutualist. Science 260:75–78

Gang G, Miedaner T, Schuhmacher U, Schollenberger M, Geiger HH (1998) Deoxynivalenol and nivalenol production by *Fusarium culmorum* isolates differing in aggressiveness toward winter rye. Phytopathology 88:879–884

Glenn AE, Bacon CW (1998) Detoxification of benzoxazinoids by *Fusarium moniliforme* and allies. Phytopathology 88:S–S32

Glenn AE, Hinton DM, Yates IE, Bacon CW (2001) Detoxification of corn antimicrobial compounds as the basis for isolating *Fusarium verticillioides* and some other *Fusarium* species from corn. Appl Environ Microbiol 67:2973–2981

Glenn AE, Gold SE, Bacon CW (2002) *Fdb1* and *Fdb2*, *Fusarium verticillioides* loci necessary for detoxification of preformed antimicrobials from corn. Mol Plant Microbe Interact 15:91–101

Glenn AE, Richardson EA, Bacon CW (2004) Genetic and morphological characterization of a *Fusarium verticillioides* conidiation mutant. Mycologia 96:968–980

Gordon WL (1952) The occurrence of *Fusarium* species in Canada. II. Prevalence and taxonomy of *Fusarium* species in cereal seed. Can J Bot 30:209–251

Hallmann J, Sikora RA (1994a) Influence of *Fusarium oxysporum*, a mutualistic fungal endophyte on *Meloidogyne incognita* infection of tomato. Pflanzenkr Pflanzenschutz 101:475–481

Hallmann J, Sikora RA (1994b) Occurrence of plant parasitic nematodes and endophytic fungi in tomato plants in Kenya and their role as mutualistic synergists for biological control of root-knot nematodes. Int J Pest Manage 40:321–325

Harris LJ, Desjardins AE, Plattner RD, Nicholson P, Butler G, Young JC, Weston G, Proctor RH, Hohn TM (1999) Possible role of trichothecene mycotoxins in virulence of *Fusarium graminearum* on maize. Plant Dis 83:954–960

Honegger R (1986) Ultrastructural studies in lichens. II. Mycobiont and photobiont cell wall surface layers and adhering crystalline lichen products in flour parmeliaceae. New Phytol 103:797–808

Hysek J, Vanova M, Sychrova E, Koutecka-Brozova J, Radova Z, Hajslova J (2000) Fusariosis of barley – the spectrum of species and the levels of mycotoxins (trichothecenes). Mitt Biol Bundesanst Land-Forstwirtsch 377:29

Kang Z, Buchenauer H (1999) Immunocytochemical localization of fusarium toxins in infected wheat spikes by *Fusarium culmorum*. Physiol Mol Plant Pathol 55:275–288

Kang Z, Buchenauer H (2000) Cytology and ultrastructure of the infection of wheat spikes by *Fusarium culmorum*. Mycol Res 104:1083–1093

Katan J (1971) Symptomless carriers of the tomato wilt pathogen. Phytopathology 61:1213–1217

Kedera CJ, Plattner RD, Desjardins AE (1999) Incidence of *Fusarium* spp and levels of fumonisin B1 in maize in western Kenya. Appl Environ Microbiol 65:41–44

Kobayashi DY, Palumbo JD (2000) Bacterial endophytes and their effects on plants and uses in agriculture. In: Bacon CW, White JF Jr (eds) Microbial endophytes. Dekker, New York, pp 199–233

Kommedahl T, Siggerirsson EI (1975) Prevalence of *Fusarium* species in roots and soil of grassland in Iceland. Mycologia 67:38–44

Kosiak B, Torp M, Thrane U (1997) The occurrence of *Fusarium* spp. in Norwegian grain – a survey. Cereal Res Commun 25:595–506

Kuldau GA, Yates IE (2000) Evidence for *Fusarium* endophytes in cultivated and wild plants. In: Bacon CW, White JF Jr (eds) Microbial endophytes. Dekker, New York, pp 85–117

Lee JC, Lobkovsky E, Pliam NB, Strobel G, Clardy J (1995) Subglutinols A and B: immunosuppressive compounds from the endophytic fungus *Fusarium subglutinans*. J Org Chem 60:7076–7077

Lemanceau P, Bakker PAHM, Dekogel WJ, Alabouvette C, Schippers, B (1993) Antagonistic effect of nonpathogenic *Fusarium oxysporum* Fo47 and psuedobactin 358 upon pathogen *Fusarium oxysporum* f. sp. *dianthi* Appl Environ Microbiol 59:74–82

Leslie JF, Pearson CAS, Nelson PE, Toussoun TA (1990) *Fusarium* spp. from corn, sorghum, and soybean fields in the central and eastern United States. Phytopathology 80:343–350

Liddell CM, Burgess LW (1985) Survival of *Fusarium moniliforme* at controlled temperature and relative humidity. Trans Br Mycol Soc 84:121–130

Logrieco A, Moretti A, Castella G, Kostecki M, Golinski P, Ritieni A, Chelkowski J (1998) Beauvericin production by *Fusarium* species. Appl Environ Microbiol 64:3084–3088

Luttrell ES (1974) Parasitism of fungi on vascular plants. Mycologia 66:1–15

Lutz MP, Feichtinger G, Defago G, Duffy B (2003) Mycotoxigenic *Fusarium* and deoxynivalenol production repress chitinase gene expression in the biocontrol agent *Trichoderma atroviride* Pl. Appl Environ Microbiol 69:3077–3084

Malalasekera RAP, Sanderson FR, Colhoun J (1973) *Fusarium* diseases of cereals. IX. Penetration and invasion of wheat seedlings by *Fusarium culmorum* and *F. navale*. Trans Br Mycol Soc 60:453–462

Manaka M, Chelkowski J (1985) Phytotoxicity and pathogenicity of *Fusarium nivale* towards cereal seedlings. Phytopathol Z 122:113–117

Marasas WFO, Nelson PE, Toussoum TA (1984) Toxigenic *Fusarium* species. Identity and mycotoxicology. The Pennsylvania State University Press, University Park, PA

Marasas WFO, Kellerman TS, Gelderblom WCA, Coetzer JAW, Thiel PG, Van der Lugt JJ (1988) Leukoencephalomalacia in a horse induced by fumonisin B1 isolated from *Fusarium moniliforme*. Onderstepoort J Vet Res 55:197–203

Marchant R (1966) Fine structure and spore germination in *Fusarium culmorum*. Ann Bot 30:821–830

Mclean M (1996) The phytotoxicity of *Fusarium* metabolites: an update since 1989. Mycopathologia 133:163–179

Mesterházy A, Bartók T, Lamper C (2003) Influence of wheat cultivar, species of *Fusarium*, and isolate aggressiveness on the efficacy of fungicides for control of Fusarium head blight. Plant Dis 87:1107–1115

Miller JD, Young CJ (1985) Deoxynivalenol in an experimental *Fusarium graminearum* infection of wheat. Can J Plant Pathol 7:132–134

Munkvold GP, Carlton WM (1997) Influence of inoculation method on systemic *Fusarium moniliforme* infection of maize plants grown from infected seeds. Plant Dis 81:211–216

Nicholson P, Simpson DR, Weston G, Rezanoor HN, Lees AK, Parry DW, Joyce D (1998) Detection and quantification of *Fusarium culmorum* and *Fusarium graminearum* in cereals using PCR assays. Physiol Mol Plant Pathol 53:17–37

Norred WP, Wang E, Yoo H, Riley RT, Merrill AH Jr (1992) In vitro toxicology of fumonisins and the mechanistic implications. Mycopathologia 117:73–78

Notz R, Maurhofer M, Dubach H, Haas D, Defago G (2002) Fusaric acid-producing stains of *Fusarium oxysjporum* alter 2.4-diacetylpholoroglucinol biosynthetic gene expression in *Psedomonas fluorescens* CHAO in vitro and in the rhizosphere of wheat. Appl Environ Microbiol 68:2229–2235

O'Donnell K, Kistler HC, McLeod L (2000) Gene genealogies reveal global phylogeographic structure and reproductive isolation among lineages of *Fusarium graminearum*, the fungus causing wheat scab. Proc Natl Acad Sci USA 97:7905–7910

Parry DW, Nicholson P (1996) Development of a PCR assay to detect *Fusarium poae* in wheat. Plant Pathol 45:383–391

Pennypacker BW (1981) Anatomical changes involved in the pathogenesis of plants by *Fusarium*. In: Nelson PE, Toussoun TA, Cook RJ (eds) *Fusarium*: diseases, biology, and taxonomy, Pennsylvania State University Press, University Park, PA, pp 400–408

Pinto LSRC, Azevedo MD, Pereiro M, Viera MLC, Labate CA (2000) Symptomless infection of banana and maize by endophytic fungi impairs photosynthetic efficiency. New Phytol 147:609–615

Plattner RD, Nelson PE (1994) Production of beauvericin by a strain of *Fusarium proliferatum* isolated from corn fodder for swine. Appl Environ Microbiol 60:3894–3896

Porter JK, Bacon CW, Wray EM, Hagler WM Jr (1995) Fusaric acid in *Fusarium moniliforme* cultures, corn, and feeds toxic to livestock and the neurochemical effects in the brain and pineal gland of rats. Nat Toxins 3:91–100

Ribichich KF, Lopez SE, Vegetti AC (2000) Histopathological spikelet changes produced by *Fusarium graminearum* in susceptible and resistant wheat cultivars. Plant Dis 84:794–802

Riley RT, An N-H, Showker JL, Yoo H-S, Norred WP, Chamberlain WJ, Wang E, Merrill AH Jr, Motelin G, Beasley VR, Haschek WM (1993) Alteration of tissue and serum sphinganine to sphingosine ratio: an early biomarker of exposure to fumonisin-containing feeds in pigs. Toxicol Appl Pharmacol 118:105–112

Schroeder HW, Christensen JJ (1963) Factors affecting resistance of wheat to scab caused by *Gibberella zeae*. Phytopathology 53:831–838

Schulz M, Wieland I (1999) Variation in metabolism of BOA among species in various field communities – biochemical evidence for co-evolutionary processes in plant communities? Chemoecology 9:133–141
Sheldon JL (1902) A corn mold (*Fusarium moniliforme.* sp.n.). Agric Exp Stn Nebraska 17:23–32
Sicker D, Frey M, Schulz M, Gierl A (2000) Role of natural benzoxazinones in the survival strategy of plants. Int Rev Cytol 198:319–346
Sieber T, Risen TK, Muller E, Fried PM (1988) Endophytic fungi in four winter wheat cultivars (*Triticum aestivum* L.) differing in resistance against *Stagonospora nodorum* (Berk.) Cast. Germ = *Septorianodorum* (Berk.) Berk. J Phytopathol 122:289–306
Stone JK, Bacon CW, White JF Jr (2000) An overview of endophytic microbes: endophytism defined. In: Bacon CW, White JF Jr (eds) Microbial endophytes. Dekker, New York, pp 3–29
Summerell BA, Leslie JF, Blackhouse D, Bryden WL, Burgess LW (2001) *Fusarium* Paul E. Nelson Memorial Symposium APS Press, St. Paul, MN
Thomas MD, Buddenhagen IW (1980) Incidence and persistence of *Fusarium moniliforme* in symptomless maize kernels and seedlings in Nigeria. Mycologia 72:882–887
Thrane U (1989) *Fusarium* species and their specific profiles of secondary metabolites. In: Chelkowski J (ed) *Fusarium*: mycotoxins, taxonomy and pathogenicity. Elsevier, Amsterdam, pp 199–225
Thrane U (2001) Developments in the taxonomy of *Fusarium* species based on secondary metabolites. In: Summerell BA, Leslie JF, Blackhouse D, Bryden WL, Burgess LW (eds) *Fusarium* Paul E. Nelson Memorial Symposium APS Press, St. Paul, MN, pp 29–49
Vesonder RF, Hesseltine CW (1981) Metabolites of *Fusarium*. In: Nelson PE, Toussoun TA, Cook RJ (eds) *Fusarium*: diseases, biology and taxonomy The Pennsylvania State University Press, University Park, PA, pp 350–364
Voorhees RK (1934) Histological studies of a seedling disease of corn caused by *Gibberella moniliformis*. J Agric Res 49:1009–1015
Voss KA, Plattner RD, Bacon CW, Norred WP (1990) Comparative studies of hepatotoxicity and fumonisin B_1 and B_2 content of water and chloroform/methanol extracts of *Fusarium moniliforme* strain MRC 826 culture material. Mycopathologia 112:81–92
Warren HL, Kommedahl T (1973) Prevalence and pathogenicity to corn of *Fusarium* species from corn roots, rhizosphere, residues, and soil. Phytopathology 63:1288–1290
Wilcoxon RD, Kommedahl T, Ozmon EA, Windels CE (1988) *Fusarium* species in scabby wheat from Minnesota and their pathogenicity to wheat. Phytopathology 78:586–589
Wildermuth GB, McNamara RB, Sparks T (1999) Different expressions of resistance to crown rot in wheat. In: Magarey RC (ed) Proceedings of the First Australasian Soilborne Disease Symposium. Bureau of Sugar Experiment Stations, Brisbane, Australia, p 79
Wong LSL, Tekauz A, Leslie D, Abramson D, McKenzie RIH (1992) Prevalence, distribution and importance of *Fusarium* head blight in wheat in Manitoba. Can J Plant Pathol 14:233–238
Xie YS, Arnason JT, Philogène BJR, Atkinson J, Morand P (1991) Distribution and variation of hydroxamic acids and related compounds in maize (*Zea mays*) root system. Can J Bot 69:677–681
Yates IE, Jaworski AJ (2000) Differential growth of *Fusarium moniliforme* relative to tissues from 'Silver Queen', a sweet maize. Can J Bot 78:472–480
Yates IE, Bacon CW, Hinton DM (1997) Effects of endophytic infection by *Fusarium moniliforme* on corn growth and cellular morphology. Plant Dis 81:723–728
Yates IE, Hiett KL, Kapczynski DR, Smart W, Glenn AE, Hinton DM, Bacon CW, Meinersmann R, Liu S, Jaworski AJ (1999) GUS transformation of the maize fungal endophyte *Fusarium moniliforme*. Mycol Res 103:129–136

9 Microbial Endophytes of Orchid Roots

Paul Bayman, J. Tupac Otero

9.1 Introduction

Orchids are interesting on many levels. Their beautiful and bizarre adaptations for pollination have fascinated many people, including Darwin (1887). Interest in orchid pollination biology has overshadowed another, equally important symbiosis: mycorrhizal relationships. In turn, interest in orchid mycorrhizae has overshadowed the relationships between orchids and endophytes. In most cases, studies on orchid root fungi have ignored all fungi not thought to be mycorrhizal. In some cases, fungi in orchid roots have been presumed to be mycorrhizal when they may in fact be endophytes. Thus the frequency, diversity and importance of orchid root endophytes remain largely unexplored.

The goal of this chapter is to review the incidence and importance of endophytes in orchid roots, and to disentangle the literature on mycorrhizal fungi from that on non-mycorrhizal endophytes.

9.2 Habits and Types of Orchid Roots

The Orchidaceae is one of the largest families of plants, with 25,000 species – close to one-tenth of all known flowering plant species (Dressler 1990). Orchids are found in all except the most extreme terrestrial environments.

Orchids are epiphytic (growing on plants), lithophytic (growing on rocks), terrestrial, or, in a few cases, some combination thereof. Epiphytic and lithophytic orchids are all tropical or subtropical; they comprise about 75% of all species (Dressler 1990). Terrestrial orchids are found worldwide; all temperate orchids are terrestrial. Most of what is known about orchid-fungal associations comes from terrestrial orchids (perhaps because

Paul Bayman: Departamento de Biologia, Universidad de Puerto Rico – Rio Piedras, PO Box 23360, San Juan, PR 00931, USA, E-mail: pbayman@upracd.upr.clu.edu

J. Tupac Otero: Universidad Nacional de Colombia-Palmira, Departamento de Ciencias Agrícolas, AA 237, Palmira, Valle del Cauca, Colombia

Soil Biology, Volume 9
Microbial Root Endophytes
B. Schulz, C. Boyle, T. N. Sieber (Eds.)

most orchidologists live in temperate countries) even though they are less speciose than epiphytes.

Roots of epiphytic/lithophytic and terrestrial orchids differ (Rasmussen 1995). Epiphytic and lithophytic roots are ecologically equivalent because in both cases the roots are exposed to light and air. Roots of epiphytic and lithophytic orchids are photosynthetic, perennial, and fairly constant throughout the year. Roots of terrestrial orchids, in contrast, are usually non-photosynthetic, live ≤ 3 years, and often show marked seasonal differences in growth and composition. They are usually buried in soil or leaf litter. Some terrestrial orchids have two morphologically distinct types of roots, one of which is mycorrhizal (Rasmussen 1995). Non-mycorrhizal roots tend to have more xylem and more amyloplasts than mycorrhizal roots.

Orchid roots have a velamen, a multiple epidermis of one to several layers of thin-walled cells. The velamen helps the root trap water and probably nutrients (Dressler 1990; Rasmusssen 1995). It has been suggested that the velamen evolved to facilitate colonization of roots by mycorrhizal fungi (Dressler 1990); this seems unlikely because pelotons are more common in the cortex than in the velamen. (Pelotons are coils of hyphae within root cells, and are characteristic of orchid mycorrhizae). Also, epiphytic orchid roots generally have more developed velamens than terrestrial orchid roots (Dressler 1990), even though the frequency of mycorrhizal infection is lower.

Terrestrial orchids are usually obligately mycorrhizal, even as adults (Rasmussen 1995). Some species are non-photosynthetic (or more accurately, myco-heterotrophic) and depend on fungi for nutrition (see Sect. 9.8). Many epiphytic orchids are facultatively mycorrhizal, at least as adults, and there is wide variation in frequency of mycorrhizal colonization; the same is probably true of epilithic orchids (Hadley and Williamson 1972; Lesica and Antibus 1990; Goh et al. 1992; Richardson et al. 1993; Zelmer et al. 1996; Currah et al. 1997; Bayman et al. 1997; Rivas et al. 1998; Otero et al. 2002; Rasmussen 2002).

9.3 Bacteria as Epiphytes and Endophytes of Orchid Roots

In general, bacterial root endophytes have been studied in an agricultural context, rather than in an ecological or biodiversity context [Chanway 1995; see Chaps. 6 (Anand et al.) and 19 (van Overbeek et al.)]. Orchids are no exception: interest in bacteria in orchid roots has focused on pathogens of cultivated plants rather than on endophytes of wild plants (Hadley et al. 1987). This oversight is unfortunate, since endophytic bacteria may be important for the health of wild plants as well as crop plants (Hallmann et al. 1997; Sturz et al. 2000).

Most knowledge of endophytic bacteria in orchid roots comes from studies in Australia. Endophytic bacteria were isolated from 12 species of terrestrial orchids in Western Australia (Wilkinson et al. 1994). The most common genus isolated was *Pseudomonas*, which varied from 23% to 73% of isolates from each orchid species. Most isolates from *Pterostylis* spp. were also Gram-negative, while most isolates from all other orchids were Gram-positive. This difference may reflect morphological differences among the orchids: in *Pterostylis* the main absorptive organs are underground stems, whereas the other orchids tested use adventitious roots. Some of these bacteria stimulated germination in vitro of seeds of *P. vittata* (Wilkinson et al. 1989).

Epiphytic bacteria on orchid roots have also been studied, mainly to determine if nitrogen-fixing bacteria can help orchids obtain nitrogen. *Arthrobacter*, *Bacillus*, *Mycobacterium*, *Pseudomonas*, *Oscillatoria* and *Nostoc* were isolated from the surfaces of roots of the terrestrial orchid *Calanthe vestita*, and *Bacillus*, *Curtobacterium*, *Flavobacterium*, *Nocardia*, *Pseudomonas*, *Rhodococcus*, *Xanthomonas* and *Nostoc* were isolated from the surface of roots of the epiphyte *Dendrobium* (Tsavkelova et al. 2001). Of these, the cyanobacteria *Oscillatoria* and *Nostoc* are capable of nitrogen fixation; they were not isolated from soil collected near the plants, suggesting some special affinity for the root surface. These and other cyanobacteria formed a biofilm on the surface of roots of epiphytic orchids in a greenhouse (Tsavkelova et al. 2003). Cyanobacteria have also often been observed within velamen cells of epiphytic orchids (Dressler 1990; Sinclair 1990). Nitrogen-fixing bacteria also occur on epiphytic *Tillandsia* plants, which often occur together with orchids (Brighigna et al. 1992). Despite the interesting implications of these studies, the transfer of nitrogen from bacteria to orchid roots has not been demonstrated (Dressler 1990; Sinclair 1990).

9.4 Orchid Endophytes or Orchid Mycorrhizal Fungi?

The focus of this chapter and volume is on endophytes, not mycorrhizal fungi. However, it is impossible to review the literature on orchid root endophytes without also discussing mycorrhizae. The study of orchid mycorrhizae and endophytes are so inextricably linked that orchids are a good example of how hard it can be to disentangle the two [see also Chaps. 14 (Cairney) and 16 (Brundrett)].

Mycorrhizae are mutualisms between plant roots and fungi, whereas endophytes are microorganisms growing inside plant tissues without causing symptoms of disease (see Chap. 1 by Schulz and Boyle). This concept of mycorrhizae is functional and describes a relationship, whereas the concept of endophytes principally describes where an organism lives, without assuming or excluding the possibility of benefit for either party. This distinction is complicated by the fact that mutualisms are part of a continuum of symbiotic relationships, and a relationship that is mutualistic may become commensalistic or parasitic, or vice versa [Bronstein et al. 2003; see Chaps. 15 (Schulz) and 16 (Brundrett)]. It is further complicated by the fact that, unlike most mycorrhizae, orchid mycorrhizae are not known to provide any benefit to the fungal partner (Andersen and Rasmussen 1996, Taylor et al. 2002); in some cases and perhaps in all, the relationship is actually parasitic rather than mutualistic. Furthermore, fungal endophytes that are not mycorrhizal in the field may stimulate orchid seed growth in culture – 'functional specificity' as opposed to 'ecological specificity', as defined by Masuhara and Katsuya (1994). This means that seed germination and seedling growth tests in vitro may not be entirely accurate in distinguishing mycorrhizal fungi from non-mycorrhizal endophytes (Rasmussen 2002). The most reliable criterion for mycorrhizae is the visual detection of pelotons in the root, but in tropical, epiphytic orchids, the pelotons observed are often degraded (Otero et al. 2002).

The distinction between orchid mycorrhizal fungi and endophytes is further complicated by the inconsistent use of terminology in the literature. Most studies of *Rhizoctonia*-like fungi assume that the relationship is mycorrhizal without demonstrating any functional benefit to the plant. Since *Rhizoctonia*-like fungi can also be plant pathogens, endophytes or saprotrophs, in some cases this may be an unwarranted assumption (Alconero 1969; Masuhara and Katsuya 1994; Carling et al. 1999; Rasmussen 2002). A more precise (but less convenient) description would be 'presumably mycorrhizal endophytes.' On the other hand, some authors are aware of this problem and err on the side of caution, preferring to call their fungi 'endophytes' for lack of functional evidence of a mycorrhizal relationship, even though they are almost certainly mycorrhizal (e.g., Hadley and Ong 1978; Ramsay et al. 1986; Currah 1991; Currah et al. 1997; Richardson et al. 1993; Otero et al. 2002). So it is hard to say how many papers with 'mycorrhiza' in the title really mean to say 'endophyte' and vice versa.

However, when the papers that focus on mycorrhizal fungi are excluded, the existing body of work on orchid endophytes is surprisingly small (Currah et al. 1997). For example, an excellent, comprehensive treatise on orchids and their mycorrhizal relationships mentions non-mycorrhizal endophytes only in passing (Rasmussen 1995).

To simplify things, in this chapter we will assume that *Rhizoctonia*-like fungi and their telomorphs (= sexual stages), including *Ceratobasidium*, *Thanatephorus*, *Tulasnella*, and *Sebacina*, are mycorrhizal in associations with orchids, that other basidiomycete fungi may be mycorrhizal in mycoheterotrophic orchids, and that members of other groups of fungi are endophytes and not mycorrhizal. Known exceptions to this generalization are mentioned where relevant.

9.5 Problems with the Taxonomy of Orchid Endophytic Fungi

Three problems complicate the study of endophytic fungi, including orchid root endophytes. First, many endophytic fungi do not sporulate in pure culture, and efforts to induce sporulation in vitro are often unsuccessful. Since traditionally fungi are classified by their spores and spore-bearing structures, nonsporulating fungi are very difficult to identify. For this reason, unidentifiable fungi are often grouped into 'morphospecies' on the basis of colony color, morphology and growth rate on agar media (Gamboa and Bayman 2001). DNA sequencing studies have shown that this technique is quite successful at grouping related fungi together (Arnold et al. 2000; Lacap et al. 2003) – but they remain unnamed, making comparisons between studies very difficult.

Second, many endophytic fungi are undescribed, and do not fit well into previously described taxa (Hawksworth and Rossman 1997; Hawksworth 1991, 2000). This complicates the job of studying them, but it also provides an incentive to do so: endophytes may be a largely unstudied reservoir of fungal biodiversity (Hawksworth and Rossman 1997; Arnold et al. 2001).

Third, some endophytes do not grow in culture. Culturing of microorganisms from plant tissues provides a skewed picture of the organisms that grow there (see Chap. 17 by Hallmann et al.). One solution to this problem is to use PCR-based methods to amplify DNA directly from orchid roots using fungal-specific primers. So far, such techniques have been used to study orchid mycorrhizal fungi (see Sect. 9.8), but not non-mycorrhizal endophytes. This approach has revealed that some endophytes of grass roots belong to previously unknown major taxa of fungi (Vandenkoornhuyse et al. 2002); it is possible that orchids also harbor important undescribed lineages of fungi However, these PCR-based approaches are also biased: specific PCR primers are needed to amplify the fungal DNA but not the plant DNA, and primers can preferentially amplify certain groups of fungi (Bruns et al. 1998); using more than one set of primers and varying amplification conditions may help reveal additional taxa.

9.6
Host Specificity of Orchid Endophytes

Host specificity of endophytes is important for estimates of global fungal biodiversity. There are many fungal species associated with each plant species, and if these fungi are host-specific, the number of endophyte species will increase linearly with the number of plant species. Ratios of fungal species to plant species are used to extrapolate fungal biodiversity from plant species richness; the most commonly cited ratio is 6:1 fungi: plants, including pathogens and endophytes (Hawksworth 2000). Endophytes are an undersampled group in terms of fungal biodiversity (Hawksworth 1991; Hawksworth and Rossman 1997; Fröhlich and Hyde 1999).There are three reasons to expect that studying orchid root endophytes may make a significant contribution to this biodiversity: (1) orchids represent almost 10% of angiosperm species; (2) orchid roots are anatomically, morphologically and ecologically different from other roots (see Sects. 9.2, 9.3) most orchids are in the tropics, probably the most undersampled area for fungal biodiversity (Fröhlich and Hyde 1999; Hawksworth 2000; Arnold et al. 2001).

There is century-old debate about the specificity of orchids for mycorrhizal fungi, (see Arditti et al.1990; Rasmussen 1995, 2002; Taylor et al. 2002 for reviews), which applies to non-mycorrhizal endophytes as well. However, few attempts have been made to compare non-mycorrhizal endophytes among orchid species using quantitative methods. Given the diversity of endophytic fungi in orchid roots and the variation in methods among studies, it is difficult to determine the levels of specificity and preference in the interaction. Taxonomic problems (see Sect. 9.5) also complicate the issue of host specificity in orchid mycorrhizal fungi: since many endophytes cannot be identified to the species level with confidence, it is difficult to determine whether different orchid species have different communities of endophytes.

9.7
Endophytic Fungi in Roots of Terrestrial, Photosynthetic Orchids

Non-mycorrhizal endophytes have been isolated from various terrestrial, photosynthetic orchids (Table 9.1), most extensively by Randall Currah and associates in Canada (Table 9.1; see Currah et al. 1997; Taylor et al. 2002). The most common and widespread endophyte isolated is *Phialocephala*, one of the 'dark septate endophytes.' These fungi are also common in many plants [Currah et al. 1987; Fernando and Currah 1995; see Chaps. 7 (Sieber and Grünig), 12 (Girlanda et al.) and 15 (Schulz)]. They may function as

Table 9.1. Non-*Rhizoctonia* fungi reported from orchid roots[a]. Basidiomycetes are in **bold**; those from myco-heterotrophic orchids have been assumed or shown to be mycorrhizal

Orchids	Fungi	Location	Reference
Terrestrial, photosynthetic orchids			
Amerorchis rotundifolia	*Phialocephala*[b]	Canada	Zelmer 1994
Amerorchis rotundifolia	*Phialocephala fortinii*	Canada	Currah et al. 1987
Blettia striata	***Favolaschia thwaitesii***	Zambia	Jonsson and Nylund 1979
Calypso bulbosa	*Leptodontidium orchidicola*, *Phialocephala fortinii*	Canada	Currah et al. 1987
Coeloglossum viride	*Dactylella* sp., *Phialocephala*	Canada	Zelmer 1994
Coeloglossum viride	*Leptodontidium orchidicola*, *Trichosporiella multisporum*	Canada	Currah et al. 1987
Cymbidium sinense	***Mycena orchidicola* sp.nov.**	China	Fan et al. 1996
Cypripedium calceolus	*Alternaria* sp., *Chaetomium* sp. *Cylindrocarpon* sp., *Epicoccum purpureum*, *Phialocephala*, *Phoma* sp.	Canada	Zelmer 1994
Cypripedium candidum	*Acremonium killense*, *Humicola* sp., *Phialocephala*	Canada	Zelmer 1994
Cypripedium montanum	*Phialocephala*	Canada	Zelmer 1994
Cypripedium passerinum	*Phialocephala*	Canada	Zelmer 1994
Cypripedium reginae	*Fusarium* sp.	Canada	Vujanovic et al. 2000
Dactylorhiza majalis	**cf. Laccaria**	Denmark	Kristiansen et al. 2001
Epipactis microphylla[c]	Tuber, other *Pezizales*	France	Selosse et al. 2004
Epipactus helleborine	*Cylindrocarpon destructans*, *Humicola fuscoatra*, *Morchella* sp., *Sordaria fimicola*	Finland	Salmia 1988
Goodyera oblongifolia	*Humicola* sp., *Phialocephala*, *Phoma sp.*, *Thermomyces verrucosus*	Canada	Zelmer 1994
Listera cordata	*Penicillium* sp., *Phialocephala*	Canada	Zelmer 1994
Piperia unalascensis	*Phialocephala*	Canada	Zelmer 1994
Piperia unalascensis	***Sistotrema* sp.**	Canada	Currah et al. 1990

Table 9.1. (continued)

Orchids	Fungi	Location	Reference
Platanthera dilatata	*Phialocephala*	Canada	Zelmer 1994
Platanthera hyperborea	*Acremonium kiliense*, *Phialocephala*, *Sporormia minima*, *Thielavia basicola*	Canada	Zelmer 1994
Platanthera hyperborea	*Leptodontidium orchidicola*, *Trichocladium opacum*	Canada	Currah et al. 1987
Platanthera obtusata	*Acremonium killense*, *Phialocephala*	Canada	Zelmer 1994
Platanthera obtusata	***Sistotrema*** sp.	Canada	Currah et al. 1990
Platanthera praeclara	*Fusarium oxysporum*, *Phialocephala*	Canada	Zelmer 1994
Spiranthes lacera	*Acremonium kiliense*, *Phialocephala*	Canada	Zelmer 1994
Spiranthes magnicamporum	*Cylindrocarpon* sp., *Papulaspora* sp., *Phialophora richardsiae*, *Ulocladium* sp.	Canada	Zelmer 1994
Spiranthes romanzoffiana	*Acremonium kiliense*, *Cylindrocarpon* sp., *Gliomastix murorum*, *Phialocephala*	Canada	Zelmer 1994
Myco-heterotrophic orchids			
Cephalanthera austinae	***Thelephora-Tomentella*** **(14 spp.)**	United States	Taylor and Bruns 1997
Corallorhiza maculata	***Armillaria melea***	United States	Campbell 1970a
Corallorhiza maculata	**Russulaceae (20 spp.)**	United States	Taylor and Bruns 1997
Corallorhiza maculata	**Russulaceae (3 spp.)**	United States	Taylor and Bruns 1999
Corallorhiza maculata	*Cylindrocarpon* sp., *Phialocephala*	Canada	Zelmer 1994
Corallorhiza maculata	*Leptodontidium orchidicola*	Canada	Currah et al. 1987
Corallorhiza striata	*Cylindrocarpon* sp., *Phialocephala*	Canada	Zelmer 1994
Corallorhiza trifida	*Phialocephala*	Canada	Zelmer 1994
Corallorhiza mertensiana	**Russulaceae (22 spp.)**	United States	Taylor et al. 2003
Corallorhiza striata	***Thelephora-Tomentella***	United States	Taylor 1997
Corallorhiza trifida	***Mycena thuja***	New Zealand	Campbell 1970a

Table 9.1. (continued)

Orchids	Fungi	Location	Reference
Corallorhiza trifida	**yellow basidiomycete w/ clamps**	Canada	Zelmer and Currah 1995
Corallorhiza trifida	***Thelephora-Tomentella***	Scotland	McKendrick et al. 2000
Danhatchia australis (= *Yoania australis*)	***Lycoperdon perlatum***	New Zealand	Campbell 1970b
Didymoplexis minor	***Marasimius coniatus***	Paleotropics	Burgeff 1959
Galeola altissima	***Erythromyces crocicreas, Ganoderma australe, Loweporus tephroporus, Microporus affinus***		Hamada and Nakamura 1963, Umata 1995
Galeola (= *Erythrorchis*) *ochobiensis*	***Hymenochaete crocicreas***		Umata 1998
Galeola (= *Erythrorchis*) *ochobiensis*	***Auricularia polytricha, Lyophyllum shimeji***		Umata 1997a, 1997b
Galeola (= *Erythrorchis*) *ochobiensis*	***Lentinula edodes***		Umata 1998
Galeola (= *Erythrorchis*) *ochobiensis*	***Lenzites betulinus, Trametes hirsuta***	Japan	Umata 1999
Galeola septentrionalis	***Armillaria mellea***	Japan	Hamada 1939, Terashita 1985
Galeola septentrionalis	***Armillaria jezoensis* sp nov.**		Cha and Igarashi 1996
Galeola sesamoides	***Fomes* sp.**	New Zealand	Campbell 1964
Gastrodia cunninghamii	***Armillaria mellea***		Campbell 1962
Gastrodia elata	***Armillaria mellea***	China	Kusano 1911
Gastrodia elata	***Armillariella mellea***	China	Lan et al. 1994
Gastrodia elata	***Mycena osmundicola***	New Zealand	Lan et al. 1996
Gastrodia minor	**brown basidiomycete w/clamps**	New Zealand	Campbell 1962
Gastrodia sesamoides	***Fomes mastoporus***		Campbell 1964
Wullschlaegelia calcarata	*Acremonium, Colletotrichum, Curvularia, Cylindrocladium, Gliocladium, Paecilomyces, Penicillium, Trichoderma, Xylaria*	Puerto Rico	J.T. Otero (unpublished)

Table 9.1. (continued)

Orchids	Fungi	Location	Reference
Epiphytic & epilithic orchids			
Campylocentrum micranthum	*Calonectria kyotensis*, *Phomopsis cf orchidophila*	Costa Rica	Richardson 1993
Catasetum maculatum	*Acrogenospora* sp., *Codinaea parva*, *Colletotrichum crassipes*, *Epicoccum andropogonis*, *Glomerella cingulata*, *Hadrotrichum* sp., *Lasiodiplodia theobromae*	Costa Rica	Richardson 1993
Catasetum maculatum	*Troposporella* sp.	Costa Rica	Richardson and Currah 1995
Dichaea standleyi	*Hadrotrichum* sp., *Nectria alata*, *Phomopsis cf. orchidophila*	Costa Rica	Richardson 1993
Dichaea trulla	*Colletotrichum crassipes*	Costa Rica	Richardson 1993
Dimerandra emarginata	*Epicoccum andropogonis*, *Hadrotrichum* sp.	Costa Rica	Richardson 1993
Dryadella pusiola	*Chaetomium homopilatum*	Costa Rica	Richardson 1993
Encyclia fragrans	*Alternaria alternata*, *Colletotrichum crassipes*, *Dactylaria* sp., *Epicoccum andropogonis*, *Glomerella cingulata*, *Hadrotrichum* sp., *Lasiodiplodia theobromae*, *Nodulisporum* sp., *Pseudallescheriaboydii*	Costa Rica	Richardson 1993
Epidendrum difforme	*Lasiodiplodia theobromae*, *Pithomyces maydicus*	Costa Rica	Richardson 1993
Epidendrum difforme	*Troposporella* sp.	Costa Rica	Richardson and Currah 1995
Epidendrum isomerum	*Nodulisporum* sp.	Costa Rica	Richardson 1993
Epidendrum nocturnum	*Epicoccum andropogonis*	Costa Rica	Richardson 1993
Epidendrum octomerioides	*Lasiodiplodia theobromae*, *Nectria ochroleuca*, *Pestalotiopsis papposa*, *Phomopsis cf. orchidophila*	Costa Rica	Richardson 1993
Epidendrum porpax	*Fusarium oxysporum*, *Guignardia* sp.	Colombia	Dreyfuss and Petrini 1984
Epidendrum schlechterianum	*Nectria haematococca*, *Periconiella* sp., *Pestalotiopsis papposa*, *Xylaria* sp.	Costa Rica	Richardson 1993

Table 9.1. (continued)

Orchids	Fungi	Location	Reference
Epidendrum stangeanum	*Fusarium oxysporum*, *Hadrotrichum* sp., *Nodulisporium* sp., *Pithomyces maydicus*	Costa Rica	Richardson 1993
Epidendrum stangeanum	*Troposporella* sp.	Costa Rica	Richardson and Currah 1995
Epidendrum spp.	*Ascochyta* sp., *Colletotrichum gloeosporioides*, *Colletotrichum* sp., *Cryptocline* sp., *Lasiodiplodia theobromae*, *Pestalotia cf. heterocornis*, *Phaeoseptoria cf. vermiformis*, *Phoma* sp., *Xylaria* sp.	Colombia/ Brazil	Dreyfuss and Petrini 1984
Gongora unicolor	*Epicoccum andropogons*, *Hypoxylon cf. unitum*, *Lasmeniella* sp., *Nodulisporium* sp., *Pestalotia poppola*, *Xylaria* sp.	Costa Rica	Richardson 1993
Hexisea imbricata	*Arthrinium* sp., *Hadrotrichum* sp., *Pestalotiopsis aquatica*	Costa Rica	Richardson 1993
Jacquinella globosa	*Colletotrichum crassipes*	Costa Rica	Richardson 1993
Lepanthes caritensis	*Penicillium*, *Trichoderma*, *Xylaria corniformis*	Puerto Rico	Tremblay et al. 1998
Lepanthes rupestris	*Acremonium*, *Colletotrichum*, *Fusarium*, *Guignardia*, *Humicola*, *Pestalotia*, *Phomposis*, *Trichoderma*, *Xylaria*	Puerto Rico	Bayman et al. 2002
Lepanthes spp.	*Aspergillus*, *Colletotrichum*, *Penicillium*, *Pestalotia*, *Xylaria arbuscula*, *X. corniformis*, *X. cf. cubensis*, *X. cf. curta multiplex*, *X. obovata*, *X. polymorpha*, *Xylaria* sp.	Puerto Rico	Bayman et al. 1997
Maxillaria confusa	*Arthrinium* sp., *Malbranchea* sp.	Costa Rica	Richardson 1993
Maxillaria endresii	*Drechslera ellisii*, *Pestalotiopsis papposa*	Costa Rica	Richardson 1993
Maxillaria neglecta	*Chaetomium subspirale*, *Colletotrichum crassipes* sp., *Drechslera australensis*, *Glomerella cingulata*, *Hadrotrichum* sp., *Humicola* sp., *Nectria haematococca*, *N. ochroleuca*, *Phomopsis cf. orchidophila*, *Xylaria* sp.	Costa Rica	Richardson 1993
Maxillaria nicaraguensis	*Pestalotiopsis gracilis*	Costa Rica	Richardson 1993
Maxillaria uncata	*Cryptosporiopsis* sp., *Hadrotrichum* sp.	Costa Rica	Richardson 1993
Maxillaria xylobiflora	*Epicoccum nigrum*	Costa Rica	Richardson 1993
Maxillaria sp.	*Colletrotrichum crassipes*, *Hadrotrichum* sp., *Xylaria* sp.	Costa Rica	Richardson 1993

Table 9.1. (continued)

Orchids	Fungi	Location	Reference
Maxillaria sp.	*Acremonium strictum, Fusarium sambucinum, Tubercularia, Xylaria* sp., *Pestalotia cf. heterocornis, Phomopsis* sp.	Colombia, Brazil	Dreyfuss and Petrini 1984
Myoxanthus scandens	*Colletotrichum crassipes*	Costa Rica	Richardson 1993
Nidema boothii	*Dactylaria* sp., *Epicoccum andropogonis, Lasiodiplodia theobromae, Leptosphaerulina australis, Pestalotiopsis papposa, Phomopsis cf. orchidophila*	Costa Rica	Richardson and Currah 1995
Nidema boothii	*Troposporella* sp.	Costa Rica	Richardson et al. 1993
Octomeria sp.	***Melanotus alpiniae***	Costa Rica	Richardson 1993
Oncidium stenotis	*Chaetomium subspirale, Colletotrichum crassipes, Humicola* sp., *Nigrospora sphaerica*	Costa Rica	Richardson 1993
Pleurothallis corniculata	*Cryptosporiopsis* sp., *Pestalotiopsis papposa*	Costa Rica	Richardson 1993
Pleurothallis guanacastensis	*Hypoxylon cf. unitum, Lasiodiplodia theobromae, Pithomyces maydicus, Xylaria* sp.	Costa Rica	Richardson 1993
Pleurothallis pantasmi	*Hadrotrichum* sp., *Humicola* sp., *Nectria peziza*	Costa Rica	Richardson 1993
Pleurothallis periodica	*Chaetomium subspirale, Cladosporium cladosporioides, Geotrichopsis* sp., *Hadrotrichum* sp., *Xylaria* sp.	Costa Rica	Richardson 1993
Pleurothallis phyllocardioides	*Lasiodiplodia theobromae, Nectria haematococca*	Costa Rica	Richardson 1993
Pleurothallis uncinata	*Nectria haematococca, Pithomyces maydicus*	Costa Rica	Richardson 1993
Pleurothallis verecunda	*Epicoccum andropogonis*	Costa Rica	Richardson 1993
Pleurothallis sp.	*Chaetomium funicola, Pyrenochaeta cf. rubi-idaei, Ramichloridium cf. subulatum*	Costa Rica	Richardson 1993
Pleurothallis sp.	*Cryptocline* sp., *Xylaria* sp.	Colombia	Dreyfuss and Petrini 1984
Polystachya foliosa	*Hadrotrichum* sp., *Xylaria* sp.	Costa Rica	Richardson 1993
Psychilis kraenzlinii	*Colletotrichum, Curvularia, Mucor, Paecilomyces, Pestalotia, Xylaria*	Puerto Rico	J.T. Otero (unpublished)

Table 9.1. (continued)

Orchids	Fungi	Location	Reference
Psychilis krugii	*Colletotrichum*, *Curvularia*, *Mucor*, *Paecilomyces*, *Pestalotia*, *Xylaria*	Puerto Rico	J.T. Otero (unpublished)
Rodriguezia compacta	*Colletotrichum crassipes*, *Epicoccum andropogonis*, *Hadrotrichum* sp., *Leptosphaerulina australis*, *Pestalotiopsis aquatica*	Costa Rica	Richardson 1993
Scaphyglottis cf. prolifera	*Nectria ochroleuca*	Costa Rica	Richardson 1993
Scaphyglottis gracilis	*Nectria haematococca*, *N. ochroleuca*	Costa Rica	Richardson 1993
Scaphyglottis minutiflora	*Arthrinium* sp.	Costa Rica	Richardson 1993
Sobralia cf. mucronata	*Arthrinium* sp., *Lasiodiplodia theobromae*	Costa Rica	Richardson 1993
Sobralia mucronata	*Pseudallescheria boydii*	Costa Rica	Richardson 1993
Sobralia powellii	*Xylaria* sp.	Costa Rica	Richardson 1993
Sobralia sp.	*Epicoccum nigrum*, *Glomerella cingulata*, *Lasiodiplodia theobromae*, *Nectria haematococca*	Costa Rica	Richardson 1993
Stelis endresii	*Curvularia cymbopogonis*, *Nectria haematococca*, *Phomopsis cf. orchidophila*	Costa Rica	Richardson 1993
Stelis sp.	*Alternaria alternata*, *Arthrinium* sp., *Chloridium virescens*, *Colletotrichum crassipes*, *Cryptosporiopsis* sp., *Epicoccum andopogensis*, *Hadrotrichum* sp., *Hypoxylon cf. unitum*, *Nectria alata*, *Nectria ochroleuca*, *N. radicicola*, *Neoplaconema napelli*, *Pithomyces maydicus*, *Xylaria* spp.	Costa Rica	Richardson 1993
Trichosalpinx blaisdellii	*Lasiodiplodia theobromae*	Costa Rica	Richardson 1993
Trichosalpinx orbicularis	*Chaetosticta cf. perforata*, *Colletotrichum crassipes*, *Nectria alata*	Costa Rica	Richardson 1993
Trichosalpinx sp.	*Hadrotrichum* sp.	Costa Rica	Richardson 1993
Trigonidium egertonianum	*Chaetosticta cf. perforata*, *Epicoccum andopogensis*	Costa Rica	Richardson 1993
Trigonidium riopalaquense	*Chaetomium aureum*, *Colletotrichum acutatum*, *C. crassipes*, *Nectria haematococca*, *Nodulisporium* sp., *Periconiella* sp.	Costa Rica	Richardson 1993

Table 9.1. (continued)

Orchids	Fungi	Location	Reference
Unidentified	*Acremonium pteridii*, *Anthostomella aracearum*, *Ascochyta* sp., *Aureobasidium caulivorum*, *Chaetosphaeria endophytica*, *Colletotrichum* spp., *Coniothyrium* sp., *Cryptocline* spp., *Cryptosporiopsis* sp., *Curvularia pallescens*, *Cytogloeum* sp., *Fusarium oxysporum*, *Gelatinosporium* spp., *Gliocladium roseum*, *Glomerella cingulata*, *Kaskaskia* sp., *Lasiodiplodia theobromae*, *Melanconium* sp., *Microascus cinereus*, *Microcyclus* sp., *Nodulisporium gregarium*, *Nodulisporium* spp., *Pestalotia adusta*, *Phialaspora* sp., *Phoma* sp., *Phomatospora berkeleyi*, *Phomopsis orchidophila*, *Phomopsis* sp., *Phyllosticta capitalensis*, *P. colocasiicola*, *Ramichloridium apiculatum*, *Verticillium lecanii*	French Guayana	Petrini and Dreyfuss 1981

[a] The following fungal genera are presumed to be mycorrhizal and are not included in this table: *Ceratobasidium*, *Oliveonia Sebacina*, *Serendipita*, *Thanatephorus*, *Tulanella* and *Ypsilonidium* (teleomorphs); *Ceratobasidium* (anamorphs) (Roberts 1999). These fungi can be found in tables published by Currah et al. (1997), Rasmussen (2002) and Taylor et al. (2002)

[b] Also called *Mycelium radicis atrovirens* (MRA)

[c] This species can be either myco-heterotrophic or photosynthetic

mycorrhizae in some plants (Fernando and Currah 1996; Jumpponen 2001), but their role in orchids is still unclear (Rasmussen 2002).

One of the most ubiquitous and interesting groups of endophytes is *Fusarium* and its teleomorphs. *Fusarium moniliforme* (= *F. verticilloides*) was isolated from leaves and roots of *Cypripedium reginae* (Peschke and Volz 1978). Inoculation of spores induced disease symptoms in hybrid orchids, suggesting that the fungus was a potential pathogen (Peschke and Volz 1978). This fungus is a ubiquitous endophyte and pathogen of all organs of various hosts, including maize (Leslie et al. 1990; Kuldau and Yates 2000; see Chap. 8 by Bacon and Yates).

On the other hand, the ability of *Fusarium* to stimulate orchid seed germination has been known for 100 years (Bernard 1909). An unidentified *Fusarium* strain isolated from a germinating seed of *Cypripedium reginae* induced germination of *C. reginae* seeds in vitro (Vujanovic et al. 2000). Although this is functional rather than ecological specificity (Masuhara and Katsuya 1994), it raises the question of whether a single *Fusarium* isolate can be endophytic, pathogenic and mycorrhizal under different circumstances.

9.8 Endophytic Fungi in Roots of Myco-Heterotrophic Orchids

Non-photosynthetic (or more precisely, myco-heterotrophic) orchids have been extensively studied because of their interesting relationships with fungi (Leake 1994). Unable to assimilate their own carbon, these orchids are parasitic on fungi. Several myco-heterotrophic orchids are very specific for certain fungi, which is interesting because orchid mycorrhizal relationships are generally considered to be non-specific (Taylor et al. 2002). In most cases the myco-heterotrophic orchids are parasitizing a mycorrhizal partner of a nearby photosynthetic plant, which means that they are indirectly parasitizing the plant as well. However, the amount of carbon taken by the orchid is probably insignificant to the plant host (McKendrick et al. 2000; Sanders 2003). An excellent review of mycorrhizal specificity in myco-heterotrophic plants is available (Taylor et al. 2002).

Fungal DNA has been amplified directly from roots or pelotons of myco-heterotrophic orchids using fungal-specific (or in some cases, basidiomycete-specific) primers. This approach has been used on several orchids in North America: *Cephalanthera* (Taylor and Bruns 1997), *Corallorhiza* spp. (Taylor and Bruns 1997, 1999) and *Hexalectris* (Taylor et al. 2003). It has also been used on *Dactylorhiza* in Denmark (Kristiansen et al. 2001) and *Neottia* in the United Kingdom, Germany (McKendrick et al. 2002) and in France (Selosse et al. 2002). In all these plants, the only fungi amplified from peletons belonged to taxa that are considered ectomycorrhizal. No other endophytic fungi were reported, which may suggest that secondary colonization of pelotons by non-mycorrhizal endophytic fungi is uncommon. No such studies have been done with the aim of identifying non-mycorrhizal fungi in orchids, e.g., using ascomycete-specific PCR primers.

Myco-heterotrophic orchids tend to associate with ectomycorrhizal fungi rather than the *Rhizoctonia*-like fungi typical of orchids, presumably because ectomycorrhizal fungi are more reliable suppliers of photosynthate. In other cases, the presumed mycorrhizal fungi are basidiomycetes that are wood decomposers not known to form ectomycorrhizae, e.g., *Armillaria*, *Fomes*, *Ganoderma* and *Phellinus* sp. (for references, see Rasmussen 2002; Taylor et al. 2002). However, most such studies have not demonstrated functional relationships between the fungus and the myco-heterotrophic orchid, so this is another area where the boundary between mycorrhizal fungi and endophytes is unclear.

Endophytes have been reported from only one tropical myco-heterotrophic orchid, *Wullschlaegelia*. Forty-one morphospecies of endophytes were isolated from plants of *W. calcarata* in Puerto Rico (J.T. Otero, unpublished; Table 9.1). Common endophytes included *Xylaria*, *Trichoderma*, *Colletotrichum* and various dematiaceous hyphomycetes. These genera were not randomly distributed among sites (χ^2 = 31.84, df = 17, P = 0.004). A morphospecies accumulation curve suggested that most of the culturable endophytic fungi in the roots were isolated (Fig. 9.1). It is likely that temperate myco-heterotrophic orchids contain a mycoflora as diverse as *W. calcarata*, but the peloton isolation technique used in many of the above studies excluded most of the non-mycorrhizal endophytes.

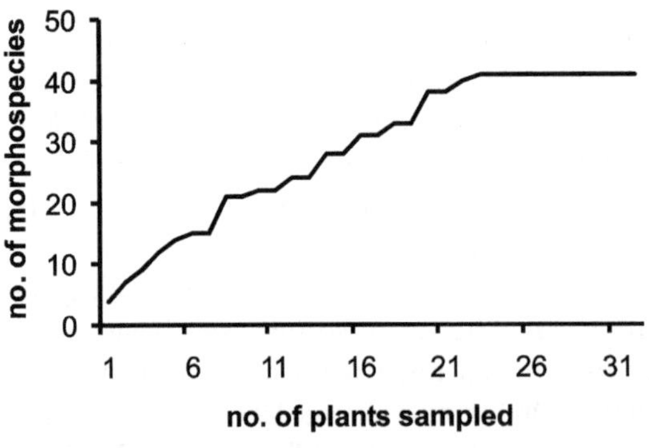

Fig. 9.1. Species accumulation curve for endophytic fungi isolated from roots of *Wullschlaegelia calcarata*, a myco-hetrotrophic orchid. A total of 32 plants were collected from eight populations in El Verde, Puerto Rico. Morphospecies were identified by morphology in culture; the identified fungi are listed in Table 9.1

9.9 Endophytic Fungi in Roots of Epiphytic and Lithophytic Orchids

Reports on non-mycorrhizal endophytes in roots of epiphytic and lithophytic orchids have mostly come from the neotropics. These reports have focused on identifying the fungi rather than exploring their relationship with the orchids. Descriptions of some common endophytes that have been published will facilitate further studies (Richardson et al. 1993; Richardson and Currah 1995; Currah et al. 1997).

South America Three taxa of Ascomycetes, 5 of Hyphomycetes and 13 of Coelomycetes were isolated from epiphytic orchid roots in French Guiana; many of these fungi were also isolated from roots of aroids and bromeliads, suggesting that the fungi are generalists (Table 9.1; Petrini and Dreyfuss 1981). Some of the same fungi were also found in orchid roots from the Colombian Amazon (Dreyfuss and Petrini 1984).

Costa Rica The most extensive sampling of epiphytic orchid roots for endophytes was done in La Selva, Costa Rica (Richardson et al. 1993; Richardson and Currah 1995). Of 59 species of epiphytic orchids sampled in La Selva, Costa Rica, mycorrhizal pelotons were observed in the roots of 23 (= 39%) (Richardson et al. 1993), a fairly low infection frequency that agrees with other studies. Basidiomycetes comprised only 3% of the fungi isolated, suggesting that mycorrhizal fungi were much less common than non-mycorrhizal endophytes or pathogens. *Hadrotrichum*, *Colletotrichum*, *Epicoccum*, *Lasiodiplodia* and *Phomopsis* were the most common genera of deuteromycetes and ascomycetes (Richardson and Currah 1995). Relative proportions of ascomycetes, hyphomycetes and coelomycetes were comparable to those reported by Petrini and Dreyfuss (1981).

Puerto Rico We have sampled endophytic fungi from roots of various epiphytic orchids in Puerto Rico (Table 9.1). The most common endophytic fungi have been fairly consistent from study to study. A variety of endophytic fungi were isolated from nine species of epiphytic orchids in Puerto Rico, including *Xylaria*, *Pestalotia* and *Colletotrichum* (Otero et al. 2002). *Rhizoctonia*-like fungi were isolated at lower frequency, from about 20% of the samples.

Fifty-five fungi (in 26 morphospecies) were isolated from roots of *Tolumnia*, from both juvenile and adult plants (J.T. Otero, unpublished). Thirty-one of these strains (in 13 morphospecies) were tested for potential mycorrhizal activity with *T. variegata* seeds. Thirteen strains (in 6 morphospecies) had a positive effect on seed germination in vitro, all of which were *Rhizoctonia*-like fungi; others were parasitic on seeds. These data suggest *Rhizoctonia* are the principal mycorrhizal fungi of *T. variegata*, but the sample did not include *Fusarium* (see Sect. 9.9).

A comparison of two sympatric populations of *Psychilis* and *P. krugii* found 26 morphospecies of endophytic fungi (Table 9.1; J.T. Otero, unpublished). There was no apparent difference between the fungal communities of the two orchids. The dominant species were *Xylaria* spp.; *Rhizoctonia*-like fungi were much less common than in the orchids cited above, and few pelotons were seen in root sections.

Endophytic fungi were isolated from roots and leaves of six species of *Lepanthes* in Puerto Rico (Bayman et al. 1997). *Xylaria* and *Rhizoctonia* were the most common genera isolated from roots and leaves. At least nine species of *Xylaria* were isolated, up to four species occurring in a single plant. Frequency of *Xylaria* species and *Rhizoctonia* did not differ significantly between roots and leaves, which is perhaps not surprising given that roots of epiphytes are exposed to light and air. Frequency of both *Rhizoctonia* and *Xylaria* differed significantly among *Lepanthes* species. However, there was also significant variation among different roots of a single plant, which means that studies that wish to compare levels of infection should sample intensively. In another study of *L. rupestris*, endophytic fungi were isolated from roots of lithophytic plants (Bayman et al. 2002). The most common genera were *Guignardia* (isolated from 22% of root pieces), *Colletotrichum* (10%) and *Xylaria* (7%). *Rhizoctonia*-like fungi (which include the presumed mycorrhizal fungi) were isolated at much lower frequencies (3%).

Is distribution of fungi a limiting factor in distribution of orchids? Tremblay et al. (1998) isolated fungi from roots of *Lepanthes caritensis*, an orchid species whose distribution is limited to a few trees along a single river. Fungi isolated from orchid roots were compared to fungi isolated from bark of host trees, and of conspecific trees without orchids (Tremblay et al. 1998). The most common fungi in *L. caritensis* roots were *Xylaria* spp, particularly *X. corniformis*. *Penicillium*, *Trichoderma* and *Rhizoctonia* were also isolated from roots, and were the most common fungi isolated from tree bark. However, *Xylaria* sp. were not common on tree bark, suggesting that they had a particular affinity for *Lepanthes* roots. The number of other, unidentified fungi was significantly higher on bark of trees without orchids than on trees with orchids. This may suggest that the presence of certain fungi inhibits the establishment of orchids.

In general, there is little evidence that non-mycorrhizal endophytes in orchids are specific to orchids. The most common groups of orchid endophytes (Table 9.1) are fungi that are ubiquitous in soil and as endophytes of other plants. There are marked differences between terrestrial and epiphytic orchids: in terrestrials, *Phialocephala* is the most frequently isolated group; in epiphytes, *Hadrotrichum*, *Epicoccum*, *Lasiodiplodia*, *Xylaria* and *Pestalotiopsis* are most frequent. However, these differences reflect differences in temperate vs. tropical mycofloras and are not particular to orchid-associated fungi. Several new species have been described from orchid endophytes (e.g., *Mycena* sp. nov. (Fan et al. 1996), *Armillaria jezoensis* sp. nov. (Cha and Igarashi 1996), *Leptodontidium* (Currah et al. 1987)), but in most cases their distribution is not sufficiently known to claim a special affinity for orchids.

9.10 Endophytic Fungi in Epiphytic Orchid Roots: Importance to Plant Hosts

There are two reasons to believe that the presence of endophytes could affect mycorrhizal fungi, and vice versa. First, orchids may produce phytoalexins such as orchinol when challenged by a fungus; these phytoalexins may then limit the ability of other fungi to colonize the plant – a type of induced resistance. These phytoalexins inhibit growth of a broad range of fungi, though bacteria are less susceptible (Gäumann et al. 1960). According to Rasmussen (1995), "...all underground parts of terrestrial orchids must either accommodate the endophyte (i.e., mycorrhizal fungus) or actively reject it." Second, endophytes and mycorrhizal fungi could compete for nutrients in the root, or could actively inhibit each other by production of secondary metabolites. Nutrient translocation from mycorrhizal fungi to orchids has been demonstrated repeatedly (see Rasmussen 1995; Bidartondo et al. 2004), but it is unknown how non-mycorrhizal endophytes might affect this transfer.

Several studies have asked whether the presence of endophytic fungi was associated with the presence of other endophytes or of mycorrhizal fungi. Pieces of *Lepanthes* roots colonized by *Colletotrichum* had a significantly lower rate of infection with *Xylaria* than would be expected from the frequency of each genus alone (Bayman et al. 2002). Presence of *Colletotrichum* also showed a significant, negative correlation with *Guignardia*. This suggests there may be competition or antagonism between these genera; alternatively, their colonization or growth could be favored by different environmental conditions. Also, mycorrhizal fungi in *Vanilla* often occur together with other, presumably pathogenic, fungi (Alconcero 1969; Porras-Alfaro and Bayman 2003). Roots of *Vanilla* plants in Puerto Rico were often colonized by both the pathogen *Fusarium oxysporum* and by *Rhizoctonia solani*, which was both mycorrhizal and pathogenic to *Vanilla* (Alconcero 1969).

Fungicides were applied to plants of *L. rupestris* to see if the costs of harboring fungi outweighed the benefits (Bayman et al. 2002). Propiconazole significantly reduced the number of total fungal colonies isolated from roots (but not from leaves), increased plant mortality, and increased loss of leaves, as compared to control plants. Benomyl significantly reduced the number of fungi isolated from leaves (but not from roots) and decreased plant mortality, but did not significantly effect plant growth. These data suggest that fungi have both positive and negative effects on growth and survival of orchid plants in the field: benomyl, which affects mainly ascomycete fungi, may have reduced pathogens and endophytes, not harming plants, whereas propiconazole, which also affects basidiomycetes, may have reduced mycorrhizal fungi as well. However, these results are very difficult to interpret: a single plant may have simultaneous infections by endophytes, mycorrhizal fungi and pathogens, and the positive effects of one may mask the negative effects of another.

The interaction between plants and endophytic fungi may be viewed as a 'balanced antagonism' (Schulz et al. 1999). The plant produces secondary metabolites that are sufficiently toxic to restrict the growth of the endophyte without being able to kill it; the endophyte in turn produces enzymes and metabolites that allow it to colonize the plant but are not sufficient to cause pathogenicity. This balance becomes still more complex if the endophyte and its activities also affect other fungi, either endophytic or mycorrhizal, as these data from *Lepanthes* suggest.

9.11 Conclusions

About 90% of the cells that comprise a human body are microbial – we are walking communities (Hamilton 1999). Most of these microorganisms are poorly understood: pathogenic microorganisms are intensively studied, but much less is known about non-pathogenic, commensal microorganisms, which are more common than pathogens. Their importance becomes obvious when an antibiotic designed to kill a pathogen affects the community of commensals as well, causing a secondary condition – for example, vaginal yeast infections in women who are taking antibiotics for bladder infections. Cross-talk between microorganisms and the host may have other roles as well; for example, it may be necessary for the proper development of the immune system (Hooper et al. 1998). Even well-studied pathogens may have complex ecological roles: *Helicobacter* was considered a serious pathogen, but recent studies suggest that it is a commensal that sometimes turns pathogenic (Pütsep et al. 1999). Our ignorance of the non-pathogenic microflora is a limiting factor in medicine.

The same situation exists with orchid root endophytes. Interest in orchid mycorrhizal fungi has overshadowed work on other orchid endophytes, which may have important interactions with mycorrhizal fungi and with the orchids themselves. In some cases the distinction between endophytes and mycorrhizal fungi is unclear, and it is likely that, as with *H. pylori* in humans, a single organism can behave as an endophyte, mycorrhizal symbiont or pathogen, depending on the environment and the health of the host plant. The study of orchid endophytes is not much more advanced than the study of orchid mycorrhizae was in 1900 – interesting observations, intriguing ideas, but little information about functional significance.

References

Alconero R (1969) Mycorrhizal synthesis and pathology of *Rhizoctonia solani* in *Vanilla* orchid roots. Phytopathology 59:526–530

Andersen TF, Rasmussen HN (1996) The mycorrhizal species of *Rhizoctonia*. In: Sneh B, Jabaji-Hare S, Neate S, Dijst G (eds) Rhizoctonia species: taxonomy, molecular biology, ecology, pathology and disease control. Kluwer, Dordrecht, pp 379–390

Arditti J, Ernst R, Yam TW, Glabe C (1990) The contribution of orchid mycorrhizal fungi to seed germination: a speculative review. Lindleyana 5:249–255

Arnold AE, Maynard Z, Gilbert G, Coley PD, Kursar TA (2000) Are tropical fungal endophytes hyperdiverse? Ecol Lett 3:267–274

Arnold AE, Maynard Z, Gilbert GS (2001) Fungal endophytes in dicotyledonous neotropical trees: patterns of abundance and diversity. Mycol Res 105:1502–1507

Bayman P, Lebrón LL, Tremblay RL, Lodge DJ (1997) Fungal endophytes in roots and leaves of *Lepanthes* (Orchidaceae). New Phytol 135:143–149

Bayman P, González EJ, Fumero JJ, Tremblay RL (2002) Are fungi necessary? How fungicides affect growth and survival of the orchid *Lepanthes rupestris* in the field. J Ecol 90:1002–1008

Bernard N (1909) L'évolution dans la symbiose. Le orchidées et leurs champignons commensaux. Ann Sci Nat Bot 9:1–196

Bidartondo MI, Burghardt B, Gebauer G, Bruns TD, Read DJ (2004) Changing partners in the dark: isotopic and molecular evidence of ectomycorrhizal liaisons between forest orchids and trees. Proc R Soc Lond B 271:1799–1806

Brighigna L, Montaini P, Favilli F, Carabez-Trejo A (1992) Role of the nitrogen-fixing bacterial microflora in the epiphytism of *Tillandsia* (Bromeliaceae). Am J Bot 79:723–727

Bronstein JL, Wilson WG, Morris WF (2003) Ecological dynamics of mutualist/antagonist communities. Am Nat 162:S24–S39

Bruns TD, Szaro TM, Gardes M, Cullings KW, Pan JJ, Taylor DL, Horton TR, Kretzer A, Garbelotto M, Li Y (1998) A sequence database for the identification of ectomycorrhizal basidiomycetes by phylogenetic analysis. Mol Ecol 7:257–272

Burgeff H (1959) Mycorrhiza of orchids. In: Withner CL (ed) The orchids: a scientific survey. Ronald, New York, pp 361–395

Campbell EO (1962) The mycorrhiza of *Gastrodia cunninghamii* Hook. Trans R Soc N Z 1:289–296

Campbell EO (1964) The fungal association in a colony of *Gastrodia sesamoides*. R. Br Trans R Soc N Z 2:237–246

Campbell EO (1970a) Morphology of the fungal association in three species of *Corallorhiza* in Michigan. Mich Bot 9:108–113
Campbell EO (1970b) The fungal association of *Yoania australis*. Trans R Soc N Z Biol Sci 12:5–12
Carling DE, Pope EJ, Brainard KA, Carter DA (1999) Characterization of mycorrhizal isolate of *Rhizoctonia solani* from orchid, including AG-12, a new anastomosis group. Phytopathology 89:942–946
Cha JY, Igarashi T (1996) *Armillaria jezoensis*, a new symbiont of *Galeola septentrionalis* (Orchidaceae) in Hokkaido. Mycoscience 37:21–24
Chanway CP (1995) Endophytes: they're not just fungi! Can J Bot 74:321–233
Currah RS (1991) Taxonomic and developmental aspects of the fungal endophytes of terrestrial orchid mycorrhizae. Lindleyana 6:211–213
Currah RS, Sigler L, Hambleton S (1987) New records and taxa of fungi from the mycorrhizae of terrestrial orchids of Alberta. Can J Bot 65:2473–2482
Currah RS, Smreciu EA, Hambleton S (1990) Mycorrhizae and mycorrhizal fungi of boreal species of *Platanthera* and *Coeloglossum* (Orchidaceae). Can J Bot 68:1171–1181
Currah RS, Zelmer CD, Hambleton S, Richardson KA (1997) Fungi from orchid mycorrhizas. In: Arditti J, Pridgeon A (eds) Orchid biology: reviews and perspectives, VII. Kluwer, Lancaster, pp 117–170
Darwin C (1887) The various contrivances by which orchids are fertilized by insects, 2nd edn. William Clowes and Sons, London
Dressler RL (1990) The orchids: natural history and classification. Harvard University Press, Cambridge MA
Dreyfuss M, Petrini O (1984) Further investigations on the occurrence and distribution of endophytic fungi in tropical plants. Bot Helv 94:33–40
Fan L, Guo S, Cao W, Xiao P, Xu J, Fan L, Guo SX, Cao WQ, Xiao PG, Xu JT (1996) Isolation, culture, identification and biological activity of *Mycena orchidicola* sp. nov. in *Cymbidium sinense* (Orchidaceae). Acta Mycol Sin 15:251–255
Fernando AA, Currah RS (1995) *Leptodontidium orchidicola* (*Mycelium radicis atrovirens* complex): aspects of its conidiogenesis and ecology. Mycotaxon 54:287–294
Fernando AA, Currah RS (1996) A comparative study of the effects of the root endophytes *Leptodontidium orchidicola* and *Phialocephala fortinii* (Fungi Imperfecti) on the growth of some subalpine plants in culture. Can J Bot 74:1071–1078
Fröhlich J, Hyde KD (1999) Biodiversity of palm fungi in the tropics: are global fungal diversity estimates realistic? Biodivers Conserv 8:977–1004
Gamboa MA, Bayman P (2001) Communities of endophytic fungi in leaves of a tropical timber tree (*Guarea guidonia*: Meliaceae). Biotropica 33:352–360
Gäumann E, Nüesch J, Rimpau RH (1960) Weitere Untersuchungen über die chemischen Abwehrreaktionen der Orchideen. Phytopathol Z 38:274–308
Goh CJ, Sim AA, Lim G (1992) Mycorrhizal associations in some tropical orchids. Lindleyana 7:13–17
Hadley G, Williamson B (1972) Features of mycorrhizal infection in some Malayan orchids. New Phytol 71:1111–1118
Hadley G, Ong SH (1978) Nutritional requirements of orchid endophytes. New Phytol 81:561–569
Hadley G, Arditti M, Arditti J (1987) Orchid diseases: a compendium. In: Arditti J (ed) Orchid biology: reviews and perspectives, vol IV. Cornell University Press, Ithaca, NY, pp 261–325
Hallmann J, Quadt-Hallmann, Mahaffee WF, Kloepper JW (1997) Bacterial endophytes in agricultural crops. Can J Microbiol 43:895–914

Hamada M (1939) Studien über die Mykorrhiza von *Galeola septentrionalis* Reichb. f. neuer Fall der Mykorrhiza-Bildung durch intraradicale Rhizomorpha. Jpn J Bot 10:151–211

Hamada M, Nakamura SI (1963) Wurzelsymbiose von *Galeola altissima* Reichb. F., einer chlorophyllfreien Orchidee, mit dem holzzerstörenden Pilz *Hymenochate crocicreas.* Berk Et Br. Sci Rep Tohoku Univ Ser IV (Biol) 29:227–238

Hamilton G (1999) Insider trading. New Scientist 6:42–46

Hawksworth DL (1991) The fungal dimension of biodiversity: magnitude, significance and conservation. Mycol Res 95:641–655

Hawksworth DL (2000) How many fungi are there? Mycol Res 104:4–5

Hawksworth DL, Rossman AY (1997) Where are all the undescribed fungi? Phytopathology 87:888–891

Hooper LV, Bry L, Falk PG, Gordon JI (1998) Host-microbial symbiosis in the mammalian intestine: exploring an internal ecosystem. BioEssays 20:336–343

Jonsson L, Nylund JE (1979) *Favolaschia dybowskiana* (Singer) Singer (Aphyllophorales), a new orchid mycorrhizal fungus from tropical Africa. New Phytol 83:121–128

Jumpponen A (2001) Dark septate endophytes - are they mycorrhizal? Mycorrhiza 11:207–211

Kristiansen KA, Taylor DL, Kjoller R, Rasmussen HN, Rosendahl S (2001) Identification of mycorrhizal fungi from single pelotons of *Dactylorhiza majalis* (Orchidaceae) using single-strand conformation polymorphism and mitochondrial ribosomal large subunit DNA sequences. Mol Ecol 10:2089–2093

Kuldau GA, Yates IE (2000) Evidence for *Fusarium* endophytes in cultivated and wild plants. In: Bacon CW, White JF (eds) Microbial endophytes, Dekker, New York, pp 85–117

Kusano S (1911) *Gastrodia elata* and its symbiotic association with *Armillaria mellea.* J Coll Agric Jpn 9:1–73

Lacap DC, Hyde KD, Liew ECY (2003) An evaluation of the fungal "morphotype" concept based on ribosomal DNA sequences Fungal Divers 12:55–63

Lan J, Xu JT, Li JS (1994) Study on symbiotic relation between *Gastrodia elata* and *Armillariella mellea* by autoradiography. Acta Mycol Sin 13:219–222

Lan J, Xu JT, Li J (1996) Study on infecting process of *Mycena osmundicola* on *Gastrodia elata* by autoradiography. Acta Mycol Sin 13:219–222

Leake JR (1994) Tansley Review No. 69: the biology of myco-heterotrophic ("saprophytic") plants. New Phytol 127:171–216

Leslie JF, Pearson CAS, Nelson PE, Tousson TA (1990) *Fusarium* spp. from corn, sorghum and soybean fields in the central and eastern United States. Phytopathology 80:343–350

Lesica P, Antibus RK (1990) The occurrence of mycorrhizae in vascular epiphytes of two Costa Rican rain forests. Biotropica 22:250–258

Masuhara G, Katsuya K (1994) In situ and in vitro specificity between *Rhizoctonia* spp. and *Spiranthes sinensis* (Persoon) Ames. var. *amoena* (M. Bieberstein) Hara (Orchidaceae). New Phytol 127:711–718

McKendrick SL, Leake JR, Read DJ (2000) Symbiotic germination and development of myco-heterotrophic plants in nature: transfer of carbon from ectomycorrhizal *Salix repens* and *Betula pendula* to the orchid *Corallorhiza trifida* through shared hyphal connections. New Phytol 145:539–548

McKendrick SL, Leake JR, Taylor DL, Read DJ (2002) Symbiotic germination and development of the myco-heterotrophic orchid *Neottia nidus-avis* in nature and its requirement for locally distributed *Sebacina* spp. New Phytol 154:233–247

Otero JT, Ackerman JD, Bayman P (2002) Diversity and host specificity of mycorrhizal fungi from tropical orchids. Am J Bot 89:1852–1858

Peschke HC, Volz PA (1978) *Fusarium moniliforme* Sheld association with species of orchids. Phytologia 40:347–356

Petrini O, Dreyfuss M (1981) Endophytische Pilze in epiphytischen Araceae, Bromeliaceae und Orchidaceae. Sydowia 34:135–148

Porras-Alfaro A, Bayman P (2003) Mycorrhizal fungi of *Vanilla*: root colonization patterns and fungal identification. Lankesteriana 7:147–150

Pütsep K, Brändén CI, Boman HG, Normark S (1999) Antibacterial peptide from *H. pylori*. Nature 398:671–672

Ramsay RR, Dixon KW, Sivasithamparam K (1986) Patterns of infection and endophytes associated with Western Australian orchids. Lindleyana 1:203–214

Rasmussen HN (1995) Terrestrial orchids from seed to mycotrophic plant. Cambridge University Press, Cambridge UK

Rasmussen HN (2002) Recent developments in the study of orchid mycorrhiza. Plant Soil 244:149–163

Richardson KA (1993) Endophytic fungi from Costa Rican orchids. MSc Thesis, University of Alberta, Edmonton

Richardson KA, Currah RS (1995) The fungal community associated with the roots of some rainforest epiphytes of Costa Rica. Selbyana 16:49–73

Richardson KA, Currah RS, Hambleton S (1993) Basidiomycetous endophytes from the roots of neotropical epiphytic Orchidaceae. Lindleyana 8:127–137

Rivas M, Warner J, Bermúdez M (1998) Presence of mycorrhizas in orchids of a neotropical botanical garden. Rev Biol Trop 46:211–216

Roberts P (1999) Rhizonctonia-forming fungi. A taxonomic guide. The trustees of the Royal Botanical Gardens, Kew, London

Salmia A (1988) Endomycorrhizal fungus in chlorophyll-free and green forms of the terrestrial orchid *Epipactis helleborine*. Karstenia 28:3–18

Sanders IR (2003) Preference, specificity and cheating in the arbuscular mycorrhizal symbiosis. Trends Plant Sci 8:143–154

Schulz B, Römmert AK, Dammann U, Aust HJ, Strack D (1999) The endophyte-host interaction: a balanced antagonism? Mycol Res 103:1275–1283

Selosse M-A, Weiß M, Jany JL, Tillier A (2002) Communities and populations of sebacinoid basidiomycetes associated with the achlorophyllous orchid *Neottia nidus-avis* and neighboring tree ectomycorrhizae. Mol Ecol 11:1831–1844

Selosse M-A, Faccio A, Scappaticci G, Bonfante P (2004) Chlorophyllous and achlorophyllous specimens of *Epipactis microphylla* (Neottieae, Orchidaceae) are associated with ectomycorrhizal septomycetes, including truffles. Microb Ecol 47:416–426

Sinclair R (1990) Water relations in orchids. In: Arditti J (ed) Orchid biology, reviews and perspectives, vol V. Timber, Portland OR, pp 61–115

Sturz AV, Christie BR, Nowak J (2000) Bacterial endophytes: potential role in developing sustainable systems of crop protection. Crit Rev Plant Prot 19:1–30

Taylor DL, Bruns TD (1997) Independent, specialized invasions of ectomycorrhizal mutualism by two nonphotosynthetic orchids. Proc Natl Acad Sci USA 94:4510–4515

Taylor DL, Bruns TD (1999) Population, habitat and genetic correlates of mycorrhizal specialization in the 'cheating' orchids *Corallorhiza maculata* and *C. mertensiana*. Mol Ecol 8:1719–1732

Taylor DL, Bruns TD, Leake JR, Read DJ (2002) Mycorrhizal specificity and function in myco-heterotrophic plants. In: van der Heijden MGA, Sanders IR (eds) Mycorrhizal ecology Ecological studies vol. 157 Springer, Berlin Heidelberg New York, pp 375–414

Taylor DL, Bruns TD, Szaro TM, Hodges SA (2003) Divergence in mycorrhizal specialization within *Hexalectris spicata* (Orchidaceae), a nonphotosynthetic desert orchid. Am J Bot 90:1168–1179

Terashita T (1985) Fungi inhabiting wild orchids in Japan (III). A symbiotic experiment with *Armillariella mellea* and *Galeola septentrionalis*. Trans Mycol Soc Jpn 26:47–53

Tremblay RL, Zimmerman J, Lebrón LL, Bayman P, Sastre I, Axelrod F, Alers-García J (1998) Host specificity and low reproductive success in the rare endemic Puerto Rican orchid*Lepanthes caritensis*. Biol Conserv 85:297–304

Tsavkelova EA, Cherdyntseva TA, Lobakova ES, Kolomeitseva GL, Netrusov AI (2001) Microbiota of the orchid rhizoplane. Mikrobiologiya 70:567–573

Tsavkelova EA, Lobakova ES, Kolomeitseva GL, Cherdyntseva TA, Netrusov AI (2003) Associative cyanobacteria isolated from the roots of epiphytic orchids. Mikrobiologiya 72:105–10

Umata H (1995) Seed germination of *Galeola altissima*, an achlorophyllous orchid, with aphyllophorales fungi. Mycoscience 36:369–372

Umata H (1998) A new biological function of shiitake mushroom, *Lentinula edodes*, in a myco-heterotrophic orchid, *Erythrorchis ochobiensis*. Mycoscience 38:355–357

Umata H (1997a) Formation of endomycorrhizas by an achlorophyllous orchid, *Erythorchis ochobiensis*, and *Auricularia polytricha*. Mycoscience 36:369–372

Umata H (1997b) In vitro germination of *Erythrorchis ochobiensis* (Orchidaceae) in the presence of *Lyophyllum shimeji*, an ectomycorrhizal fungus. Mycoscience 38:335–339

Umata H (1999) Germination and growth of *Erythrorchis ochobiensis* (Orchidaceae) accelerated by monokaryons and dikaryons of *Lenzites betulinus* and *Trametes hirsuta*. Mycoscience 40:367–371

Vandenkoornhuyse P, Baldauf SL, Leyval C, Straczek J, Young JPW (2002) Extensive fungal diversity in plant roots. Science 295:2051

Vujanovic V, St.-Arnaud M, Barabé D, Thibeault G (2000) Viability testing of orchid seed and the promotion of colouration and germination. Ann Bot 86:79–86

Wilkinson KG, Dixon KW, Sivasithamparam K (1989) Interaction of soil bacteria, mycorrhizal fungi and orchid seed in relation to germination of Australian orchids. New Phytol 112:429–435

Wilkinson KG, Sivasithamparam K, Dixon KW, Fahy PC, Bradley JK (1994) Identification and characterisation of bacteria associated with Western Australian orchids. Soil Biol Biochem 26:137–142

Zelmer CD (1994) Interactions between northern terrestrial orchids and fungi in nature. MSc Thesis, University of Alberta, Edmonton, Alberta, Canada

Zelmer CD, RS Currah (1995) Evidence of fungal liaison between *Corallorhiza trifida* (Orchidaceae) and *Pinus contorta* (Pinaceae). Can J Bot 73:862–866

Zelmer CD, Cuthbertson L, Currah RS (1996) Fungi associated with terrestrial orchid mycorrhizas, seeds and protocorms. Mycoscience 37:439–448

10 Fungal Endophytes in Submerged Roots

Felix Bärlocher

10.1 Introduction

It has long been known that plants harbour fungal endophytes, and it was suspected that systemic grass endophytes, primarily clavicipitaceous fungi, are associated with toxicity to grazing livestock (Saikkonen et al. 1998). This connection was firmly established in the 1970s (Bacon et al. 1977). The early emphasis on grasses and their endophytes have led some authors to consider the term endophyte as being synonymous with mutualist. However, many fungal pathogens may be latent in grasses without causing disease or long before the outbreak of disease symptoms (Petrini 1991; Fisher and Petrini 1993). The first systematic surveys of plants other than grasses were stimulated by the observation that many common phyllosphere fungi invade stomatal cavities of Douglas fir needles within their first year. Bernstein and Carroll (1977) demonstrated that with increasing age, all needle segments become infected with endophytes. The presence of primarily non-balansiaceous endophytes was extended to other conifers (Carroll et al. 1977) and has since been documented in every tree, shrub and herb that has been examined (Carroll 1995; Sridhar and Raviraja 1995; Saikkonen et al. 1998). Generally, a large number of species can be isolated from a given host, yet only four to five are common and likely to be host specific (Fisher and Petrini 1993). Community ordination analyses have generally shown that endophyte assemblages are specific at the host species level, and may be impoverished outside the host's natural range. While a few of these associations provide clear benefits to the plant by fungal interference with herbivores or microbial pathogens, others eventually cause damage to the plant, while some are essentially neutral. A widely accepted definition of an endophyte is as an agent of a currently asymptomatic infection, without specifying the role of the agent in the host or its development at a later stage (Petrini 1991; Fisher and Petrini 1993; Schulz et al. 1998). However, Schulz et al. (1999) showed that even in infections without visible symptoms, colonisation led to the synthesis of higher concentrations of

Felix Bärlocher: 63B York Street, Department of Biology, Mount Allison University, Sackville, New Brunswick, E4L 1G7, Canada, E-mail: fbaerlocher@mta.ca

Soil Biology, Volume 9
Microbial Root Endophytes
B. Schulz, C. Boyle, T. N. Sieber (Eds.)

potentially antimicrobial compounds. In vitro, endophytic fungi produce more herbicidally active substances than soil fungi. Schulz et al. (1999) therefore hypothesise that the host-endophyte interaction is a case of balanced antagonism: pathogens overcome the host's defences to the extent that they cause visible damage, whereas endophytic virulence is only sufficient to be able to infect and colonise without causing visible damage. If the balance shifts, the endophyte may turn pathogenic.

Much of the current interest in endophytes is based on the hope of finding unique secondary metabolites and enzymes affecting plants, herbivores and microbes, with potential applications in medicine and agriculture (Petrini et al. 1992). The continuum of endophyte interactions with plants also provides interesting case studies for the evolution of mutualism and pathology, and for understanding how environmental factors might favour one or the other (Carroll 1988). Until recently, it has commonly been believed that the first fungi were saprotrophs from which necrotrophs and biotrophs evolved, but there are convincing arguments supporting an alternative view (Parbery 1996). A recent comparison of 1,551 ribosomal sequences of the two sister groups of chitinous fungi, the Glomeromycota and the Dikaryomycota, both of which have symbiotic life-styles, suggests that the symbiosis between fungi and green plants was present before the colonisation of land by plants (Tehler et al. 2003). Finally, endophytes may not have been seriously taken into account when assessing diversity (Hawksworth 1991, 2001); Dreyfuss and Chapela (1994) concluded that over 1.3 million species of fungal endophytes remain to be discovered and described.

10.2 Aquatic Hyphomycetes

Up to 99% of the energy available to stream communities consists of terrestrial plant detritus (leaves, needles, twigs; Allan 1995). Aquatic hyphomycetes, a heterogeneous group of aquatic fungi, are an indispensable link in the food web between this detritus and stream invertebrates (Bärlocher 1992). The annual fungal production per stream bed area falls within the same order of magnitude as that of bacteria and invertebrates (Suberkropp 1997). Aquatic hyphomycetes disperse from leaf to leaf by producing conidia, whose shapes are predominantly tetraradiate (four arms) or sigmoid. Both types have been shown to increase the conidium's likelihood of settling and germinating on new leaves; they are clearly the result of convergent evolution (Webster 1987).

In temperate streams, the number of conidia in the water column declines from up to $30{,}000\,\mathrm{l}^{-1}$ in late fall to almost nil during summer, undoubtedly a response to the seasonal availability of terrestrial leaves

(Bärlocher 1992, 2000). Combined with the unidirectional displacement of substrates and spores in running water, this raises the question of how aquatic hyphomycetes can maintain themselves within a given reach of a stream and avoid being washed downstream. Potential solutions include (1) the fact that the fungi also colonise woody substrates, which can persist for several years in a stream, (2) the presence of teleomorphs in some species with ascospores that may be dispersed aerially, (3) dispersal of fungal-colonised leaves or conidia by animals, (4) the occurrence of the fungi in terrestrial habitats, e.g. as plant pathogens or endophytes (Bärlocher 1992). For example, Hartig (1880) first described a parasite of maple seedlings as *Cercospora acerina*. Later, its identity as *Centrospora acerina* was established, and it is now known as *Mycocentrospora acerina* (Hartig) Deighton. It is a remarkably widespread and versatile species: it is a well-known plant pathogen, has been implicated in human infections, and is a common stream fungus. Morphologically, there is no difference among various strains. Iqbal and Webster (1969) showed that strains they isolated from a stream were pathogenic to carrots and parsnips. Nemec (1969) isolated *Anguillospora longissima* (Sacc. & Syd.) Ingold and *Tetracladium marchalianum* de Wild., which had been isolated from aqueous habitats, from the roots of diseased strawberry plants. Several other species have also been isolated from terrestrial root surfaces of apparently healthy plants (Waid 1954; Taylor and Parkinson 1965; Parkinson and Thomas 1969; Watanabe 1975). These observations suggested that some aquatic hyphomycetes might be root endophytes.

10.3 Fungi in Submerged Roots

Fisher and Petrini (1989) were the first to demonstrate an endophytic phase of two aquatic hyphomycete species. They examined terrestrial roots of *Alnus glutinosa* (L.) Gaertner on the banks of Exeter Canal (Exeter, Devon, UK). Only 1.7 and 0.7% of 300 root segments were colonised by the aquatic hyphomycetes *Tricladium splendens* and *Campylospora purvula*, respectively, compared to the 19% that were colonised by the most common endophyte *Cylindrocarpon destructans* In a later study, Fisher et al. (1991) compared aquatic and terrestrial alder roots along the banks of the River Dart (Devon, UK) They separated roots into bark and xylem (decorticated roots), and found more endophytic aquatic hyphomycetes in the former. Mean frequency of occurrence of aquatic species in submerged roots was as high as 30%, compared to 12% on terrestrial roots. In addition to typical aquatic hyphomycetes, they also found species of the genera *Fusarium* and *Cylindrocarpon*. Members of these two taxa are often found on leaves in decaying streams. Cluster and correspondence analyses suggested that aquatic and soil root samples are colonised by two distinct endophyte populations, indicating that the external environment may have a greater influence on endophyte communities of roots than those of leaves (Fisher and Petrini 1993). Three previously unknown species (*Fontanospora fusiramos*, and two species belonging to *Filosporella*) were subsequently isolated and described from submerged alder roots (Marvanová and Fisher 1991; Marvanová et al. 1992, 1997).

The host range of endophytic aquatic hyphomycetes was extended by Sridhar and Bärlocher (1992a). They found additional species in spruce (*Picea glauca* [Moench] Voss), birch (*Betula papyrifera* Marsh) and maple (*Acer spicatum* Lam.). Again, fungal endophytes were more common in the bark, suggesting that roots are colonised by fungi settling on surfaces and growing toward the interior. In addition to plating out surface-sterilised root fragments, Sridhar and Bärlocher (1992a) aerated them in distilled water and were able to observe release of typical tetraradiate or sigmoid conidia. However, aeration had to continue for 4 days (compared to the usual 1–2 days) before spores were detected (any superficial mycelia that may have been present were killed by surface sterilisation). Spore production per unit mass was less than 1 mg^{-1}, compared to 100–150 mg^{-1} from dead submerged branches, and up to 8,000 mg^{-1} on dead leaves (Gessner et al. 2003). Nevertheless, the root biomass in streams is considerable and its turnover rapid (Waid 1974), suggesting that it may be an important secondary resource for aquatic hyphomycetes. Their existence as endophytes may provide them with a head start in the use of root detritus, a possible advantage of the endophytic life style that has also been suggested for leaf-decomposing saprobes (Fisher and Petrini 1993).

On spruce roots, the number aquatic hyphomycetes in the xylem was highest in 4- to 5-year-old segments (Sridhar and Bärlocher 1992b). Total fungal biomass, estimated by ergosterol, amounted to 0.002 to 0.2% of root biomass. This compares to values exceeding 15% on decaying leaves (Gessner et al. 2003).

Iqbal et al. (1994) reported 17 species of endophytic aquatic hyphomycetes from tree roots along canal banks in Pakistan (*Mangifera indica* L., *Populus hybrida* Reichb., *Salix babylonica* L.). Two plantation crops (*Coffea arabica* Linn., *Hevea brasiliensis* M.) and four ferns (*Diplazium esculentum* (Retz) Sw., *Macrothelypteris torresiana* (Gaudich.) Ching., *Angiopteris evecta* (Forst) Hoffin., *Christela dentata* Brownsey & Jermy), all from India, were also shown to harbour aquatic endophytes in their submerged roots (Raviraja et al. 1996). Again, their incidence was higher in the bark than in the xylem of tree roots. Conidium release per root biomass upon aeration was much higher from the fern *C. dentata* (12,900 g^{-1}) than from the other plants (36–410 g^{-1}).

Permanently or periodically submerged roots are common in mangrove swamps. Ananda and Sridhar (2002) examined fungal epiphytes in prop roots or pneumatophores (which cycle between exposure to air and immersion in salt or brackish water) of *Avicennia officinalis* L., *Rhizophora mucronata* Lamk. and *Sonneratia caseolaris* (L.) Engl.. Aquatic hyphomycetes were represented by *Mycocentrospora acerina* in roots of *A. officinalis* L., and *Triscelophorus acuminatus* in roots of *R. mucronata* and *S. caseolaris*. In addition, various marine and terrestrial fungi were found.

Overall, 35 aquatic hyphomycete species (plus seven taxa identified to genus) have been reported from submerged roots of 13 plants, including Angiosperms, Gymnosperms and ferns in eight studies (Table 10.1). This corresponds to roughly 10% of the total number of described species (L. Marvanová, personal communication). Clearly, submerged roots can provide a stationary refuge for aquatic hyphomycetes, which may help them maintain their presence in a given stream reach despite the unidirectional flow of water.

Table 10.1. Endophytic aquatic hyphomycetes recovered from roots, submerged in saltwater (*) or freshwater (all others). Root sections: *R* Entire root, *B* bark, *X* xylem (decorticated root)

Fungus	Substrate	Root section	References
Anguillospora filiformis Greath.	*Acer spicatum*	B, X	Raviraja et al. 1996; Sridhar and Bärlocher 1992b
	Betula papyrifera	B, X	Sridhar and Bärlocher 1992a
	Picea glauca	B	Sridhar and Bärlocher 1992a
A. longissima (de Wild.) Ingold	*Mangifera indica*	B, X	Iqbal et al. 1995
	Populus hybrida	B, X	Iqbal et al. 1995
	Salix babylonica	B, X	Iqbal et al. 1995
Articulospora antipodea Roldán	*Picea glauca*	B, X	Sridhar and Bärlocher 1992a
A. atra Descals	*Alnus glutinosa*	B	Fisher et al. 1991
	Picea glauca	B, X	Sridhar and Bärlocher 1992a
A. tetracladia Ingold	*Acer spicatum*	B, X	Fisher et al. 1991
	Alnus glutinosa	B, X	Fisher et al. 1991; Sridhar and Bärlocher 1992a, 1992b
	Picea glauca	B, X	Sridhar and Bärlocher 1992a
A. proliferata Jooste, Radon & Merwe	*Mangifera indica*	B, X	Iqbal et al. 1995
	Populus hybrida	B, X	Iqbal et al. 1995
	Salix babylonica	B, X	Iqbal et al. 1995
Bacillispora inflata Iqbal & Bhatty	*Mangifera indica*	B	Iqbal et al. 1995
	Populus hybrida	B	Iqbal et al. 1995
	Salix babylonica	B	Iqbal et al. 1995

Table 10.1. (continued)

Fungus	Substrate	Root section	References
Campylospora chaetocladia Ranzoni	*Salix babylonica*	B	Iqbal et al. 1995
Clavariopsis aquatica de Wild.	*Alnus glutinosa*	B, X	Fisher et al. 1991
	Picea glauca	X	Sridhar and Bärlocher 1992a
	Populus hybrida	B	Iqbal et al. 1995
	Salix babylonica	B	Iqbal et al. 1995
C. azlanii Nawawi	*Mangifera indica*	B	Iqbal et al. 1995
Cylindrocarpon aquaticum (Nils.) Marvanová & Descals	*Acer spicatum*	B, X	Sridhar and Bärlocher 1992a
	Mangifera indica	B, X	Iqbal et al. 1995
	Picea glauca	B	Sridhar and Bärlocher 1992a
	Populus hybrida	B, X	Iqbal et al. 1995
	Salix babylonica	B, X	Iqbal et al. 1995
Filosporella sp.	*Alnus glutinosa*	B	Fisher et al. 1991
F. fistucella Marvanová & Fisher	*Alnus glutinosa*	B	Marvanová and Fisher 1991
F. versimorpha Marvanová et al.	*Alnus glutinosa*	B	Marvanová et al. 1992
Flagellospora curvula Ingold	*Mangifera indica*	B	Iqbal et al. 1995
	Populus hybrida	B	Iqbal et al. 1995
	Salix babylonica	B, X	Iqbal et al. 1995
F. fusarioides Iqbal	*Mangifera indica*	B, X	Iqbal et al. 1995
	Populus hybrida	B, X	Iqbal et al. 1995
	Salix babylonica	B, X	Iqbal et al. 1995
F. penicillioides Ingold	*Mangifera indica*	B, X	Iqbal et al. 1995
	Populus hybrida	B, X	Iqbal et al. 1995
	Salix babylonica	B, X	Iqbal et al. 1995
Fontanospora fusiramosa Marvanova et al.	*Alnus glutinosa*	R	Marvanová et al. 1997
Geniculospora sp.	*Picea glauca*	B, X	Sridhar and Bärlocher 1992b
Heliscus lugdunensis Sacc. & Therry	*Acer spicatum*	B, X	Sridhar and Bärlocher 1992a
	Alnus glutinosa	B, X	Iqbal et al. 1995
	Betula papyrifera	B, X	Sridhar and Bärlocher 1992a
	Picea glauca	B, X	Sridhar and Bärlocher 1992a, 1992b
	Salix babylonica	B,X	Iqbal et al. 1995
Lunulospora curvula Ingold	*Alnus glutinosa*	B	Fisher et al. 1991
	Angiopteris evecta	R	Raviraja et al. 1996
	Christela dentata	R	Raviraja et al. 1996
	Coffee arabica	B, X	Raviraja et al. 1996

Table 10.1. (continued)

Fungus	Substrate	Root section	References
	Hevea brasiliensis	X	Raviraja et al. 1996
	Mangifera indica	B	Iqbal et al. 1995
	Populus hybrida	B	Iqbal et al. 1995
	Salix babylonica	B	Iqbal et al. 1995
Mycocentrospora sp. 1	*Alnus glutinosa*	B	Fisher et al. 1991
Mycocentrospora sp. 2	*Acer spicatum*	B, X	Sridhar and Bärlocher 1992a
	Picea glauca	B, X	Sridhar and Bärlocher 1992a
Mycocentrospora sp. 3	*Coffee arabica*	B	Raviraja et al. 1996
	Diplazium esculentum	R	Raviraja et al. 1996
	Hevea brasiliensis	X	Raviraja et al. 1996
	Macrothelypteris torresiana	R	Raviraja et al. 1996
M. acerina (Hartig) Deighton	**Avicennia officinalis*	R	Ananda and Sridhar 2002
M. clavata Iqbal	*Betula papyrifera*	B, X	Sridhar and Bärlocher 1992a
	Picea glauca	B, X	Sridhar and Bärlocher 1992a
M. iqbalii sp. ind. F. Bareen	*Mangifera indica*	B, X	Iqbal et al. 1995
	Salix babylonica	B, X	Iqbal et al. 1995
Phalangispora constricta Nawawi & Webster	*Picea glauca*	B, X	Sridhar and Bärlocher 1992b
Pseudoanguillospora sp.	*Alnus glutinosa*	B	Fisher et al. 1991
Tetrabrachium elegans Nawawi & Kuthubutheen	*Acer spicatum*	B, x	Sridhar and Bärlocher 1992a
	Betula papyrifera	B, X	Sridhar and Bärlocher 1992a
	Picea glauca	B	Sridhar and Bärlocher 1992a
Tetracladium sp	*Angiopteris evecta*	R	Raviraja et al. 1996
T. furcatum Descals	*Angiopteris evecta*	R	Raviraja et al. 1996
T. marchalianum de Wild.	*Mangifera indica*	B	Iqbal et al. 1995
	Populus hybrida	B	Iqbal et al. 1995
	Salix babylonica	B	Iqbal et al. 1995
T. setigerum (Grove) Ingold	*Picea glauca*	B	Sridhar and Bärlocher 1992b
Tricladium chaetocladium Ingold	*Alnus glutinosa*	B	Fisher et al. 1991
	Alnus glutinosa	B	Fisher et al. 1991
Tricellula aquatica Webster	*Mangifera indica*	B	Iqbal et al. 1995
Triscelophorus acuminatus Nawawi	*Angiopteris evecta*	R	Raviraja et al. 1996
	Christela dentata	R	Raviraja et al. 1996
	Coffea arabica	B	Raviraja et al. 1996
	Diplazium esculatum	R	Raviraja et al. 1996
	Hevea brasiliensis	B, X	Raviraja et al. 1996

Table 10.1. (continued)

Fungus	Substrate	Root section	References
	Macrothelypteris torresiana	R	Raviraja et al. 1996
	**Rhizophora mucronata*	R	Ananda and Sridhar 2002
	**Sonneratia caseolaris*	R	Ananda and Sridhar 2002
T. konajensis Sridhar & Kaveriappa	*Antipteris evecta*	R	Raviraja et al. 1996
	Christela dentata	R	Raviraja et al. 1996
	Coffea arabica	B	Raviraja et al. 1996
	Macrothylpteris torresiana	R	Raviraja et al. 1996
T. monosporus Ingold	*Angiopteris evecta*	R	Raviraja et al. 1996
	Christela dentata	R	Raviraja et al. 1996
	Coffea arabica	B, X	Raviraja et al. 1996
	Diplazium esculatum	R	Raviraja et al. 1996
	Macrothelypteris torresiana	R	Raviraja et al. 1996
	Mangifera indica	B	Iqbal et al. 1995
	Populus hybrida	B	Iqbal et al. 1995
	Salix babylonica	B	Iqbal et al. 1995
Tumularia aquatica (Ingold) Marvanová & Descals	*Alnus glutinosa*	B	Fisher et al. 1991
Varicosporium elodeae Kegel	*Alnus glutinosa*	B	Fisher et al. 1991
	Picea glauca	B, X	Sridhar and Bärlocher 1992a, 1992b
V. giganteum Crane	*Picea glauca*	B, X	Sridhar and Bärlocher 1992a, 1992b

10.4 Conclusions and Outlook

Work on submerged roots, primarily in fresh water, has been dominated by a very specific objective: to evaluate their role as habitat for aquatic hyphomycetes. Other aspects of the plant-endophyte relationship include the potential production of unique secondary metabolites allowing the fungi to live within the plant without overt symptoms, and which might be toxic to potential pathogens or herbivores. Several observations suggest that aquatic fungi can produce diffusible antibiotics. For example, *Massarina aquatica*, the teleomorph of *Tumularia aquatica*, releases antifungal substances (Fisher and Anson 1983). Similar observations on other species have been reported by Asthana and Shearer (1990) and Poch et al. (1992). Chamier et al. (1984) demonstrated inhibition of bacteria by aquatic hyphomycetes in field experiments. Isolation and characterisation of antimicrobial compounds from *Anguillospora longissima* and *A. crassa* resulted in the discovery of novel metabolites (Harrigan et al. 1995). Two surveys of aquatic hyphomycetes and ascomycetes demonstrated that antibacterial and antifungal substances are produced by about one-half of the species tested (Gulis and Stephanovich 1999; Shearer and Zare-Maivan 1988). Lignicolous aquatic ascomycetes and hyphomycetes were generally more antagonistic than foliicolous species, possibly because long-lasting substrata, such as wood, favour colonisation by species capable of defending captured resources (Shearer 1992). It is currently unknown how such compounds affect the root's susceptibility toward pathogens or herbivores.

Sexual and asexual reproduction of the endophyte are often initiated upon the death of the host tissue (Fisher et al. 1986). Sridhar and Bärlocher (1992a) reported that *Heliscus lugdunensis* produced a teleomorph upon subculturing; aquatic hyphomycetes endophytic in roots may therefore be useful in establishing additional anamorph-teleomorph connections (Webster 1992; Sivichai and Jones 2003).

It is generally accepted that aquatic hyphomycetes and ascomycetes had terrestrial ancestors, and several have indeed close terrestrial relatives (Kong et al. 2000; Liew et al. 2002). Shearer (1993) suggested that when terrestrial plants invaded freshwater habitats, they brought with them fungal pathogens, endophytes and saprobes. Alternatively, plant detritus, precolonised by fungal biotrophs or saprotrophs may have fallen into streams. Some of these fungi may subsequently have adapted to dispersal and reproduction in water. The most comprehensive analysis of fungal gene sequences suggests that the biotrophic lifestyle was a synapomorphic trait (i.e. present in a common ancestor; Tehler et al. 2003). The first fungi that colonised the terrestrial habitat most likely did so while closely associated with plants. The question remains whether such fungi first evolved into terrestrial saprobes, and then into aquatic hyphomycetes, or whether there was a direct transition from terrestrial biotrophs to aquatic saprobes. Were the presumed ancestors restricted to specific plant organs, e.g. aerial twigs and leaves, or roots? Upon their death, roots submerged in streams may have released propagules of fungal endophytes, some of which settled on other types of imported terrestrial detritus, such as leaves. Over time, this may have favoured adaptation to life in running water. Or, terrestrial leaves infected with endophytes were shed and landed in streams. Even waterfilms on soil or between layers of terrestrial leaf layer may have selected for tetraradiate spore shapes (Bandoni 1975) and predisposed some fungi for their eventual evolution into aquatic hyphomycetes. There are several reports of aquatic hyphomycete conidia in rainwater dripping from trees

(Ando and Tubaki 1984; Bärlocher 1992; Czeczuga and Orlowska 1999), and Widler and Müller (1984) isolated two undescribed species of *Gyoerffyella* and *Varicosporium* from green twigs. Iqbal et al. (1995) reported 14 species from submerged green leaves.

It will be of considerable interest to investigate the common route of invasion by aquatic hyphomycetes: do they first colonise the roots, and then spread through the rest of the plant, or do some invade aerial parts? A thorough study of endophytes in aerial and subterranean plant parts at various ages, and molecular data of such strains may eventually allow us to reconstruct the origins of aquatic hyphomycetes.

References

Allan JD (1995) Stream ecology. Chapman and Hall, London

Ananda K, Sridhar KR (2002) Diversity of endophytic fungi in the roots of mangrove species on the west coast of India. Can J Microbiol 48:871–878

Ando K, Tubaki K (1984) Some undescribed hyphomycetes in the rain drops from intact leaf-surface. Trans Mycol Soc Jpn 25:21–37

Asthana A, Shearer CA (1990) Antagonistic activity of *Pseudohalonectria* and *Ophioceras*. Mycologia 82:554–561

Bacon CW, Porter JK, Robbins JD, Luttrell ES (1977) *Epichloë typhina* from toxic tall fescue grasses. Appl Environ Microbiol 34:576–581

Bandoni RJ (1975) Significance of the tetraradiate form in dispersal of terrestrial fungi. Rep Tottori Mycol Inst 12:105–113

Bärlocher F (1992) Research on aquatic hyphomycetes: historical background and overview. In: Bärlocher F (ed) The ecology of aquatic hyphomycetes Springer, Berlin Heidelberg New York, pp 1–15

Bärlocher F (2000) Water-borne conidia of aquatic hyphomycetes: seasonal and yearly patterns in Catamaran Brook, New Brunswick, Canada. Can J Bot 78:157–167

Bernstein ME, Carroll G (1977) Internal fungi in old-growth Douglas fir foliage. Can J Bot 55:644–653

Carroll GC (1988) Fungal endophytes in stems and leaves: from latent pathogen to mutualistic symbiont. Ecology 69:2–9

Carroll GC (1995) Forest endophytes: pattern and process. Can J Bot 73 [Suppl 1]:S1316–S1324

Carroll FE, Müller E, Sutton BC (1977) Preliminary studies on the incidence of needle endophytes in some European conifers. Sydowia 29:87–103

Chamier A-C, Dixon P, Archer SA (1984) The spatial distribution of fungi on decomposing alder leaves in a freshwater stream. Oecologia 64:92–103

Czeczuga B, Orlowska M (1999) Hyphomycetes in rain water, melting snow and ice. Acta Mycol 34:181–200

Dreyfuss MM, Chapela ICH (1994) Potential of fungi in the discovery of novel, low-molecular weight pharmaceuticals. In Gullo VP (ed) The discovery of natural products with therapeutic potential. Butterworth-Heinemann, Boston, pp 49–80

Fisher PJ, Anson AE (1983) Antifungal effects of *Massarina aquatica* growing on oak wood. Trans Br Mycol Soc 81:523–527

Fisher PJ, Petrini O (1989) Two aquatic hyphomycetes as endophytes in *Alnus glutinosa* roots. Mycol Res 92:367–368

Fisher PJ, Petrini O (1993) Ecology, biodiversity and physiology of endophytic fungi. Curr Top Bot Res 1:271–279

Fisher PJ, Anson AE, Petrini O (1986) Fungal endophytes in *Ulex europaeus* and *Ulex gallii.* Trans Brit Mycol Soc 86:153–156

Fisher PJ, Petrini O, Webster J (1991) Aquatic hyphomycetes and other fungi in living aquatic and terrestrial roots of *Alnus glutinosa.* Mycol Res 95:543–547

Gessner MO, Bärlocher F, Chauvet E (2003) Qualitative and quantitative analyses of aquatic hyphomycetes in streams. Fungal Div Res Ser 10:127–157

Gulis VI, Stephanovich AI (1999) Antibiotic effects of some aquatic hyphomycetes. Mycol Res 103:111–115

Harrigan GG, Armentrout GL, Gloer JB, Shearer CA (1995) Anguillosporal, a new antibacterial and antifungal metabolite from the freshwater fungus *Anguillospora longissima.* J Nat Prod 58:1467–1469

Hartig R (1880) Der Ahornkeimlingspilz, *Cercospora acerina.* Untersuch Forstbot Inst München 1:58–61

Hawksworth DL (1991) The fungal dimension of biodiversity: magnitude, significance and conservation. Mycol Res 95:641–655

Hawksworth DL (2001) The magnitude of fungal diversity: the 1.5 million species estimate revisited. Mycol Res 105:1422–1431

Iqbal SH, Webster J (1969) Pathogenicity of aquatic isolates of *Centrospora acerina* to carrots and parsnips. Trans Br Mycol Soc 53:486–490

Iqbal SH, Firdaus-e-Bareen, Yousaf N (1994) Freshwater hyphomycete communities in a canal. 1. Endophytic hyphomycetes of submerged roots of trees sheltering a canal bank. Can J Bot 73:538–543

Iqbal SH, Akhtar G, Firdaus-e-Bareen (1995) Endophytic freshwater hyphomycetes of submerged leaves of some plants lining a canal bank. Pak J Plant Sci 1:239–254

Kong RY, Chan JY, Mitchell JI, Vrijmoed LLP, Jones EBG (2000) Relationships of *Halosarpheia, Lignincola* and *Nais* inferred from partial 18S rDNA. Mycol Res 103:1399–1403

Liew ECY, Aptroot A, Hyde KD (2002) An evaluation of the monophyly of *Massarina* based on ribosomal DNA sequences. Mycologia 94:803–813

Marvanová L, Fisher PJ (1991) A new endophytic hyphomycete from alder roots. Nova Hedwigia 52:33–37

Marvanová L, Fisher PJ, Aimer R, Segedin BC (1992) A new *Filosporella* from alder roots and from water. Nova Hedwigia 54:151–158

Marvanová L, Fisher PJ, Descals E, Bärlocher F (1997) *Fontanospora fusiramosa* sp. nov., a hyphomycete from live tree roots and from stream foam. Czech Mycol 50:3–11

Nemec S (1969) Sporulation and identification of fungi isolated from root-rot in diseased strawberry plants. Phytopathology 59:1552–1553

Parbery DG (1996) Trophism and the ecology of fungi associated with plants. Biol Rev 71:473–527

Parkinson D, Thomas S (1969) Studies on fungi in the root region. Plant Soil 31:299–310

Petrini O (1991) Fungal endophytes of tree leaves. In: Andrews J, Hirano S (eds) Microbial ecology of leaves. Springer, Berlin Heidelberg New York, pp 179–197

Petrini O, Sieber TH, Toti L, Viret O (1992) Ecology, metabolite production, and substrate utilization in endophytic fungi. Nat Toxins 1:185–196

Poch GK, Gloer JB, Shearer CA (1992) New bioactive metabolites from a freshwater isolate of the fungus *Kirschsteiniothelia* sp. J Nat Prod 55:1093–1099

Raviraja NS, Sridhar KR, Bärlocher F (1996) Endophytic aquatic hyphomycetes of roots from plantation crops and ferns from India. Sydowia 48:152–160

Saikkonen K, Faeth SH, Helander M, Sullivan TJ (1998) Fungal endophytes: a continuum of interactions with host plants. Annu Rev Ecol Syst 29:319–343
Schulz B, Guske S, Dammann U, Boyle C (1998) Endophyte-host interactions. II. Defining symbiosis of the endophyte-host interaction. Symbiosis 25:213–227
Schulz B, Römmert A-K, Damman U, Aust H-J, Strack D (1999) The endophyte-host interaction: a balanced antagonism? Mycol Res 103:1275–1283
Shearer CA (1992) The role of woody debris. In: Bärlocher F (ed) The ecology of aquatic hyphomycetes. Springer, Berlin Heidelberg New York, pp 77–98
Shearer CA (1993) The freshwater ascomycetes. Nova Hedwigia 56:1–33
Shearer CA, Zare-Maivan H (1988) In vitro hyphal interactions among wood- and leaf-inhabiting ascomycetes and fungi imperfecti from freshwater habitats. Mycologia 80:31–37
Sivichai S, Jones EBG (2003) Teleomorphic-anamorphic connections of freshwater fungi. Fung Div Res Ser 10:259–272
Sridhar KR, Bärlocher F (1992a) Endophytic aquatic hyphomycetes of roots from spruce, birch and maple. Mycol Res 96:305–308
Sridhar KR, Bärlocher F (1992b) Aquatic hyphomycetes in spruce roots. Mycologia 84:580–584
Sridhar KR, Raviraja NS (1995) Endophytes – a crucial issue. Curr Sci 69:570–571
Suberkropp K (1997) Annual production of leaf-decaying fungi in a woodland stream. Freshwater Biol 38:169–178
Taylor GS, Parkinson D (1965) Studies in the root region. IV. Fungi associated with the roots of *Phaseolus vulgaris* L. Plant Soil 22:1–20
Tehler A, Little DP, Farris JS (2003) The full-length phylogenetic tree from 1551 ribosomal sequences of chitinous fungi. Mycol Res 107:901–916
Waid JS (1954) Occurrence of aquatic hyphomycetes upon the root surfaces of beech grown in woodland soils. Trans Br Mycol Soc 37:420–421
Waid JS (1974) Decomposition of roots. In: Dickinson CH, Pugh GJF (eds) Biology of plant litter decomposition. Academic, New York, pp 175–211
Watanabe T (1975) *Tetracladium setigerum*, an aquatic hyphomycete associated with gentian and strawberry roots. Trans Mycol Soc Jpn 16:348–350
Webster J (1987) Convergent evolution and the functional significance of spore shape in aquatic and semi-aquatic fungi. In: Rayner ADM, Brasier CM, Moore D (eds) Evolutionary biology of the fungi. Cambridge University Press, Cambridge, pp 191–201
Webster J (1992) Anamorph-teleomorph relationships. In: Bärlocher F (ed) The ecology of aquatic hyphomycetes. Springer, Berlin Heidelberg New York, pp 99–117
Widler B, Müller E (1984) Untersuchungen über endophytische Pilze von Arctostaphylos uva-ursi (L.) Sprengel (Ericaceae). Bot Helv 94:307–337

11 Nematophagous Fungi as Root Endophytes

Luis V. Lopez-Llorca, Hans-Börje Jansson,
José Gaspar Maciá Vicente, Jesús Salinas

11.1 Introduction

Nematophagous fungi constitute a group of fungal antagonists to nematodes. The latter are small roundworms living in soil and water. Most nematodes are saprotrophic, but many species are parasites of plants and animals (Poinar 1983). The nematophagous fungi have been suggested as promising candidates for biological control of parasitic nematodes (Stirling 1991), but so far no successful commercial products have been presented. Many of the previous studies on these organisms have been concerned with the ecology and physiology of interactions between nematophagous fungi and nematodes. More recently, molecular techniques have been employed (Jansson and Lopez-Llorca 2001). Nematophagous fungi also have the ability to infect and colonise other organisms, including other fungi and plant roots (Jansson and Lopez-Llorca 2004). In the current review we will briefly describe the nematophagous fungi, with special emphasis on their interactions with plant roots.

11.2 Nematophagous Fungi

Nematophagous fungi, or nematode-destroying fungi, have the capacity to infect, kill and digest living stages of their nematode hosts (eggs, juveniles and adults). These fungi are ubiquitous soil inhabitants found in most parts of the world and in all climate types (Barron 1977). Many of the

Luis V. Lopez-Llorca: Departamento de Ciencias del Mar y Biología Aplicada, Universidad de Alicante, Apdo. 99, 03080 Alicante, Spain, E-mail: lv.lopez@ua.es

Hans-Börje Jansson: Departamento de Ciencias del Mar y Biología Aplicada, Universidad de Alicante, Apdo. 99, 03080 Alicante, Spain, E-mail: hb.jansson@ua.es

José Gaspar Maciá Vicente: Departamento de Ciencias del Mar y Biología Aplicada, Universidad de Alicante, Apdo. 99, 03080 Alicante, Spain, E-mail: jgmv@ua.es

Jesús Salinas: Departamento de Ciencias del Mar y Biología Aplicada, Universidad de Alicante, Apdo. 99, 03080 Alicante, Spain, E-mail: jesus.salinas@ua.es

Soil Biology, Volume 9
Microbial Root Endophytes
B. Schulz, C. Boyle, T. N. Sieber (Eds.)

nematophagous fungi are facultative parasites and can grow saprophytically in soil. In the presence of hosts they can change from a saprophytic to a parasitic stage and form infection structures, e.g. trapping organs, hyphal coils or appressoria. These infection structures vary depending on the type of host–nematode, fungus or plant.

Entomopathogenic fungi, e.g. *Lecanicillium lecanii*, have the capacity to infect both nematode eggs (Meyer 1998) and other fungi. Furthermore, species closely related to nematophagous fungi, e.g. *Arthrobotrys ferox*, can infect other small soil animals like springtails (Rubner 1996), but are generally not known to infect nematodes.

11.2.1 Nematode Parasites

The nematophagous fungi can be divided into four groups depending on their mode of attacking their hosts. The first three groups infect vermiform nematodes (juveniles and adults), whereas the fourth group infects nematode females and eggs (Jansson and Lopez-Llorca 2001). *Nematode-trapping fungi* use various types of trapping organs formed on their hyphae, e.g. adhesive networks, adhesive knobs (Fig. 11.1a) or constricting rings, and these fungi are facultative parasites to various extents. The nematodes are captured in the traps formed by the fungi either by adhesion or mechanical function. In the *endoparasitic fungi*, the spores (conidia, zoospores) function as infection structures, which either adhere to the nematode cuticle or are ingested. These fungi are generally obligate parasites of nematodes (Fig. 11.1b). The *toxin-producing fungi*, comprising for instance the common wood-decomposing oyster mushroom, intoxicate their nematode victims before penetrating them. The *egg- and female parasitic* attack mature females of cyst- and root-knot nematodes and the eggs they contain (Fig. 11.1c). Infection usually takes place via appressoria. Common to all types of nematophagous fungi is that after contact with the nematode cuticle, or egg shell, penetration takes place followed by digestion of the contents resulting in formation of new fungal biomass inside, and later outside, the nematode.

Taxonomy

The nematophagous fungi are found in most fungal taxa (Dackman et al. 1992). In the *Basidiomycetes*, nematophagous fungi such as the oyster mushroom (*Pleurotus ostreatus*) and *Hohenbuehelia* spp. (teleomorph of *Nematoctonus* spp.) can be found. Many of the nematode-trapping fungi belong to the *Deuteromycetes* or mitosporic fungi, e.g. *Arthrobotrys* spp. and *Monacrosporium* spp., but the *Arthrobotrys* spp. have been found

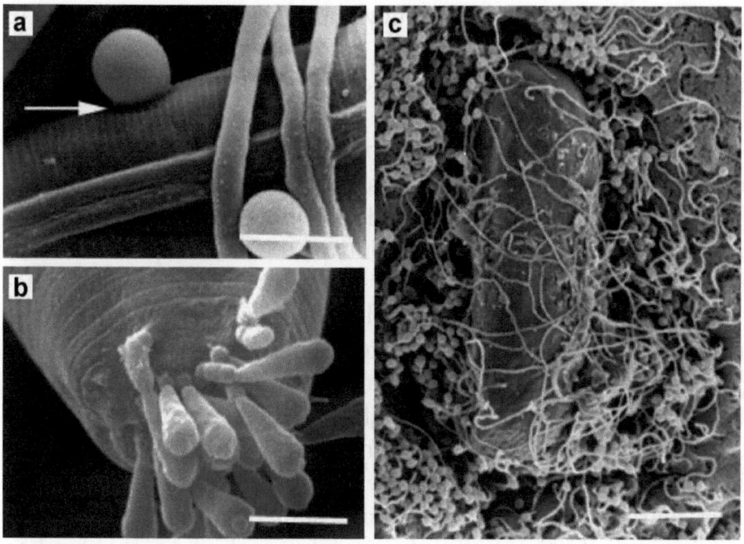

Fig. 11.1. **a** Adhesive knob trap of *Monacrosporium haptotylum* adhering to the nematode cuticle. Note adhesive pad between trap and nematode (*arrow*). **b** Conidia of the endoparasitic fungus *Drechmeria coniospora* adhering to the mouth region of a nematode. (From Jansson and Nordbring-Hertz 1983, courtesy of Society of General Microbiology). **c** A nematode egg infected by the egg-parasitic fungus *Pochonia rubescens*. (From Lopez-Llorca and Claugher 1990, courtesy of Elsevier). *Bars* **a**, **b** 5 μm; **c** 4 μm

to be *Ascomycetes* with their teleomorph in *Orbilia* spp. (Pfister 1997). Other nematode-trapping fungi, e.g. *Stylopage* and *Cystopage* spp. are *Zygomycetes*, and some endoparasitic fungi, such as *Catenaria anguillulae*, are zoosporic *Chytridiomycetes*. The main egg-parasitic fungi are now placed in the new genus *Pochonia* (formerly *Verticillium*) (Gams and Zare 2003). Therefore, the various nematophagous fungi seem to have acquired their nematophagous ability independently during the course of evolution.

The few phylogenetic studies that have been presented (Ahrén et al. 1998; Hagedorn and Scholler 1999) show that the orbiliaceous nematode-trapping fungi are closely related, and that the type of trap, rather than traditional spore morphology, is more determinate on the species level.

Biology

We will focus on two types of nematophagous fungi: the nematode-trapping *Arthrobotrys oligospora* and the egg parasite *Pochonia chlamydosporia*. These fungi are common soil inhabitants living both saprophytically and parasitically and, as we will show, also endophytically.

Arthrobotrys oligospora forms three-dimensional adhesive network traps in the presence of nematodes (Nordbring-Hertz 1977). Apart from a low nutrient status, small peptides, e.g. phenylalanyl-valine, can induce trap formation (Nordbring-Hertz 1973). When traps are formed, and even before traps are fully developed, nematodes can be captured in the adhesive covering the traps. The adhesive has been partially characterised and appears to be a polymer complex containing proteins, neutral sugars and uronic acids (Tunlid et al. 1991). The adhesive changes properties, from an amorphous stage to directed fibrils, after contact with the nematode cuticle (Veenhuis et al. 1985). This is in contrast to the endoparasitic fungus *Drechmeria coniospora*, where the fibrils appear directed whether nematodes are present or not (Jansson and Nordbring-Hertz 1988). After adhesion, the fungus penetrates the nematode cuticle from the trap, probably using both mechanical and enzymatic means. Since the nematode cuticle contains mainly proteinaceous material (Bird and Bird 1991), extracellular proteolytic enzymes involved in cuticle penetration have been studied. The major protease appears to be subtilisin PII. This serine protease has been characterised and genomically cloned (Åhman et al. 1996). After penetration, an infection bulb is formed, from which trophic hyphae grow out and digest the contents of the nematode. New hyphae and traps are then formed outside the nematode corpus to start a new infection cycle.

Pochonia spp. adhere to nematode egg shells by means of an appressorium formed at the hyphal tip (Lopez-Llorca and Claugher 1990, Lopez-Llorca et al. 2002b). An extracellular material (ECM) probably functions as adhesive, but possibly also seals the perforation in the egg shell caused by the penetration hypha beneath the fungal appressorium. This extracellular material can be labelled with the lectin Concanavalin A, indicating that the ECM contains mannose/glucose moieties probably on the side chains of glycoproteins (Lopez-Llorca et al. 2002b). Most ECMs of fungal hyphae consist of proteins and carbohydrates (Nicholson 1996). The nematode egg shell consists mainly of proteins and chitin (Bird and Bird 1991) and therefore proteases and chitinases would be important for fungal penetration of the egg shell. Serine proteases have been isolated from *Pochonia rubescens*, P32 (Lopez-Llorca 1990), *P. chlamydosporia*, VcP1 (Segers et al. 1994) and *Paecilomyces lilacinus*, PL (Bonants et al. 1995). In some cases these have been immunolocalised in infected hosts, their peptides sequenced, and the coding genes cloned. Recently, the chitinolytic system of *P. rubescens* and *P. chlamydosporia* has been studied (Tikhonov et al. 2002). For both species, among other enzymes, a similar major 43 kDa endochitinase (CHI43) was purified and characterised. A combination of protease P32 and chitinase CHI43, or the enzymes individually, caused removal of egg shell layers of the potato cyst nematode *Globodera pallida* (Tikhonov et al. 2002). Following penetration of the egg shell, the fungus digests the contents of the egg, proliferates inside and later grows outside the egg to penetrate neighbouring eggs in nematode cysts or egg masses; alternatively it can grow saprophytically.

11.2.2 Mycoparasites

Mycoparasitism is a common feature of fungi (Jeffries 1997). The ability of nematophagous fungi to attack other fungi was first described by Tzean and Estey (1978). Nematode-trapping fungi such as *A. oligospora* attack their host fungi, e.g. *Rhizoctonia solani*, in a manner similar to that of the well known mycoparasite *Trichoderma* spp. (Chet et al. 1981). The mycoparasitic behaviour of *A. oligospora* takes place by coiling of the hyphae of the nematode-trapping fungi around the host hyphae, which, in contrast to *Trichoderma* spp., results in disintegration of the host cell cytoplasm without penetration of the host (Persson et al. 1985). It has been shown using radioactive phosphorous tracing that nutrient transfer takes place between the nematode-trapping fungus *A. oligospora* and its host *R. solani* (Olsson and Persson 1994). Although this phenomenon has never been observed in soil, it may increase the fitness of the nematode-trapping fungi in soil by reducing competition and providing nutrients. Moreover, it may extend the biocontrol capability of nematophagous fungi as biocontrol agents to fungal parasites as well as nematodes. Furthermore, *P. chlamydosporia* has been described as being able to infect propagules of important plant pathogens, such as uredospores of rust fungi (Leinhos and Buchenauer 1992), and oospores of *Phytophthora* and other *Oomycetes* (Sneh et al. 1977).

11.2.3 Root Endophytes

Most work on the root biology of nematophagous fungi has concerned external root colonisation (ectorhizosphere). Lately, colonisation in the root tissues has also been reported (endorhizosphere). Some of these studies will be discussed in this chapter.

Ectorhizosphere

Since plant-parasitic nematodes generally attack plant roots it has been an important task to study the rhizosphere biology of nematophagous fungi – the root zone is an area with an abundant supply of the nematode prey. Not surprisingly, the nematode-trapping fungi have been found to be more frequent in the rhizosphere than in the bulk soil (Peterson and Katznelson 1965; Gaspard and Mankau 1986; Persmark and Jansson 1997).

Egg-parasitic fungi were also found to be more abundant in the rhizosphere (Bourne et al. 1996; Kerry 2000).

External root colonisation varies between plant species. For instance, Persmark and Jansson (1997) studied the presence and frequency of nematode-trapping fungi in field soils planted with barley, pea or white mustard. The pea rhizosphere harboured by far the highest frequency of nematode-trapping fungi: 19 times higher than in the root free soil. The number of species of nematode-trapping fungi was also higher in the pea rhizosphere, with *A. oligospora* as being the most common species (Persmark and Jansson 1997). In an investigation on chemotropic growth towards roots of pea, barley and white mustard by seven species of nematophagous fungi, only isolates of *A. oligospora* were attracted to the roots of all plants, but this was confined to the 2 mm closest to the roots (Bordallo et al. 2002). In a pot experiment, the colonisation of tomato roots by several nematophagous fungi was followed for 3 months. *Monacrosporium ellipsosporum* and *Arthrobotrys dactyloides* were especially competent in colonising the roots (Persson and Jansson 1999). Several nematode-trapping fungi are able to form so-called "conidial traps" in response to roots and root exudates (Persmark and Nordbring-Hertz 1997). The conidial traps are capture organs formed directly on the conidia without hyphal growth, and give the fungi an extra advantage by spreading the fungi and capturing nematodes, similar to most endoparasitic fungi, which infect nematodes entirely with adhesive conidia.

External root colonisation by the egg-parasite *Pochonia chlamydosporia* also varied with plant species. For instance, kale and cabbage had a density of the fungus twice as high as that on soya bean and tomato (Bourne et al. 1996). Furthermore, rhizosphere colonisation by *P. chlamydosporia* was increased when plants were infected with the root-knot nematode *Meloidogyne incognita*. This effect is possibly due to increased leakage of root exudates after damage to the root surface by the nematodes (Bourne et al. 1996). In none of the investigations mentioned above was the fungal colonisation of internal root tissues examined.

Endorhizosphere

Using axenic barley and tomato plants inoculated with the nematophagous fungi *P. chlamydosporia* or *A. oligospora*, we found that both fungi have the capacity to colonise epidermis and root cortex of barley and the epidermis of tomato (Lopez-Llorca et al. 2002a, Bordallo et al. 2002).

In these experiments roots were sequentially sampled, cryo-sectioned, and observed under light- or cryo-scanning electron microscopes. Both fungi grew inter- and intra-cellularly and formed appressoria (Figs. 11.2a,b) when penetrating plant cell walls of epidermis and cortex cells, but never

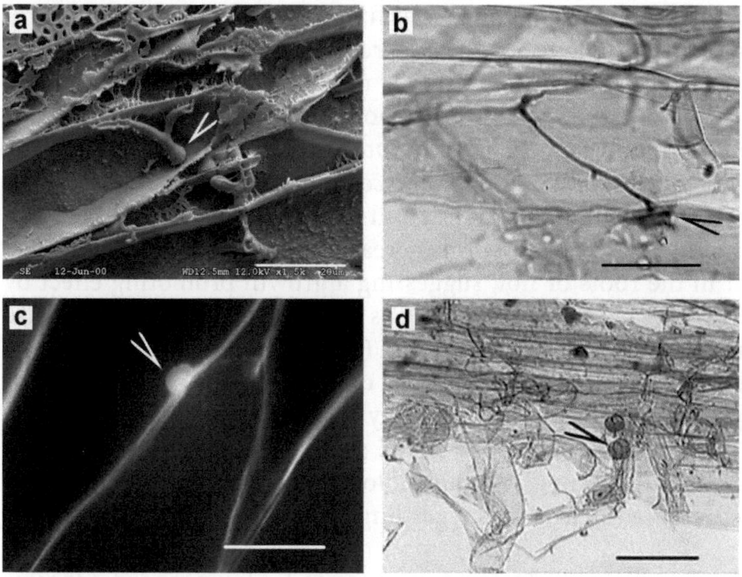

Fig. 11.2. **a** Formation of appressoria (*arrow*) by *Arthrobotrys oligospora* during penetration of the epidermis of barley roots. (From Bordallo et al. 2002, courtesy of Blackwell Science). **b** Colonisation of tomato roots by *Pochonia chlamydosporia*. Note appressorium and cell wall protein apposition (*arrow*). (From Bordallo et al. 2002, courtesy of Blackwell Science). **c** Colonisation of barley roots by *A. oligospora*. Callose deposit in papillae (*arrow*) stained with Sirofluor. (From Bordallo et al. 2002, courtesy of Blackwell Science). **d** Colonisation of tomato roots by *Pochonia chlamydosporia*. Note externally produced chlamydospores (*arrow*). (From Bordallo et al. 2002, courtesy of Blackwell Science). *Bars* **a**, **c** 20 μm; **b**, **d** 30 μm

entered vascular tissues. In contrast to *Pochonia* spp., appressoria had previously never been observed in *A. oligospora*. Using histochemical stains, we could show plant defence reactions, e.g. papillae, lignitubers and other cell wall appositions induced by nematophagous fungi, but these never prevented root colonisation. Callose deposits in papillae induced by *A. oligospora* in barley roots are shown in Fig. 11.2c. Nematophagous fungi grew extensively, especially in monocotyledonous plants, producing abundant mycelia, conidia and chlamydospores (*P. chlamydosporia*) (Fig. 11.2d). Necrotic areas of the roots were observed at initial stages of colonisation by *A. oligospora*, but were never seen at later stages even when the fungus proliferated in epidermal and cortical cells. Roots colonised by *P. chlamydosporia* displayed higher proteolytic activity than non-inoculated control roots using immunochemical techniques. The significance of this fact for biological control of root pathogens is under investigation in our laboratory.

The growth of the two nematophagous fungi in plant roots appears to resemble that of an endophyte, i.e. the host remains asymptomatic. Whether this endophytic growth induces systemic resistance to nematodes and/or plant pathogens in plants is as yet unknown, but worth further investigation. We have found that *P. chlamydosporia* could reduce growth of the plant-pathogenic fungus *Gaeumannomyces graminis* var. *tritici* (take-all fungus, Ggt) in dual culture Petri dish and in growth tube experiments. In pot experiments, *P. chlamydosporia* increased plant growth whether Ggt was present in the roots or not, suggesting a growth promoting effect by *P. chlamydosporia* (Monfort et al. 2005), as has also been found in the case of colonisation by other endophytic fungi (see Chap. 15 by Schulz).

In a recent screening in our laboratory on the capacity of various types of nematophagous fungi to grow endophytically in roots, we have shown that fungi other than *A. oligospora* and *P. chlamydosporia* had this ability (Table 11.1). With these fungi, all four ecological groups of nematophagous fungi are represented. The results regarding root colonisation are shown in Table 11.2.

Hirsutella rhossiliensis, which infects nematodes by means of adhesive conidia (Jaffee and Zehr 1982), behaves ecologically as an endoparasitic fungus ("obligate" parasite), although it can grow in the laboratory on artificial media. The fungus reacts in a density-dependent manner (Jaffee et al. 1992) and, in spite of its low efficiency as a nematode antagonist, it suppresses plant-parasitic nematodes in agroecosystems with little human disturbance, such as old peach orchards in the United States (Stirling 1991). *H. rhossiliensis*, unlike *A. oligospora* and *P. chlamydosporia*, does not seem to colonise barley roots endophytically. Three weeks after inoculation, cortex and epidermal cells were free from hyphal colonisation (Table 11.2). However, the fungus seems to colonise the rhizoplane abundantly, where it forms viable conidiophores (Fig. 11.3a).

Nematoctonus (teleomorph *Hohenbuehelia*) is a peculiar genus of basidiomycetous nematophagous fungi. It comprises about 15 species, some of

Table 11.1. Endophytic root colonisation of barley by the four ecological groups of nematophagous fungi

Fungus	Type	Colonisation	Reference
Pochonia chlamydosporia	Egg-parasite	+	Lopez-Llorca et al. 2002a
Arthrobotrys oligospora	Nematode-trapping	+	Bordallo et al. 2002
Arthrobotrys dactyloides	Nematode-trapping	+	This chapter
Nematoctonus robustus	Nematode-trapping	+	This chapter
Nematoctonus pachysporus	Endoparasite	–	This chapter
Hirsutella rhossiliensis	Endoparasite	–	This chapter
Pleurotus djamor	Toxin producing	+	This chapter

Table 11.2. Semiquantitative endophytic barley root colonisation by nematophagous fungi (unpublished results). In these experiments barley seeds were sterilised, germinated and planted together with the appropriate fungus in culture tubes containing sterilised vermiculite and water according to Bordallo et al. (2002). After 21 days the roots were harvested, surface sterilised to remove external hyphal growth, cut into 1-cm fragments and plated on corn meal agar. Half of the roots were not surface sterilised. After approx 7 days the root fragments were examined and fungal growth was recorded. Data are based on three roots, with six fragments of each ($n = 18$)

Fungus	Type	Root fragments with fungus (%) Non-surface sterilised	Surface sterilised
Arthrobotrys dactyloides	Nematode-trapping	77.8	5.6
Nematoctonus robustus	Nematode-trapping	100.0	22.0
Nematoctonus pachysporus	Endoparasite	50.0	0
Hirsutella rhossiliensis	Endoparasite	100.0	0
Pleurotus djamor	Toxin producing	77.8	11.1

which fit the endoparasitic behaviour, infecting nematodes by means of adhesive conidia, whereas others capture nematodes with adhesive traps, yet others share both types of behaviour (Thorn and Barron 1986). This diverse nematophagous behaviour was also reflected in root colonisation by the two species studied. Whereas the nematode-trapping *N. robustus* penetrated and colonised barley roots (Fig. 11.3b) and formed typical clamp-connections (Fig. 11.3c) as soon as 1 week after inoculation, the endoparasite *N. pachysporus* did not enter the roots but colonised the root surface as did *H. rhossiliensis* (Table 11.2). It thus appears that, in general, the endoparasitic fungi may not be endophytic root colonisers, but this may also be a reflection of their slow growth rate and their mode of infecting nematodes.

The genus *Pleurotus* also belongs to the *Basidiomycetes.* It forms basidiocarps and its standard means of living is saprophytic growth on decaying wood. The most common species, *P. ostreatus*, is a commercially cultivated edible mushroom. The fungus compensates for the lack of nitrogen in wood, its natural substrate, with its nematophagous habit (Thorn and Barron 1984). In fact, the nematophagous habit has been described for several species of *Pleurotus* (Thorn and Barron 1984). The fungus immobilises nematodes with a toxin (Kwok et al. 1992) prior to infection and digestion of its prey (Nordbring-Hertz et al. 1995). The strong relation with plant tissues, as a natural wood decomposer, was also confirmed in roots, when we found that, just as *N. robustus*, *P. djamor* penetrated early, and extensively colonised barley roots (Fig. 11.3d). The fungus was in fact very aggressive since some parts of the root appeared to be decorticated 3 weeks

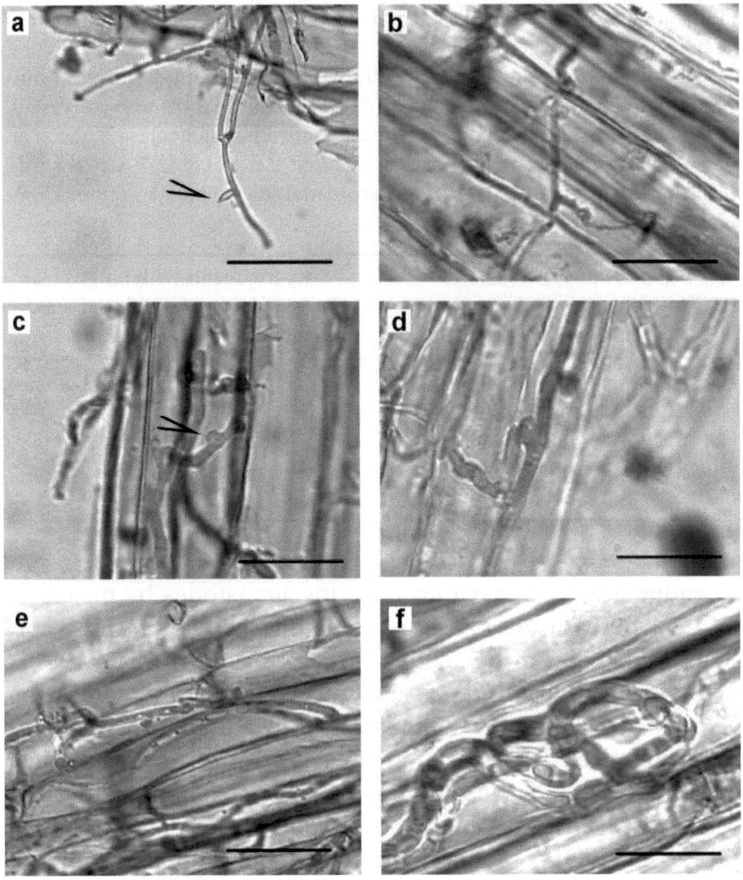

Fig. 11.3. **a** Rhizoplane colonisation of barley roots by the endoparasitic fungus *Hirsutella rhossiliensis* 3 weeks after inoculation, forming conidiophore with phialide (*arrow*). **b** Colonisation of cortex of barley roots by the nematode-trapping basidiomycete *Nematoctonus robustus* 2 weeks after inoculation. **c** Colonisation of cortex and epidermis of barley roots by the nematode-trapping basidiomycete *Nematoctonus robustus* 1 week after inoculation. Note clamp-connection (*arrow*). **d** Colonisation of cortex of barley roots by *Pleurotus djamor* 10 days after inoculation. **e** Colonisation of cortex of barley roots by the nematode-trapping fungus *Arthrobotrys dactyloides* 2 weeks after inoculation. **f** Colonisation of cortex of barley roots by the nematode-trapping fungus *Arthrobotrys dactyloides* 2 weeks after inoculation showing coiling structure. *Bars* **a, b** 30 μm; **c–f** 15 μm

after inoculation, as was *A. oligospora* in previous experiments (Bordallo et al. 2002).

Arthrobotrys dactyloides is a nematode-trapping fungus that captures nematodes by means of constricting rings (Barron 1977). The fungus was shown to form functional traps in soil (Jansson et al. 2000) and on the surface of tomato roots infected with root-knot nematodes (Riekert and Tiedt

1994), and has also been used in biological control experiments of plant parasitic nematodes (Stirling and Smith 1998). In our recent experiments, *A. dactyloides* was an active, and early, root coloniser. We found evidence of epidermal cell penetration and colonisation (Fig. 11.3e) 1 week after inoculation. Like *P. chlamydosporia*, the fungus formed coiling structures in barley root cells (Fig. 11.3f) and extensively colonised the roots. Such structures are also formed by other root endophytes, e.g. *Piriformospora indica* (Varma et al. 1999; Chap. 15 by Schulz), and presumably improve the exchange of metabolites.

These preliminary studies lack the most important component, the nematode. With trapping and endoparasitic nematophagous fungi, such experiments are easy to perform axenically, since the nematophagous habit can easily be triggered by free-living nematodes. Such experiments are underway in our laboratory to address the question whether the root endophytic behaviour of nematophagous fungi is functional in their nematophagous habit, i.e. if the mycelium growing in roots is able to develop active trapping organs or adhesive spores.

In the case of fungal nematode egg parasites the inclusion of a plant parasitic nematode is technically more difficult. This is because the natural targets of nematophagous *Pochonia* spp. are cyst and root knot nematodes. These phytopathogenic nematodes have an endoparasitic behaviour and a life span of at least a month – too long for our axenic system. Although these nematodes can be multiplied in axenic systems based on *Agrobacterium*-transformed roots on a tissue culture medium (Verdejo-Lucas 1995), these extremely rich conditions are incompatible for studying root colonisation by nematophagous fungi. An alternative, which we have explored (unpublished) is the use of migratory endoparasitic nematodes such as *Pratylenchus* spp., some species of which have broad host specificity and can infect cereals such as barley under our conditions. We have encountered two problems that have prevented further development. One is that *Pratylenchus* spp. is not a host of *Pochonia* spp. (or has not been described yet). Since the nematode lays eggs in root tissue, one could expect to find egg infection. However, in our experiments we found that the axenic conditions are too favourable for the nematode, which multiplies much faster than the fungus. This may be a question of optimising inoculum. We are trying to adapt our methods to include endoparasitic nematodes such as *Meloidogyne* spp., and the final goal of our studies is to develop biological control strategies to such severe plant pathogens. Development of alternative control methods, e.g. biological control, is especially important since the phasing out of the most widespread method of plant parasitic nematode control, fumigation with methyl bromide, is to be enforced.

Endophytic rhizobacteria that reduce plant-parasitic nematodes have also been described [Hallman et al. 2001; Chaps. 4 (Berg and Hallmann) and 3 (Kloepper and Ryu)], as well as arbuscular mycorrhizal fungi that reduce root knot nematodes (Waecke et al. 2001). If this is also the case for nematophagous fungi this will open up a new area of biocontrol using these fungi. The internal root colonisation by egg-parasitic fungi, e.g. *Pochonia* spp., may give the fungi an opportunity to infect nematode eggs in egg sacks of root-knot nematodes inside the roots and reduce subsequent spread and infection of roots by the second stage juveniles. Structures resembling trapping organs were observed in epidermal cells colonized by *A. oligospora* (Bordallo et al. 2002), and these may serve the purpose of trapping newly hatched juveniles escaping the roots. The ability to colonise plant roots may also be a survival strategy of these fungi and could explain soil suppressiveness to plant-parasitic nematodes in nature. Root inoculation of nematophagous fungi may help us to circumvent the lack of receptivity of soil (even sand) to inoculum of nematophagous fungi (Monfort et al. 2006), which has hampered the capability of fungi such as *P. chlamydosporia* to control agronomically important nematodes such as *Meloidogyne* spp. (Verdejo-Lucas et al. 2003). This, despite the fact that the nematode is naturally found to be infected by *P. chlamydosporia* and other egg-parasitic fungi in similar Mediterranean agroecosystems (Verdejo-Lucas et al. 2002; Olivares-Bernabeu and Lopez-Llorca 2002). The root colonisation of plant roots is a new area of research that deserves in-depth investigation, particularly for biocontrol purposes.

11.3 Concluding Remarks

The role of the host plant in the tritrophic relationship between nematophagous fungi, plants and phytopathogenic nematodes has largely been neglected. In this chapter we have collected and drawn conclusions from the data accumulated in the literature. We have also presented our own data, which indicate that the outcome of the interaction between nematophagous fungi and plants depends both on the host plant (mono- vs di-cotyledon) and the fungal species, but also on the ecological groups of nematophagous fungi (trapping, endoparasitic, egg-parasite, toxin producer).

The plant host is, after all, the most important living entity in an agroecosystem and is, of course, the target of any approach to disease control. We would like to stress the biological – theoretical – importance of the endophytic behaviour of nematophagous fungi, for instance, as a likely means to explain soil suppressiveness to plant parasitic nematodes. Another theoretical component is the possible discovery of a new mechanism of the mode of action of nematophagous fungi, that of interaction with the plant host. This may function in two ways. We have initial evidence of plant growth promotion. The other function could be modulation of plant defences. Similar activities have been found for antagonists (bacterial and fungal) of other plant pathogens, but also for other endophytic fungi that are not necessarily antagonists [see Chaps. 3 (Kloepper and Ryu), 4 (Berg and Hallmann) and 14 (Schulz)].

Basic discoveries in the field of pathogen–plant host interaction have led to new developments and approaches to plant disease control. For instance, the recent proliferation of compounds modulating plant defence responses as "magic bullets" to control a wide array of biotic and abiotic stresses in plants had its origin in the study of the molecular basis of virulence/avirulence in the plant pathogenic bacteria–plant interaction. There are similar examples, such as the effect of oligosaccharides or other molecules on plant defence systems. A future possibility for enhancing control of plant pathogens may be to devise a way to obtain synergies with non-chemical means of control. Perhaps we are in the dawn of a new era of biological control of nematodes that may circumvent the difficulties found in the mere application of nematophagous fungi to soil.

References

Åhman J, Ek B, Rask L, Tunlid A (1996) Sequence analysis and regulation of a gene encoding a cuticle-degrading serine protease from the nematophagous fungus *Arthrobotrys oligospora*. Microbiology 142:1605–1616

Ahrén D, Ursing BM, Tunlid A (1998) Phylogeny of nematode trapping fungi based on 18S rDNA sequences. FEMS Microbiol Lett 158:179–184

Barron GL (1977) The nematode-destroying fungi. Topics in mycobiology No. 1 Canadian Biological Publications, Guelph

Bird AF, Bird J (1991) The structure of nematodes. Academic, San Diego

Bonants PJM, Fitters PFL, Thijs H, Den Belder E, Waalwijk C, Willem J, Henfling DM (1995) A basic serine protease from *Paecilomyces lilacinus* with biological activity against *Meloidogyne hapla* eggs. Microbiology 141:775–784

Bordallo JJ, Lopez-Llorca LV, Jansson H-B, Salinas J, Persmark L, Asensio L (2002) Effects of egg-parasitic and nematode-trapping fungi on plant roots. New Phytol 154:491–499

Bourne JM, Kerry BR, De Leij FAAM (1996) The importance of the host plant on the interaction between root-knot nematodes (*Meloidogyne* spp.) and the nematophagous fungus, *Verticillium chlamydosporium* Goddard. Biocontrol Sci Technol 6:539–548

Chet I, Harman GER, Baker R (1981) *Trichoderma hamatum*: its hyphal interaction with *Rhizoctonia solani* and *Pythium* spp. Microb Ecol 7:29–38

Dackman C, Jansson H-B, Nordbring-Hertz B (1992) Nematophagous fungi and their activities in soil. In: Bollag J-M, Stotzky G (eds) Soil biochemistry, vol 7 Decker, New York, pp 95–130

Gams W, Zare R (2003) A taxonomic review of the clavicipitaceous anamorphs parasitizing nematodes and other microinvertebrates. In: White JF, Bacon CW, Hywel-Jones NL, Spatafora JW (eds). Clavicipitalean fungi: evolutionary biology, chemistry, biocontrol, and cultural impacts Dekker, New York, pp 17–73

Gaspard JT, Mankau R (1986) Nematophagous fungi associated with *Tylenchulus semipenetrans* and the citrus rhizosphere. Nematologica 32:359–363

Hagedorn G, Scholler M (1999) A reevaluation of predatory orbiliaceous fungi. I. Phylogenetic analysis using rDNA sequence data. Sydowia 51:27–48

Hallmann J, Quadt-Hallmann A, Miller WG, Sikora RA, Lindow SE (2001) Endophytic colonization of plants by the biocontrol agent *Rhizobium etli* G12 in relation to *Meloidogyne incognita* infection. Phytopathology 91:415–422

Jaffee BA, Zehr EI (1982) Parasitism of the nematode *Criconemella xenoplax* by the fungus *Hirsutella rhossiliensis*. Phytopathology 72:1378–1381

Jaffee B, Phillips R, Muldoon A, Mangel M (1992) Density-dependent host pathogen dynamics in soil microcosms. Ecology 73:495–506

Jansson H-B, Lopez-Llorca LV (2001) Biology of nematophagous fungi. In: Misra JK, Horn BW (eds). Trichomycetes and other fungal groups. Science Publishers, Enfield, pp 145–173

Jansson H-B, Lopez-Llorca LV (2004) Control of nematodes by fungi. In: Arora DK (ed) Fungal biotechnology in agriculture, food, and environmental applications Dekker, New York, pp 205–215

Jansson H-B, Nordbring-Hertz B (1983) The endoparasitic nematophagous fungi *Meria coniospora* infects nematodes specifically at the chemosensory organs. J Gen Microbiol 129:1121–1126

Jansson H-B, Nordbring-Hertz B (1988) Infection events in the fungus-nematode system. In: Poinar GO, Jansson H-B (eds) Diseases of nematodes, vol 2. CRC, Boca Raton, pp 59–72

Jansson H-B, Persson C, Odselius R (2000) Growth and capture activities of nematophagous fungi in soil visualized by low temperature scanning electron microscopy. Mycologia 92:10–15

Jeffries P (1997) Mycoparasitism. In: Dicklow TD, Söderström B (eds). The mycota IV. Environmental and microbiological relationships, Springer, Berlin Heidelberg New York, pp 149–164

Kerry BR (2000) Rhizosphere interactions and the exploitation of microbial agents for the biological control of plant-parasitic nematodes. Annu Rev Phytopathol 38:423–441

Kwok OCH, Plattner R, Weisleder D, Wicklow DT (1992) A nematicidal toxin from *Pleurotus ostreatus* NRRL 3526. J Chem Ecol 18:127–136

Leinhos GME, Buchenauer H (1992) Hyperparasitism of selected fungi on rust fungi on cereals. Z Pflanzenkr Pflanzenschutz 99:482–498

Lopez-Llorca LV (1990) Purification and properties of extracellular proteases produced by the nematophagous fungus *Verticillium suchlasporium*. Can J Microbiol 36:530–537

Lopez-Llorca LV, Claugher D (1990) Appressoria of the nematophagous fungus *Verticillium suchlasporium*. Micron Microsc Acta 21:125–130

Lopez-Llorca LV, Bordallo JJ, Salinas J, Monfort E, Lopez-Serna ML (2002a) Use of light and scanning electron microscopy to examine colonisation of barley rhizosphere by the nematophagous fungus *Verticillium chlamydosporium*. Micron 33:61–67

Lopez-Llorca LV, Olivares-Bernabeu C, Salinas J, Jansson H-B (2002b) Pre-penetration events in fungal parasites of nematode eggs. Mycol Res 106:499–506

Meyer SLF (1998) Evaluation of *Verticillium lecanii* strains applied in root drenches for suppression of *Meloidogyne incognita* on tomato. J Helminthol 65:82–86

Monfort E, Lopez-Llorca LV, Jansson H-B, Salinas J, Park J-O, Sivasithamparam K (2005) Colonisation of seminal roots of wheat and barley by egg-parasitic nematophagous fungi and their effects on *Gaeumannomyces graminis* var. tritici and development of root-rot. Soil Biol Biochem 37:1229–1235

Monfort E, Lopez-Llorca LV, Jansson H-B, Salinas J (2006) In vitro soil receptivity assays to egg-parasitic nematophagous fungi. Mycol Prog 5:18–23
Nicholson RL (1996) Adhesion of fungal propagules. In: Nicole M, Gianinazzi-Pearson V (eds) Histology, ultrastructure and molecular cytology of plant-microorganism interactions. Kluwer, Amsterdam, pp 117–134
Nordbring-Hertz B (1973) Peptide-induced morphogenesis in the predacious fungus *Arthrobotrys oligospora*. Physiol Plant 29:223–233
Nordbring-Hertz B (1977) Nematode-induced morphogenesis in the predacious fungus *Arthrobotrys oligospora*. Nematologica 23:443–451
Nordbring-Hertz B, Jansson H-B, Friman E, Persson Y, Dackman C, Hard T, Poloczek E, Feldman R (1995) Nematophagous Fungi Institut für den Wissenschaftlichen Film, Göttingen. Film No C 1851
Olivares-Bernabeu C, Lopez-Llorca LV (2002) Fungal egg-parasites of plant-parasitic nematodes from Spanish soils. Rev Iberoam Micol 19:104–110
Olsson S, Persson Y (1994) Transfer of phosphorus from *Rhizoctonia solani* to the mycoparasite *Arthrobotrys oligospora*. Mycol Res 98:1065–1068
Persmark L, Jansson H-B (1997) Nematophagous fungi in the rhizosphere of agricultural crops. FEMS Microbiol Ecol 22:303–312
Persmark L, Nordbring-Hertz B (1997) Conidial trap formation of nematode-trapping fungi in soil and soil extracts. FEMS Microbiol Ecol 22:313–323
Persson C, Jansson H-B (1999) Rhizosphere colonization and control of *Meloidogyne* spp. by nematode-trapping fungi. J Nematol 31:164–171
Persson Y, Veenhuis M, Nordbring-Hertz B (1985) Morphogenesis and significance of hyphal coiling by nematode-trapping fungi in mycoparasitic relationships. FEMS Microbiol Ecol 31:283–291
Peterson EA, Katznelson H (1965) Studies on the relationships between nematodes and other soil microorganisms. IV. Incidence of nematode-trapping fungi in the vicinity of plant roots. Can J Microbiol 11:491–495
Pfister DH (1997) Castor, Pollux and life histories of fungi. Mycologia 89:1–23
Poinar GO (1983) The natural history of nematodes. Prentice-Hall, Englewood Cliffs
Riekert HF, Tiedt LR (1994) Scanning electron microscopy of *Meloidogyne incognita* juveniles entrapped in maize roots by a nematode-trapping fungus *Arthrobotrys dactyloides*. S Afr J Zool 29:189–191
Rubner A (1996) Revision of predacious hyphomycetes in the *Dactylella-Monacrosporium* complex. Stud Mycol 39:1–134
Segers R, Butt TM, Kerry BR, Peberdy F (1994) The nematophagous fungus *Verticillium chlamydosporium* Goddard produces a chymoelastase-like protease which hydrolyses host nematode proteins in situ. Microbiology 140:2715–2723
Sneh B, Humble SJ, Lockwood JL (1977) Parasitism of oospores of *Phytophthora megasperma* var. sojae, *P. cactorum*, *Pythium* and *Aphanomyces euteiches* by oomycetes, chytridomycetes, hyphomycetes, actinomycetes and bacteria. Phytopathology 67:622–628
Stirling GR (1991) Biological control of plant parasitic nematodes. Progress, problems and prospects. CAB International, Wallingford
Stirling GR, Smith LJ (1998) Field tests of formulated products containing either *Verticillium chlamydosporium* or *Arthrobotrys dactyloides* for biological control of root-knot nematodes. Biol Contr 11:231–239
Thorn RG, Barron GL (1984) Carnivorous mushrooms. Science 224:76–78
Thorn RG, Barron GL (1986) *Nematoctonus* and the Tribe resupinatae in Ontario, Canada. Mycotaxon 25:321–453

Tikhonov VE, Lopez-Llorca LV, Salinas J, Jansson H-B (2002) Purification and characterization of chitinases from the nematophagous fungi *Verticillium chlamydosporium* and *V. suchlasporium*. Fungal Gen Biol 35:67–78

Tunlid A, Johansson B, Nordbring-Hertz B (1991) Surface polymers of the nematode-trapping fungus *Arthrobotrys oligospora*. J Gen Microbiol 137:1231–1240

Tzean SS, Estey RH (1978) Nematode-trapping fungi as mycopathogens. Phytopathology 68:1266–1270

Varma A, Verma S, Sudha, Sahay N, Bütehorn B, Franken P (1999) *Piriformospora indica*, a cultivable plant-growth-promoting root endophyte. Appl Environ Microbiol 65:2741–2744

Veenhuis M, Nordbring-Hertz B, Harder W (1985) An electron-microscopical analysis of capture and initial stages of penetration of nematodes by *Arthrobotrys oligospora*. Antonie van Leeuwenhoek 51:385–398

Verdejo-Lucas S (1995) Dual cultures: nematodes. In: Singh RP, Singh US (eds) Molecular methods in plant pathology. CRC, Boca Raton, pp 301–312

Verdejo-Lucas S, Ornat C, Sorribas FJ, Stchiegel A (2002) Species of root-knot nematodes and fungal egg parasites recovered from vegetables in Almeria and Barcelona, Spain. J Nematol 34:405–408

Verdejo-Lucas S, Sorribas FJ, Ornat C, Galeano M (2003) Evaluating *Pochonia chlamydosporia* in a double-cropping system of lettuce and tomato in plastic houses infested with *Meloidogyne javanica*. Plant Pathol 52:521–528

Waecke JW, Waudo SW, Sikora R (2001) Suppression of *Meloidogyne hapla* by arbuscular mycorrhiza fungi (AMF) on pyrethrum in Kenya. Int J Pest Manage 47:135–140

12 Molecular Diversity and Ecological Roles of Mycorrhiza-Associated Sterile Fungal Endophytes in Mediterranean Ecosystems

Mariangela Girlanda, Silvia Perotto, Anna Maria Luppi

12.1 Introduction

Under natural conditions, plant roots sustain considerable fungal diversity (Vandenkoornhuyse et al. 2002). In healthy plants, colonisation of root tissues is not restricted to mycorrhizal symbionts; other fungi can also grow asymptomatically in the roots, occurring either in the presence or absence of ecto- or endo-mycorrhizal mycobionts. Isolation from surface-sterilised roots usually yields a great proportion of fungi with dematiaceous hyphae and ascomycetous septa, which are sterile in culture, and are referred to as "dark septate endophytes" (DSE) or "dark sterile mycelia" (DSM) (Schild et al. 1988; Summerbell 1989; Read 1991; Holdenrieder and Sieber 1992; Girlanda and Luppi-Mosca 1995; Ahlich and Sieber 1996; Ahlich et al. 1998; Jumpponen and Trappe 1998a; Schadt et al. 2001; see Chap. 7 by Sieber and Grünig). These mycelia can also be directly observed to grow inter- and intra-cellularly in the root cortex, producing peculiar structures, such as microsclerotia that fill cortical cells (Jumpponen and Trappe 1998a; Barrow and Aaltonen 2001; Yu et al. 2001; Kovacs and Szigetvari 2002; Ruotsalainen et al. 2002; Barrow 2003). By virtue of their constant association with roots, they can be qualified as true root symbionts (see Chap. 1 by Schulz and Boyle).

Although the presence of DSE in roots was noted early in the last century (Melin 1922, 1923; Peyronel 1924), their abundant, regular, and ubiquitous occurrence was given prominence only recently, and is suggestive of a significant role in natural ecosystems. As a group, these fungi colonise a broad range of hosts, being reported from nearly 600 plant species, representing about 320 genera and 114 families (Jumpponen and Trappe 1998a). However, DSE represent a heterogeneous assemblage of ascomycetous taxa.

Mariangela Girlanda: Dipartimento di Biologia Vegetale and IPP - Torino, Viale PA Mattioli 25, 10125 Torino, Italy, E-mail: mariangela.girlanda@unito.it

Silvia Perotto: Dipartimento di Biologia Vegetale and IPP - Torino, Viale PA Mattioli 25, 10125 Torino, Italy

Anna Maria Luppi: Dipartimento di Biologia Vegetale, Viale PA Mattioli 25, 10125 Torino, Italy

Soil Biology, Volume 9
Microbial Root Endophytes
B. Schulz, C. Boyle, T. N. Sieber (Eds.)

Although these fungi are mostly sterile when brought into culture, sporulation has been occasionally induced under particular incubation conditions (Richard and Fortin 1973; Wang and Wilcox 1985; Fernando and Currah 1995; Ahlich and Sieber 1996), leading to recognition of distinct conidial forms (Gams 1963; Deacon 1973; Richard and Fortin 1973; Wang and Wilcox 1985; Currah et al. 1987). In persistently sterile isolates, which are unidentifiable with conventional criteria, systematic heterogeneity is indicated by different cultural macro- and micro-scopic morphologies and by polymorphisms revealed by molecular markers (Stoyke et al. 1992; Harney et al. 1997; Jumpponen and Trappe 1998a; Schadt et al. 2001; Addy et al. 2001; Grünig et al. 2001, 2002b; see Chap. 7 by Sieber and Grünig). For instance, sequencing of the 18S nuclear ribosomal DNA (rDNA) region has shown polyphyletic placement of DSE fungi within Ascomycetes, with representatives of distinct orders such as Pleosporales, Leotiales, and Pezizales (Lobuglio et al. 1996; Jumpponen and Trappe 1998a; Schadt et al. 2001; Girlanda et al. 2002). As a consequence of such heterogeneity, actual host specificity might vary within the DSE group. Specificity is an important attribute of fungus-plant associations, and understanding this aspect of the association is crucial to clarifying the functional nature of the interactions with the host plants, and hence the ecological role of DSE associates. Indeed, inoculation experiments of different plants with different DSE strains have indicated that plant growth response may depend on the particular combination of the fungus and plant being tested (Jumpponen and Trappe 1998a; Jumpponen 2001).

To date, most of the knowledge available on the specific identity of DSE fungi, as assessed by molecular analyses of internal transcribed spacer (ITS) regions of nuclear rDNA, derives from isolates obtained from subantarctic and northern temperate forests or from Arctic areas in Europe and Canada (Stoyke et al. 1992; Harney et al. 1997; Jumpponen 1999; Addy et al. 2001; Grünig et al. 2001, 2002a, 2002b; Schadt et al. 2001; see Chap. 7 by Sieber and Grünig); other biomes remain largely unexplored in DSE research. Extending investigations to different environments offers the opportunity of assessing both host and habitat specificity patterns for these fungi. Mediterranean ecosystems are especially interesting in this respect. They develop in temperate-warm climates (mean annual temperature generally ranging from 14° to 20°C), with precipitation of 350–1,000 mm/year, characterised by hot, dry summers and mild, wet winters ("winter-rain and summer dry" climates). Such climatic conditions occur in five distinct regions of the world, i.e. the Mediterranean basin, California, Central Chile, the Cape Province in South Africa, and Southern Australia (Di Castri and Mooney 1973). Although other regions may exhibit similar mean annual temperatures and rainfall, they differ in rain distribution over the year (Japan, for instance, lacks summer drought). Vegetation in Mediterranean climate areas has different regional expressions (such as "maquis" and "garrigue" in the Mediterranean basin, "chaparral" in California, "matorral" in Chile, "fynbos" and "veld" in capensic flora, "kwongan" and "mallee" in Australia), which, however, display striking convergence in life forms as an adaptation to the distinctive climatic regime (Pignatti et al. 2002). A characteristic, shared feature of Mediterranean vegetation is a high diversity of plants and their associated mycorrhizal types: sclerophyllous and evergreen shrubs, and small trees bearing arbuscular, ericoid, arbutoid mycorrhiza and ectomycorrhiza, which coexist and may take up equal size and dominance (Allen 1991). These environments offer therefore an interesting scenario for comparative studies of root-fungus associations in plants with different mycorrhizal status.

We have been investigating molecular diversity and the possible ecological roles of DSE associates of host pairs in Mediterranean ecosystems in Northern Italy (Liguria). Neighbouring, healthy-looking individuals of the ectomycorrhizal *Pinus halepensis* Mill. (Halep pine) and *Quercus ilex* L. (holm oak) and the endomycorrhizal *Rosmarinus officinalis* L. (arbuscular mycorrhiza; rosemary) and *Erica arborea* L. (ericoid mycorrhiza; Mediterranean heather) were selected for isolation of DSE fungi. Using both molecular approaches and synthesis experiments the fungi were characterised for their range of diversity and for possible ecophysiological traits involved in interactions with different plant hosts.

12.2 Diversity of DSE Associates of Ecto- and Endo-Mycorrhizal Plants in Mediterranean Ecosystems in Northern Italy

To assess the diversity of DSE isolates obtained from surface-sterilised mycorrhizal roots of the two different host pairs (*Pinus halepensis* / *Rosmarinus officinalis*, *Quercus ilex* / *Erica arborea*), morphotypes were recognised based on macro- and micro-scopic somatic features and assessed for consistency through internal transcribed spacer-restriction fragment length polymorphism (ITS-RFLP); further molecular characterisation was carried out on morphotypes repeatedly obtained from both hosts (Girlanda et al. 2002; Bergero et al. 2000, 2003). Such shared DSE groups occurred regularly and at high frequency. In the *P. halepensis* / *R. officinalis* study, for instance, such groups were isolated in all the collection periods, spanning the course of 11 years. Ribosomal DNA sequences obtained for representative isolates from each morphotype were used as queries for BLAST searches in public DNA databases as well as for phylogenetic reconstructions carried out on datasets comprising both named alignable sequences retrieved from BLAST searches and sequences of closely related taxa.

In the investigation on the *P. halepensis* / *R. officinalis* host pair, carried out at Varigotti (Girlanda et al. 2002), taxonomic affiliation through sequence analysis of rDNA regions has revealed a peculiar spectrum of taxa, distinct from those recognised to date in other hosts. In some cases, ITS (ITS1–5.8S–ITS2) sequence matches with sequences in GenBank were poor (<80% sequence identity), being limited mostly to the 5.8S gene region. These morphotypes would therefore represent fungi having no close relatives in GenBank or hitherto undescribed fungi. They could therefore be given taxonomic placement only at a high level based on 18S rDNA sequence data, and were assigned to the Dothideomycetidae and Chaetothyriomycetidae. Interestingly, one such morphotype displayed significant ITS identity with a DSE morphotype obtained from *E arborea* roots and mature *Q ilex* woodland and garrigue soil in the *Q. ilex* / *E. arborea* investigations (Bergero et al. 2000, 2003; see below). Better sequence matches referred to the entire ITS region, (94–95% sequence identity), and were obtained for a different morphotype with *Diaporthe*/*Phomopsis* sequences. *Diaporthe* is a genus belonging to the Valsaceae (Diaporthales, Sordariomycetidae) that has anamorphs primarily in *Phomopsis* (Kirk et al. 2001). Results from phylogenetic analyses based on ITS, 18S and elongation factor-1 alpha (EF1-α) gene sequences indicate that this morphotype, which is likely identifiable as a *Diaporthe/Phomopsis* species, would, however, be distinct from those currently represented in GenBank.

A single morphotype could be identified with confidence to the species level as *Rhizopycnis vagum* D.F. Farr, based on a nearly identical ITS sequence (approx. 99% identity over 500 bp, i.e. the entire ITS region). Such an in silico identification is consistent with colony features and morphology of somatic structures, including characteristic, tuberculate dark chlamydospores. *R. vagum* is a recently described coelomycete (*gen. et sp. nov.*; Farr et al. 1998), which was originally known to contribute to 'vine decline' of cucurbit crops in several parts of the world. It was first reported from cantaloupe in the Rio Grande Valley of Texas and was subsequently isolated from melon, watermelon and cucumber in Guatemala, Honduras, California, Spain and Central Italy (Mertely et al. 1991; Miller et al. 1996; Bruton and Miller 1997a, 1997b, 1997c; Gwinne et al. 1997; Aegerter et al. 2000; Armengol et al. 2000, 2003; Montuschi 2001). In Central and Southern Italy *R. vagum* has also been isolated from tomato roots exhibiting typical corky root symptoms (Porta-Puglia et al. 2001). Teleomorphic connections are unknown for *R. vagum*, but phylogenetic ITS-based analyses placed this fungus in a sister-group to *Massarina walkeri* Shoemaker, C.E. Babc. & J.A.G. Irwin, a teleomorphic species belonging to Lophiostomataceae, Pleosporales (Dothideomycetidae; Girlanda et al. 2002; Armengol et al. 2003). *Massarina walkeri* is the teleomorph of *Acrocalymma medicaginis*, a coelomycete exhibiting morphological resemblance to *R. vagum*

(Shoemaker et al. 1990), causing a root and crown rot disease of lucerne in Australia, characterised by symptoms that are somewhat reminiscent of those induced by *R. vagum* on cucurbit hosts (Alcorn and Irwin 1987).

None of the DSE morphotypes investigated, therefore, is identifiable with any of the taxa known to date as DSE sporulating forms (such as *Phialocephala* spp., *Phialophora* spp., *Cadophora finlandia* (Wang & Wilcox) Harrington & McNew, *Chloridium paucisporum* C.J.K. Wang & H.E. Wilcox, *Leptodontidium orchidicola* Sigler & Currah; Gams 1963; Deacon 1973; Richard and Fortin 1973; Wang and Wilcox 1985; Currah et al. 1987; Harrington and McNew 2003). Among these, *Phialocephala fortinii* C.J.K. Wang & Wilcox appears as the major component of DSE assemblages in northern alpine and subalpine forest ecosystems, occurring with no apparent host specificity in North America, Europe, and Japan (Wang and Wilcox 1985; Currah et al. 1987; Stoyke and Currah 1991, 1993; Stoyke et al. 1992; Currah and Tsuneda 1993; O'Dell et al. 1993; Ahlich and Sieber 1996; Dahlberg et al. 1997; Hambleton and Currah 1997; Harney et al. 1997; Addy et al. 2001; Grünig et al. 2001, 2002a, 2002b; see Chap. 7 by Sieber and Grünig). Although the occurrence of these taxa in the Mediterranean ecosystem we have studied cannot be ruled out entirely, they certainly do not appear among the dominant DSE in such an environment, where *P. halepensis* and *R. officinalis* represent the most abundant plant species.

In the *Q. ilex* (holm oak) / *E. arborea* (Mediterranean heather) studies (Bergero et al. 2000, 2003), investigations were carried out at different stages in the evolution of the plant community at a site near Borgio Verezzi. If left undisturbed, Mediterranean vegetation in Northern Italy develops into pure *Q. ilex* woodland, characterised by an extremely reduced understorey vegetation, due to the disappearance of several plant species of earlier stages of succession. Later stages are dominated by ectomycorrhiza, in contrast to maquis and garrigue vegetation, where ectomycorrhizal plants often co-dominate with endomycorrhizal hosts, such as *E. arborea*, a typical ericoid Mediterranean shrub. In the climax woodland, disturbances such as human activities and fire, which play a key role in shaping Mediterranean vegetation, open the way for recolonisation by pioneer plant species, thus initiating secondary successions. Investigations were carried out both within a pure woodland where *Q. ilex* establishment had caused the disappearance of most pioneer species, including *E. arborea*, and in post-cutting clearings where the latter species had become re-established.

Two DSE morphotypes (Sd2 and Sd9) were isolated from both holm oak and heather roots and characterised in detail. When tested in dual axenic culture trials on *E. arborea*, isolates assigned to these morphotypes were found to be able to form typical intracellular hyphal coils characteristic of ericoid mycorrhizal infection (see below). Morphotype Sd2 was also isolated from soil of the thickest part of the mature holm oak woodland (under adult trees), using *E. arborea* as a bait plant. No significant ITS sequence identity was found in GenBank for this morphotype. However, based on more conserved sequences such as the 5′ end of 28S rDNA, it was found to cluster with ericoid fungi from *Gaultheria shallon* in a clade comprising *Capronia* species (Chaetothyriomycetidae; Allen et al. 2003). In contrast, morphotype Sd9 displayed significant matches over the ITS region with nameless fungi of Helotialean affinities represented in GenBank. This morphotype apparently belongs to an undefined complex within the Helotiales, comprising DSE, ericoid endomycorrhizal and ectomycorrhizal fungi from ericoid (*Gaultheria shallon*), epacrid (*Astroloma pinifolium, Woollsia pungens*), tree and sedge hosts (*Tsuga* sp., *Pinus sylvestris, Betula pubescens, Kobresia myosuroides*) from Australia, Norway, Canada and Colorado (Monreal et al. 1999; McLean et al. 1999; Vrålstad et al. 2002a; Bergero et al. 2003; and unpublished sequences). In particular, Sd9 shared high sequence identity of approx. 99% over ca. 440 bp with an ectomycorrhizal fungus from the sedge *Kobresia myosuroides* from Colorado (Fig. 12.1).

Genetic relatedness between sterile mycorrhizal isolates obtained from ericoid and epacrid hosts in the northern and southern hemispheres, respectively, has been shown by several authors, in accordance with the current view of these plants as members of the single family Ericaceae (Crayn and Quinn 2000), and suggests the monophyletic origin of their endomycorrhizal associations (see Chap. 14 by Cairney). Such mycorrhizal fungi of Ericaceae from both hemispheres exhibit phylogenetic affinities to the Helotiales. Their precise placement within this order, however, remains unclear: nameless mycorrhizal fungi are either part of a *Hymenoscyphus ericae*-related group or of separate, unspecified groups often receiving poor bootstrap support (McLean et al. 1999; Monreal et al. 1999; Chambers et al. 2000; Sharples et al. 2000; Cairney and Ashford 2002; see Chap. 14 by Cairney). Genetic relatedness has also been demonstrated between *Hymenoscyphus ericae* and the fungal symbiont forming *Piceirhiza bicolorata* ectomycorrhizae, belonging to a monophyletic aggregate possibly comparable with a generic unit (Vrålstad et al. 2000). It is thus accepted that

▶ **Fig. 12.1.** Neighbour-joining tree for nuclear rDNA ITS (ITS1–5.8S–ITS2) sequences of DSE morphotypes Sd1, Sd9, Sm1, Sm2, Sm3, Sm5 and Sm8 and other alignable sequences from BLAST searches. Morphotypes were isolated from *Erica arborea* roots (Sd1, Sd9 and Sm1), *Quercus ilex* roots (Sd1 and Sd9), pure woodland soil where *Q. ilex* establishment had caused the disappearance of *E. arborea* (Sd1, Sm1, Sm2, Sm5 and Sm8), soil of post-cutting clearings where the latter species had re-established (Sd1, Sd9, Sm1, Sm2 and Sm3), and were found to be able to form typical mycorrhizal hyphal coils when tested in dual axenic culture trials on *E. arborea* (Bergero et al. 2000, 2003). The Kimura-2-parameter model was used for pairwise distance measurement. Bootstrap values above 50% are indicated (1,000 replicates). The *bar* indicates 0.1 bp changes. The tree was rooted automatically

Hymenoscyphus ericae AJ319078
Hymenoscyphus sp. AF081435
Phialophora finlandia AF011327
Phialocephala fortinii AF214579
Microglossum viride AY144534
Morphotype Sm3
Epacrid root endophyte AF098291
Calyptrozyma arxii AJ133432
Morphotype Sm1
Morphotype Sm2
Leotiales sp. AY011014
Leaf litter ascomycete AF502801
Leaf litter ascomycete AF502870
Cadophora sp. (mitosporic Helotiales) AY371513
Leaf litter ascomycete AF502651
Neofabraea alba AY359236
Neofabraea alba AY359235
Mycorrhizal fungal sp. from hemlock AY394891
Mycorrhizal fungal sp. from hemlock AY394920
Mycorrhizal fungal sp. from hemlock AY394916
Mycoarthris corallinus AF128440
Morphotype Sm8
Epacrid root endophyte AF099089
Morphotype Sd1
Mycorrhizal sp. from Gaultheria shallon AF081442
Mycorrhizal fungal sp. from hemlock AY394901
Mycorrhizal fungal sp. from hemlock AY394888
Epacrid root endophyte sp. AY279185
Epacrid root endophyte sp. AY279184
Morphotype Sd9
Ectomycorrhizal sp. from Kobresia myosuroides AY046959
Ectomycorrhizal sp. from Kobresia myosuroides AY046961
Ectomycorrhizal sp. from Kobresia myosuroides AY046960
Morphotype Sm5
Helotiales sp. AJ292198
Axenic ectomycorrhizal isolate (Helotiales) AJ430412
Ectomycorrhizal isolate (Helotiales) AJ430411
Ectomycorrhizal isolate (Helotiales) AJ430414
Helotiales sp. AJ292197
0.1

Helotiales encompass fungi with different strategies of association with roots, including ecto-, ectoendo-, ericoid endo-mycorrhizal and DSE fungi (LoBuglio et al. 1996; Monreal et al. 1999; Sharples et al. 2000; Vrålstad et al. 2000, 2002a). However, subgrouping within the order, and hence the possible occurrence of strict ericoid-, ecto-, and DSE lineages (Read 2000), is still unclear, since the boundaries between intra- and inter-specific ITS variation are presently uncertain for these fungi. More detailed molecular comparisons, using sequence data from further loci, as well as other genetic markers are needed before relationships within this group can be resolved with further detail (Cairney and Ashford 2002). In the case of morphotype Sd9, however, identical random amplified polymorphic DNA (RAPD) profiles were obtained for isolates from both *E. arborea* and *Q. ilex*, indicating that not only the same taxon, but also the same genotype of that taxon is shared between an ericoid and an ectomycorrhizal host (Bergero et al. 2000).

Database searches using ITS sequences from Mediterranean DSEs have allowed comparisons over geographically distinct areas and even biomes. Some taxa of Helotialean affinities appear to recur across unrelated biomes, suggesting that such fungi find a suitable niche within root tissues, independently of the precise host and environment. The apparent absence in the Mediterranean habitats investigated of *Phialocephala fortinii*, the dominating DSE inhabitant of roots in northern temperate forests, and, by contrast, the presence of *Rhizopycnis vagum* (see above) suggests the existence of some environmental selection on distribution of these fungi. Within a given biome, while ITS data leave the question of actual host specificity unresolved, there is evidence from RAPD data on Mediterranean DSE that some DSE genotypes may actually associate with both ecto- and endomycorrhizal plants. Such a capacity is a prerequisite for these fungi to play a role in interactions between different plant hosts.

12.3 Ecological Relationships with Conventional Mycorrhizal and Pathogenic Symbionts

The functional significance of DSE associations is variously described in the literature. A range of enzymatic activities has been reported for DSE, conferring potential ability to utilise some of the major organic detrital nutrient pools (Bååth and Söderström 1980; Haselwandter 1983; Currah and Tsuneda 1993; Fernando and Currah 1995; Caldwell et al. 2000). By using axenically grown *E arborea* as a bait plant, we have shown that soil from mature *Q ilex* forest from which the ericoid host had disappeared at least 10 years previously, maintains a high and diverse inoculum of

DSE fungi capable of associating with the ericaceous plant (Bergero et al. 2003). DSE fungi have also been isolated from soil from northern temperate forests (Jumpponen and Trappe 1998a). However, it remains uncertain whether DSE fungi are endowed with "competitive saprophytic ability" (Garrett 1950, 1956), i.e. are actually efficient at saprotrophically decomposing organic debris in the complex soil environment, or whether they exist primarily as root associates of different plants, including ectomycorrhizal hosts (see Chap. 13 by Rice and Currah). Enzymatic activity may assist penetration into root tissues (Jumpponen and Trappe 1998a; Schulz et al. 2002).

Whatever their actual free-living and saprotrophic abilities, when in association with host roots DSE isolates have been shown to increase host foliar P and N concentrations and plant biomass under some experimental conditions (Haselwandter and Read 1982; Fernando and Currah 1996; Jumpponen and Trappe 1998b; Jumpponen et al. 1998; Newsham 1999). These results suggest that at least some strains of DSE may have, from a functional point of view, a relationship with their host plants not dissimilar from that of conventional mycorrhizal symbionts. It should also be considered that variation in host response to classical mycorrhizal fungi may represent a continuum, ranging from parasitism to mutualism [Jumpponen 2001; see Chaps. 16 (Brundrett) and 15 (Schulz)]. Involvement in nutrient and possibly water acquisition could be especially relevant in unfavourable environments exposed to droughts (Sengupta et al. 1989; Jumpponen et al. 1998). DSE are found extensively in xeric cold and warm environments (Read and Haselwandter 1981; Haselwandter and Read 1982; Kohn and Stasovski 1990; Barrow et al. 1997; Ruotsalainen et al. 2002), where they may be even more prevalent than conventional mycorrhizal fungi, colonising roots extensively with active structures (Barrow and Aaltonen 2001; Barrow and Osuna 2002; Barrow 2003). This suggests special adaptations to the harsh conditions of dry soil, possibly providing, when they occur concurrently with mycorrhizal symbionts, a back-up system during periods when the latter are inhibited by environmental conditions (Jumpponen and Trappe 1998a).

Inoculation experiments in *E arborea* with DSE isolates obtained from either this plant or *Q ilex* have shown that, irrespective of the host of origin, some strains are capable of forming typical intracellular hyphal coils, suggesting ericoid mycorrhizal behaviour (Bergero et al. 2000, 2003). Morphotype Sd9 isolates could also produce an unusual phenotype in the form of an additional hyphal net surrounding root epidermal cells (Fig. 12.2). While ecto- and ectoendo-mycorrhizal potential was reported for fungi of the DSE complex (Wilcox and Wang 1987a, 1987b; Ursic and Peterson 1997; Vrålstad et al. 2002b), the capacity to form endomycorrhizal structures such as coils had thus far not been described for the latter fungi. Such

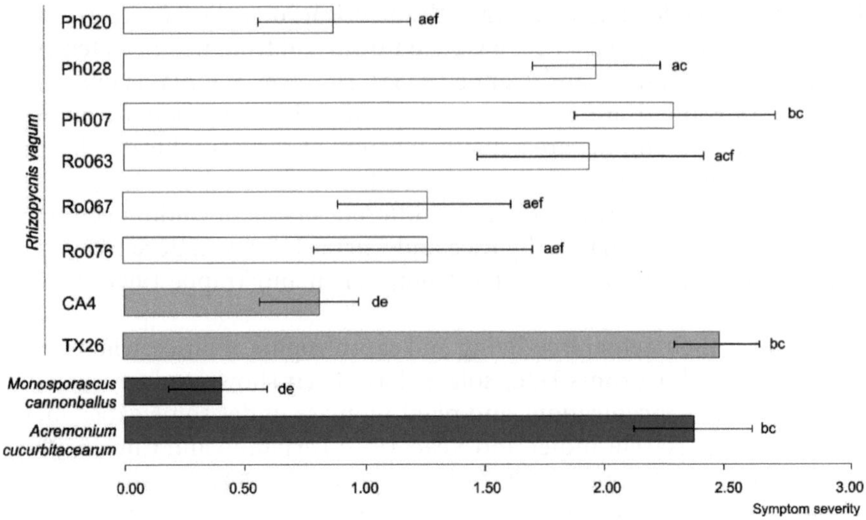

Fig. 12.2. Effects of *Rhizopycnis vagum* isolates of different origins on melon roots. Roots were rated for disease severity on the scale of Aegerter et al. (2000) [0 = no symptoms; 1 = few lesions (covering < 10% of root), secondary root rot slight; 2 = rot of secondary roots or lesions covering approximately 25% of the root; 3 = lesions covering at least 50% of the root and dead secondary roots; 4 = general root rot, most of the root affected] after inoculation with *R. vagum* (5,000 cfu/g soil) at 25–28°C for 40–50 days. *Monosporascus cannonballus* and *Acremonium cucurbitacearum*, two other vine decline pathogens, were inoculated for comparison [30 cfu/g soil and 20,000 cfu/g soil, respectively (Aegerter et al. 2000)]. *Error bars* indicate the standard deviation of the mean of ten observations in one growth chamber experiment. Different letters indicate differences significant at $P < 0.05$ (ANOVA, Tukey post-hoc test; different values have no common letter)

phenotypic plasticity in the production of distinct interfaces in different hosts, e.g. intracortical microsclerotia and hyphal coils, is not unusual and is reminiscent of the behaviour of some conventional mycorrhizal symbionts, such as fungi capable of forming ectomycorrhiza on tree hosts and endomycorrhiza on non-photosynthetic, "mycoheterotrophic" orchids (Taylor et al. 2002; see Chap. 9 by Bayman and Otero). Recently, the ability of a fungus grouping with *Cadophora finlandia* in the "*Hymenoscyphus ericae* aggregate" to simultaneously form both ectomycorrhizas in aseptic synthesis with *Pinus sylvestris* seedlings and ericoid mycorrhizas in *Vaccinium myrtillus* was demonstrated (Villarreal-Ruiz et al. 2004). This suggests that ericoid and ectomycorrhizal fungi may be part of a common guild in boreal and temperate woodland ecosystems (Vrålstad 2004). Of other fungal isolates from the same aggregate, only isolates of ericoid mycorrhizal origin formed classic ericoid mycorrhizal associations, while ectomycorrhiza formation was restricted to isolates of ectomycorrhizal origin. None of the tested isolates were able to form both kinds of mycorrhizal symbioses (Vrålstad et al. 2002b). Similarly, Sd2 and Sd9, the two morphotypes isolated from the *Q. ilex*/ *E. arborea* host pair in Mediterranean plant communities, were only able to form typical ericoid mycorrhiza in the latter host under the conditions tested, despite the fact that RAPD data for the Sd9 morphotype demonstrate the presence of the same genetic individual in both host plants (Bergero et al. 2000). Nonetheless, ectomycorrhizal behaviour on suitable hosts cannot be ruled out entirely, as suggested by high ITS sequence identity with an ectomycorrhizal symbiont of *Kobresia myosuroides*.

Whatever their specific structural association with the root, in nature the multiple association potential of DSE fungi may favour inter-plant interactions, which can in turn affect the diversity and dynamics of plant communities. Mycelia of DSE fungi might interlink roots of different hosts, which could translocate nutrients via the hyphae of such shared fungal associates, similar to what has been shown in situations involving mycorrhizal fungi (Simard et al. 1997a, 1997b). Experiments with isotope tracers are obviously required to confirm such a hypothesis. A second possibility, not necessarily implying physical integrity and continuity of mycelia colonising different hosts, is that the ectomycorrhizal host provides a "reservoir" for mycorrhizal infection of other plants. Such a role as an efficient source of mycorrhizal inoculum for newly establishing seedlings would be especially relevant in highly disturbed habitats such as occur in Mediterranean environments. Results of isolation experiments from mature *Q. ilex* woodland soil have established survival of ericoid mycorrhizal fungi in the absence of the ericoid plant, which could facilitate secondary successions.

Further experimentation is needed to verify the actual abilities of the Mediterranean DSE investigated thus far both in improving host growth and health and in forming typical mycorrhizal structures under natural conditions. In nature, these fungi face a variety of abiotic conditions as well as interactions with complex communities of microorganisms also interacting with the roots.

A different behaviour was highlighted in the Halep pine/rosemary investigation (Girlanda et al. 2002). Identification of a DSE morphotype as *Rhizopycnis vagum* was unexpected since this fungus had thus far only been known as a root pathogen involved in cucurbit crop "vine declines" under agricultural conditions. *R. vagum*-specific PCR primers designed towards the ITS region have been developed to assist disease diagnosis (Ghignone et al. 2003). Association of *R. vagum* with unrelated hosts such as cucurbit crops and wild garrigue tree and shrub plants appears to differ from situations in which crop pathogens are endophytic in weeds in affected fields (Sinclair and Cerkauskas 1996). Pathogenic association has also been reported between *Phialocephala fortinii* and some host plants, as revealed by resynthesis experiments under controlled conditions (Wilcox and Wang 1987b; Stoyke and Currah 1993; Fernando and Currah 1996). No disease, however, has so far been associated with DSE colonisation under natural conditions. Reports for *R. vagum* therefore raise several questions about the actual ecological plasticity of this fungus. Analyses of polymorphisms in single and multilocus genetic markers in pathogenic and endophytic isolates from cucurbit, *P. halepensis* and *R. officinalis* plants from Italy and Northern and Central America are currently underway to assess host-specific and geographical genetic variation within the fungus. Based on disease reaction in melon roots, the pathogenicity of endophytic *R. vagum* isolates from *P. halepensis* and *R. officinalis* was confirmed in greenhouses and growth chambers under different temperature regimes. Melon seedling inoculation demonstrated virulence on this host, establishing inherent pathogenic potential on a cucurbit plant, with at least some endophytic Italian isolates being more aggressive, under certain experimental conditions, than either of two pathogenic American isolates from diseased cucurbits (Fig. 12.2).

Pathogenic behaviour has also been reported for *Phomopsis / Diaporthe*, another DSE morphotype from Halep pine and rosemary. Isolates of these genera are among the most common of the avirulent endophytes of both the above-ground organs and the roots of many plants, including trees, while others are known as plant pathogens (although mostly from epigeous plant organs) with a widespread occurrence (Sutton 1980; Alexopoulos et al. 1996). Many fungal pathogens are often identified among the large numbers of fungi isolated from asymptomatic stems of woody hosts (Redlin and Carris 1996; Saikkonen et al. 1998). Disease symptoms were not apparent in the Halep pine and rosemary individuals sampled for DSE isolation. Endophytic fungi colonising asymptomatically plant parts may also include pathogens that have extended latency periods before development of disease, which occurs under conditions that induce host stress (Saikkonen et al. 1998, see Chap. 1 by Schulz and Boyle). Implicit in such a concept of latent pathogen is virulence, referring to the ability of the fungus to cause a disease in the particular hosts. Inoculation experiments in Petri dishes, containing mineral agar, of axenically cultured *P. halepensis* seedlings with DSE isolates identifiable as *Phomopsis / Diaporthe* did not result in disease symptoms, despite of experimental conditions not especially favourable to the plant, and the exposure to high fungal inoculum concentrations (unpublished data). Although the possibility of disease development in *P. halepensis* or *R. officinalis* under specific abiotic or biotic environmental conditions cannot be ruled out entirely, an alternative possibility is actual loss of virulence on these hosts and an inability to breach their defence barriers, while retaining some capacity of penetration into root tissues, which permits endophytic colonisation. Fungal endophytes of epigeous parts of both grasses and woody plants are closely related to pathogenic fungi, and are thought to have evolved from them via an extension of latency periods and a reduction of virulence (Schardl and Clay 1997; Saikkonen et al. 1998). DSE fungi could thus fit a definition of "saprotrophic symbionts", living in close association with their hosts but confined by saprotrophy to the dead host tissues, to diffusates or even refractory structural components of such tissues, or to exudates from living host tissues, or even to organic material made available by other root associates (Cooke and Rayner 1984; Cooke and Whipps 1987).

The results described here suggest that the outcome of the interaction between the same DSE fungus and different plant hosts, determining a conventional mycorrhizal, endophytic or even pathogenic association, may result from differences in fungal gene expression in response to the plant or differences in the ability of a plant to respond to the fungus, as has been suggested for other plant-fungus associations (Redman et al. 2001). This would fit well to a model of a finely tuned equilibrium between fungal virulence and plant defence characterising the fungal endophyte-plant host interaction, which could be described as a "balanced antagonism" (Schulz et al. 1999, 2002; see Chap. 1 by Schulz and Boyle).

12.4 Conclusions

Molecular studies based on rDNA sequences are beginning to unravel the heterogeneity of the assemblage of DSE endophytes associated with different plants in different ecosystems. Analyses of sequence data from isolates from Mediterranean environments in Italy have widened the spectrum of DSE taxa known to associate with ecto- and endo-mycorrhizal hosts, pointing to broad taxonomic boundaries of these endophytes. Sequencing of rDNA coding regions has indicated affiliation to distant taxa within Ascomycota (distinct subclasses); however, taxonomic placement of these fungi has been achieved with varying degrees of resolution. ITS-based species-level identification of sterile fungi remains elusive in most cases, either because of matches with sequences from other unidentified fungi, or because of poor matches with all available sequences in the EMBL/GenBank/DDBJ databases. In spite of the daily increase in fungal sequences available in public databases, less than 1% of the estimated 1.5 million fungal species are represented in public databases (Vilgalys 2003), with large gaping holes for most fungal groups – Ascomycota included, with many families and even orders having no sequenced members to date. Misidentifications of named published sequences (upwards of 20% of the named sequences may be attributed to incorrectly named organisms;

Vilgalys 2003), and hence their unreliability, may represent another problem restricting feasibility of sequence-based identifications (see e.g. Bridge et al. 2003, 2004; Hawksworth 2004).

Since intraspecific variation in the ITS region may attain ca. 15% in some fungal taxa (Egger and Sigler 1992; Seifert et al. 1995), conspecificity may be supposed with confidence only in cases where ITS sequence identity is quite high. For this reason, the amount of nucleotide divergence in single loci cannot in itself be used to define taxonomic rank, and phylogenetic species recognition relying on concordance between several independent gene genealogies (Taylor et al. 2000) may be a more valuable diagnostic tool. However, fungal non-ITS sequences of systematic value at the species level are scarce in databases.

Regardless of their precise identity, however, database searches for DSE ITS sequences indicate that while some DSE fungi are possibly restricted to specific environments, in other cases the same fungus, or very closely related taxa, may occur in different biomes and unrelated hosts. The possibility that taxonomically diverse fungi have evolved not only different patterns of biome and host specificity, but also diverse functional significance for the plant is not unexpected. Data from inoculation experiments under semicontrolled conditions suggest that the range of DSE associates may vary from fungi with conventional mycorrhizal potential to fungi with pathogenic attributes. Findings for Mediterranean DSE isolates adds credence to the view of a possible continuum of mycorrhizal, endophytic or pathogenic root associations, depending on phylogenetic and life history constraints, geography, interactions with other species in the community and prevailing abiotic factors (Saikkonen et al. 1998; Brundrett 2002; see Chap. 16 by Brundrett). Assessing the ecological significance of information derived from axenic culture work with isolated DSE and the influences of plant and fungal genotypes, as well as local abiotic and biotic environments, on endophyte-host interactions under natural conditions is a major challenge for the future.

Acknowledgements. Stefano Ghignone, Roberta Bergero, Cristina Mariani and Giacomo Tamietti contributed to the experimental work described in this chapter. Grants were provided by IPP-CNR Torino, UNITO 60%, CEBIOVEM.

References

Addy HD, Hambleton S, Currah RS (2001) Distribution and molecular characterisation of the root endophyte *Phialocephala fortinii* along an environmental gradient in the boreal forest of Alberta. Mycol Res 104:1213–1221

Aegerter BJ, Gordon TR, Davis RM (2000) Occurrence and pathogenicity of fungi associated with melon root rot and vine decline in California. Plant Dis 84:224–230

Ahlich K, Sieber TN (1996) The profusion of dark septate endophytic fungi in non-ectomycorrhizal fine roots of forest trees and shrubs. New Phytol 132:259–270

Ahlich K, Rigling D, Holdenrieder O, Sieber TN (1998) Dark septate hyphomycetes in Swiss conifer forest soils surveyed using Norway-spruce seedlings as bait. Soil Biol Biochem 30:1069–1075

Alcorn JL, Irwin JAG (1987) *Acrocalymma medicaginis* gen. et sp. nov. causing root and crown rot of *Medicago sativa* in Australia. Trans Br Mycol Soc 88:163–167

Alexopoulos CJ, Mims CW, Blackwell M (1996) Introductory mycology, 4th edn. Wiley, New York

Allen MF (1991) The ecology of mycorrhizae. Cambridge University Press, Cambridge

Allen TR, Millar T, Berch S, Berbee ML (2003) Culturing and direct DNA extraction find different fungi from the same ericoid mycorrhizal roots. New Phytol 160:255–272

Armengol J, Pellicer I, Vicent A, Bruton BD, García-Jiménez J (2000) *Rhizopycnis vagum* DF Farr, un nuevo Coelomycete asociato a raíces de planta de melón con síntomas de colapso en España. Boletín de Sanidad Vegetal Plagas (1° Trimestre) 26:103–112

Armengol J, Vicent A, Martínez-Culebras P, Bruton BD, García-Jiménez J (2003) Identification, occurrence and pathogenicity of *Rhizopycnis vagum* on muskmelon in Spain. Plant Pathol 52:68–73

Bååth E, Söderström B (1980) Degradation of macromolecules by microfungi from different podzolic soil horizons. Can J Bot 58:422–425

Barrow JR (2003) Atypical morphology of dark septate fungal root endophytes of *Bouteloua* in arid southwestern USA rangelands. Mycorrhiza 13:239–247

Barrow JR, Aaltonen RE (2001) A method of evaluating internal colonisation of *Atriplex canescens* (Pursh) Nutt. roots by dark septate fungi and how they are influenced by physiological activity. Mycorrhiza 11:199–205

Barrow JR, Osuna P (2002) Phosphorus solubilization and uptake by dark septate fungi in fourwing saltbush, *Atriplex canescens* (Pursh) Nutt. J Arid Environ 51:449–459

Barrow JR, Havstad KM, McCaslin BD (1997) Fungal root endophytes in fourwing saltbush, *Atriplex canescens*, on arid rangelands of southwestern USA. Arid Soil Res Rehabil 11:177–185

Bergero R, Perotto S, Girlanda M, Vidano G, Luppi AM (2000) Ericoid mycorrhizal fungi are common root associates of a Mediterranean ectomycorrhizal plant (*Quercus ilex*). Mol Ecol 9:1639–1649

Bergero R, Girlanda M, Bello F, Luppi AM, Perotto S (2003) Soil persistence and biodiversity of ericoid mycorrhizal fungi in the absence of the host plant in a Mediterranean ecosystem. Mycorrhiza 13:69–75

Bridge PD, Roberts PJ, Spooner BM, Panchal G (2003) On the unreliability of published DNA sequences. New Phytol 160:43–48

Bridge PD, Spooner BM, Roberts PJ (2004) Reliability and use of published sequence data. New Phytol 161:15–17

Brundrett MC (2002) Tansley review no. 134. Coevolution of roots and mycorrhizas of land plants. New Phytol 154:275–304

Bruton BD, Miller ME (1997a) Occurrence of vine decline diseases of muskmelon in Guatemala. Plant Dis 81:694

Bruton BD, Miller ME (1997b) Occurrence of vine decline diseases of melons in Honduras. Plant Dis 81:696

Bruton BD, Miller ME (1997c) The impact of vine declines on muskmelon production in Central America. Phytopathology 88:S120

Cairney JWG, Ashford AE (2002) Tansley review no. 135. Biology of mycorrhizal associations of epacrids (Ericaceae). New Phytol 154:275–304

Caldwell BA, Jumpponen A, Trappe JM (2000) Utilization of major detrital substrates by dark-septate root endophytes. Mycologia 92:230–232

Chambers SM, Liu G, Cairney JWG (2000) ITS rDNA sequence comparison of ericoid mycorrhizal endophytes from *Woollsia pungens*. Mycol Res 104:168–174

Cooke RC, Rayner ADM (1984) Ecology of saprotrophic fungi. Longman, London

Cooke RC, Whipps JM (1987) Saprotrophy, stress and symbiosis. In: Rayner ADM, Brasier CM, Moore D (eds) Evolutionary biology of the fungi. Cambridge University Press, Cambridge, pp 137–148

Crayn DM, Quinn CJ (2000) The evolution of the *atpβ-rbc*L intergenic spacer in the epacrids (Ericales) and its systematic and evolutionary implications. Mol Phylogenet Evol 16:238–252

Currah RS, Tsuneda A (1993) Vegetative and reproductive morphology of *Phialocephala fortinii* (Hyphomycetes, *Mycelium radicis atrovirens*) in culture. Trans Mycol Soc Japan 34:345–356

Currah RS, Sigler L, Hambleton S (1987) New records and new taxa of fungi from the mycorrhizae of terrestrial orchids of Alberta. Can J Bot 65:2473–2482

Dahlberg A, Jonsson L, Nylund J-E (1997) Species diversity and distribution of biomass above and below ground among ectomycorrhizal fungi in an old-growth Norway spruce forest in south Sweden. Can J Bot 75:1323–1335

Deacon JC (1973) *Phialophora radicicola* and *Gaeumannomyces graminis* on roots of grasses and cereals. Trans Br Mycol Soc 61:471–485

Di Castri F, Mooney H (1973) Mediterranean type ecosystems. Springer, New York Berlin Heidelberg

Egger KN, Sigler L (1992) Using molecular data to infer anamorph-teleomorph connections: the case of *Scytalidium vaccinii* and *Hymenoscyphus ericae*. In: Reynolds DR, Taylor JW (eds) The fungal holomorph: mitotic, meiotic and pleomorphic speciation in fungal systematics. CAB International, Wallingford, pp 141–146

Farr DF, Miller ME, Bruton BD (1998) *Rhizopycnis vagum* gen et sp nov, a new coelomycetous fungus from roots of melons and sugarcane. Mycologia 90:290–296

Fernando AA, Currah RS (1995) *Leptodontidium orchidicola* (*Mycelium Radicis Atrovirens* complex): aspects of its conidiogenesis and ecology. Mycotaxon 54:287–294

Fernando AA, Currah RS (1996) A comparative study of the effects of the root endophytes *Leptodontidium orchidicola* and *Phialocephala fortinii* (Fungi Imperfecti) on the growth of some subalpine plants in culture. Can J Bot 74:1071–1078

Gams W (1963) *Mycelium radicis atrovirens* in forest soils, isolation from soil microhabitats and identification. In: Doeksen J, van der Drift J (eds) Soil organisms. North-Holland Publications, Amsterdam, pp 176–182

Garrett SD (1950) Ecology of the root-inhabiting fungi. Biological Reviews 25:220–254

Garrett SD (1956) Biology of root-infecting fungi. Cambridge University Press, Cambridge

Ghignone S, Tamietti G, Girlanda M (2003) Development of specific PCR primers for identification and detection of *Rhizopycnis vagum*. Eur J Plant Pathol 109:861–870

Girlanda M, Luppi-Mosca AM (1995) Microfungi associated with ectomycorrhizae of *Pinus halepensis* Mill. Allionia 33:93–98

Girlanda M, Ghignone S, Luppi AM (2002) Diversity of root-associated fungi of two Mediterranean plants. New Phytol 155:481–498

Grünig CR, Sieber TN, Holdenrieder O (2001) Characterisation of dark septate endophytic fungi (DSE) using inter-simple-sequence-repeat-anchored polymerase chain reaction (ISSR-PCR) amplification. Mycol Res 105:24–32

Grünig CR, Sieber TN, Rogers SO, Holdenrieder O (2002a) Genetic variability among strains of *Phialocephala fortinii* and phylogenetic analysis of the genus *Phialocephala* based on rDNA ITS sequence comparisons. Can J Bot 80:1239–1249

Grünig CR, Sieber TN, Rogers SO, Holdenrieder O (2002b) Spatial distribution of dark septate endophytes in a confined forest plot. Mycol Res 106:832–840

Gwinne BJ, Davis RM, Gordon TR (1997) Occurrence and pathogenicity of fungi associated with melon vine decline in California. Phytopathology 87:S37

Hambleton S, Currah RS (1997) Fungal endophytes from the roots of alpine and boreal Ericaceae. Can J Bot 75:1570–1581

Harney SK, Rogers SO, Wang CJK (1997) Molecular characterisation of dematiaceous root endophytes. Mycol Res 101:1397–1404

Harrington TS, McNew DL (2003) Phylogenetic analysis places the *Phialophora*-like anamorph genus *Cadophora* in the Helotiales. Mycotaxon 87:141–152

Haselwandter K (1983) Pectic enzymes produced by fungal root associates of alpine plants. Phyton 28:55–64

Haselwandter K, Read DJ (1982) The significance of root-fungus associations in two *Carex* species of high-alpine plant communities. Oecologia 53:352–354

Hawksworth DL (2004) 'Misidentifications' in fungal DNA sequence databanks. New Phytol 161:13–15

Holdenrieder O, Sieber TN (1992) Fungal associations of serially washed, healthy, non-mycorrhizal roots of *Picea abies*. Mycol Res 96:151–156

Jumpponen A (1999) Spatial distribution of discrete RAPD phenotypes of a root endophytic fungus, *Phialocephala fortinii*, at a primary successional site on a glacier forefront. New Phytol 141:333–344

Jumpponen A (2001) Dark septate endophytes – are they mycorrhizal? Mycorrhiza 11:207–211

Jumpponen A, Trappe JM (1998a) Dark septate endophytes: a review of facultative biotropic root-colonising fungi. New Phytol 140:295–310

Jumpponen A, Trappe JM (1998b) Performance of *Pinus contorta* inoculated with two strains of root endophytic fungus *Phialocephala fortinii*: effects of resynthesis system and glucose concentration. Can J Bot 76:1205–1213

Jumpponen A, Mattson KG, Trappe JM (1998) Mycorrhizal functioning of *Phialocephala fortinii*: interactions with soil nitrogen and organic matter. Mycorrhiza 7:261–265

Kirk PM, Cannon PF, David JC, Stalpers JA (2001) Ainsworth and Bisby's dictionary of the fungi, 9th edn. CAB International, Wallingford

Kohn L, Stasovski E (1990) The mycorrhizal status of plants at Alexandra Fiord, Ellesmere Island, Canada, a high Arctic site. Mycologia 82:23–35

Kovacs GM, Szigetvari C (2002) Mycorrhizae and other root-associated fungal structures of the plants of a sandy grassland on the Great Hungarian Plain. Phyton-Annu Rev Bot A 42:211–223

LoBuglio KF, Berbee ML, Taylor JW (1996) Phylogenetic origins of the asexual mycorrhizal symbiont *Cenococcum geophilum* Fr. and other mycorrhizal fungi among the ascomycetes. Mol Phylogenet Evol 6:287–294

McLean CB, Cunnington JH, Lawrie AC (1999) Molecular diversity within and between ericoid endophytes from the Ericaceae and Epacridaceae. New Phytol 144:351–358

Melin E (1922) On the mycorrhizas of *Pinus sylvestris* L. and *Picea abies* Karst. A preliminary note. J Ecol 9:254–257

Melin E (1923) Experimentelle Untersuchungen über die Konstitution und Okologie der Mykorrhizen von *Pinus silvestris* und *Picea abies*. Mykologische Untersuchungen und Berichte von R. Falck 2:73–331

Merteley JC, Martyn RD, Miller ME, Bruton BD (1991) Role of *Monosporascus cannonballus* and other fungi in a root rot/vine decline of muskmelon. Plant Dis 75:1133–1137

Miller ME, Bruton BD, Farr DF (1996) Association of a *Stagonospora*-like fungus on roots of melons exhibiting vine decline symptoms. Phytopathology 86:S3

Monreal M, Berch SM, Berbee M (1999) Molecular diversity of ericoid mycorrhizal fungi. Can J Bot 77:1580–1594

Montuschi C (2001) Collasso delle cucurbitacee: una malattia da indagare. Agricoltura 30:31–33

Newsham KK (1999) *Phialophora graminicola*, a dark septate fungus, is a beneficial associate of the grass *Vulpia ciliata* ssp. *ambigua*. New Phytol 144:517–524

O'Dell TE, Massicotte HB, Trappe JM (1993) Root colonisation of *Lupinus latifolius* Agardh. and *Pinus contorta* Dougl. by *Phialocephala fortinii* Wang & Wilcox. New Phytol 124:93–100

Peyronel B (1924) Prime ricerche sulla micorize endotrofiche e sulla microflora radicola normale delle fanerogame. Rivista di Biologia 6:17–53

Pignatti E, Pignatti S, Ladd PG (2002) Comparison of ecosystems in the Mediterranean Basin and Western Australia. Plant Ecol 163:177–186

Porta-Puglia A, Pucci N, Di Gianbattista G, Infantino A (2001) First report of *Rhizopycnis vagum* associated with tomato roots in Italy. Plant Dis 85:1210

Read DJ (1991) Experimental simplicity versus natural complexity in mycorrhizal systems. In: Fontana A (ed) Funghi, piante e suolo. Consiglio Nazionale delle Ricerche, Turin, pp 75–104

Read DJ (2000) Links between genetic and functional diversity – a bridge too far? New Phytol 145:363–365

Read DJ, Haselwandter K (1981) Observations on the mycorrhizal status of some alpine plant communities. New Phytol 88:341–352

Redlin SC, Carris LM (1996) Endophytic fungi in grasses and woody plants. APS Press, St Paul, MN

Redman RS, Dunigan DD, Rodriguez RJ (2001) Fungal symbiosis from mutualism to parasitism: who controls the outcome, host or invader? New Phytol 151:705–7161

Richard C, Fortin JA (1973) The identification of *Mycelium radicis atrovirens*. Can J Bot 51:2247–2248

Ruotsalainen AL, Vare H, Vestberg M (2002) Seasonality of root fungal colonisation in low-alpine herbs. Mycorrhiza 12:29–36

Saikkonen K, Faeth SH, Helander M, Sullivan TJ (1998) Fungal endophytes: a continuum of interactions with host plants. Annu Rev Ecol Syst 29:319–343

Schadt CW, Mullen RB, Schmidt SK (2001) Isolation and phylogenetic identification of a dark-septate fungus associated with the alpine plant *Ranunculus adoneus*. New Phytol 150:747–755

Schardl CL, Clay K (1997) Evolution of mutualistic endophytes from plant pathogens. In: Carroll GC, Tudzynski P (eds) The Mycota V: Plant relationships, part B. Springer, Berlin Heidelberg New York, pp 221–238

Schild DE, Kennedy A, Stuart MR (1988) Isolation of symbiont and associated fungi from ectomycorrhizas of Sitka spruce. Eur J Forest Pathol 18:51–61

Schulz B, Römmert AK, Dammann U, Aust HJ, Strack D (1999) The endophyte-host interaction: a balanced antagonism? Mycol Res 103:1275–1283

Schulz B, Boyle C, Draeger S, Römmert AK, Krohn K (2002) Endophytic fungi: a source of novel biologically active secondary metabolites. Mycol Res 106:996–1004

Seifert KA, Wingfield BD, Wingfield MJ (1995) A critique of DNA sequence analysis in the taxonomy of filamentous Ascomycetes and ascomycetous anamorphs. Can J Bot 73:S760–S767

Sengupta A, Chakraborty DC, Chaudhuri S (1989) Do septate endophytes also have a mycorrhizal function for plants under stress? In: Mahadevan A, Raman N, Natarajan K (eds) Mycorrhizae for Green Asia. Proceedings of the 1st Asian Conference on Mycorrhizae, Nadras, India, pp 169–174

Sharples JM, Chambers SM, Meharg AA, Cairney JWG (2000) Genetic diversity of root-associated fungal endophytes from *Calluna vulgaris* at contrasting field sites. New Phytol 148:153–162

Shoemaker RA, Babcock CE, Irwin JAG (1990) *Massarina walkeri* n. sp., the teleomorph of *Acrocalymma medicaginis* from *Medicago sativa* contrasted with *Leptosphaeria pratensis, L. weimeri* n. sp. & *L. viridella*. Can J Bot 69:569–573

Simard SW, Jones MD, Durall DM, Perry DA, Myrold DD, Molina R (1997a) Reciprocal transfer of carbon isotopes between ectomycorrhizal *Betula papyrifera* and *Pseudotsuga menziesii* New Phytol 137:529–542

Simard SW, Perry DA, Jones MD, Myrold DD, Durall DM, Molina R (1997b) Net transfer of carbon between ectomycorrhizal tree species in the field. Nature 388:579–582

Sinclair JB, Cerkauskas RF (1996) Latent infection vs endophytic colonisation by fungi. In: Redlin SC, Carris LM (eds) Endophytic fungi in grasses and woody plants. APS Press, St Paul, MN, pp 3–29

Stoyke G, Currah RS (1991) Endophytic fungi from the mycorrhizae of alpine ericoid plants. Can J Bot 69:347–352

Stoyke G, Currah RS (1993) Resynthesis in pure culture of a common sub-alpine fungus-root association using *Phialocephala fortinii* and *Menziesia ferruginea* (Ericaceae). Arctic Alpine Res 25:189–193

Stoyke G, Egger KN, Currah RS (1992) Characterisation of sterile endophytic fungi from the mycorrhizae of subalpine plants. Can J Bot 70:2009–2016

Summerbell RC (1989) Microfungi associated with the mycorrhizal mantle and adjacent microhabitats within the rhizosphere of black spruce. Can J Bot 67:1085–1095

Sutton BC (1980) The Coelomycetes – Fungi Imperfecti with pycnidia, acervuli and stromata. Commonwealth Mycological Institute, Kew

Taylor JW, Jacobson DJ, Kroken S, Kasuga T, Geiser DM, Hibbett DS, Fisher MC (2000) Phylogenetic species recognition and species concepts in fungi. Fungal Genet Biol 31:21–32

Taylor DL, Bruns TD, Leake JR, Read DJ (2002) Mycorrhizal specificity and function in myco-heterotrophic plants. In: Van der Heijden MGA, Sanders I (eds) Mycorrhizal ecology. Springer, Berlin Heidelberg New York, pp 375–413

Ursic M, Peterson RL (1997) Morphological and anatomical characterisation of ectomycorrhizas and ectendomycorrhizas on *Pinus strobus* seedlings in a southern Ontario nursery. Can J Bot 75:2057–2072

Vandenkoornhuyse P, Baldauf SL, Leyval C, Straczek J, Young JPW (2002) Extensive fungal diversity in plant roots. Science 295:2051

Vilgalys R (2003) Taxonomic misidentification in public DNA databases. New Phytol 160:4–5

Villarreal-Ruiz L, Anderson IC, Alexander IJ (2004) Interaction between an isolate from the *Hymenoscyphus ericae* aggregate and roots of *Pinus* and *Vaccinium*. New Phytol 164:183–192

Vrålstad T (2004) Are ericoid and ectomycorrhizal fungi part of a common guild? New Phytol 164:7–10

Vrålstad T, Fossheim T, Schumacher T (2000) *Piceirhiza bicolorata* – the ectomycorrhizal expression of the *Hymenoscyphus ericae* aggregate? New Phytol 145:549–563

Vrålstad T, Myhre E, Schumacher T (2002a) Molecular diversity and phylogenetic affinities of symbiotic root-associated ascomycetes of the Helotiales in burnt and metal polluted habitats. New Phytol 155:131–148

Vrålstad T, Schumacher T, Taylor AFS (2002b) Mycorrhizal synthesis between fungal strains of the *Hymenoscyphus ericae* aggregate and potential ectomycorrhizal and ericoid hosts. New Phytol 153:143–152

Wang CJK, Wilcox HE (1985) New species of ectendomycorrhizal and pseudomycorrhizal fungi: *Phialophora finlandia*, *Chloridium paucisporum*, and *Phialocephala fortinii*. Mycologia 77:951–958

Wilcox HE, Wang CJK (1987a) Ectomycorrhizal and ectendomycorrhizal associations of *Phialophora finlandia* with *Pinus resinosa*, *Picea rubens*, and *Betula alleghaniensis*. Can J For Res 17:976–990

Wilcox HE, Wang CJK (1987b) Mycorrhizal and pathological associations of dematiaceous fungi in roots of 7-month-old tree seedlings. Can J For Res 17:884–899

Yu TEJ-C, Egger KN, Peterson RL (2001) Ectendomycorrhizal associations – characteristics and functions. Mycorrhiza 11:167–177

13 *Oidiodendron maius*: Saprobe in Sphagnum Peat, Mutualist in Ericaceous Roots?

Adrianne V. Rice, Randolph S. Currah

13.1 Introduction

Oidiodendron maius Barron is a hyphomycete species isolated from peat, soil, decaying organic matter, and plant roots throughout temperate ecosystems, including peatlands, forests, and heathlands (e.g. Barron 1962; Nordgren et al. 1985; Schild et al. 1988; Nilsson et al. 1992; Hambleton et al. 1998; Qian et al. 1998; Lumley et al. 2001; Thormann 2001; Thormann et al. 2001, 2004; Tsuneda et al. 2001; Rice and Currah 2002; Rice et al. 2006). In pure culture, colonies are white due to the presence of abundant arthroconidia that develop in chains at the apex of thick-walled, melanized erect conidiophores 30–500 μm tall (Fig. 13.1a). Conidia are thin-walled, subglobose to elongate or irregular, 2-5x1–2.5 μm, and have an asperulate perispore (Rice and Currah 2001) (Fig. 13.1b).

A sexual state is unknown but morphological characters indicate a close affiliation to other taxa in the Myxotrichaceae (a cleistothecial family in the Helotiales; Tsuneda and Currah 2004): six teleomorph species within the Myxotrichaceae have *Oidiodendron* states (Hambleton et al. 1998; Rice and Currah 2005) and sterile ascomata with peridial elements resembling those formed by species of *Myxotrichum* can be induced when the species is grown on autoclaved lichen (Rice and Currah 2002) (Fig. 13.1c). Molecular evidence confirms the position of *O. maius* among other species of *Oidiodendron* and their teleomorphic counterpart, *Myxotrichum* (Hambleton et al. 1998).

The distribution of *O. maius* seems to parallel that of members of the Ericaceae (blueberries, cranberries, rhododendrons, etc.), a family that often dominates the vegetation in arctic and alpine meadows, temperate heathlands, the understory in boreal forests and peatlands (Hambleton 1998; Chambers et al. 2000; Hambleton and Currah 2000), and Mediterranean ecosystems (Perotto et al. 1995). This shared distribution pattern may be

Adrianne V. Rice: Northern Forestry Centre, Canadian Forest Service, Natural Resources Canada, 5320-122 St., Edmonton, AB, T6H 3S5 Canada, E-mail: ARice@NRCan.gc.ca

Randolph S. Currah: Department of Biological Sciences, University of Alberta, Edmonton, AB, T6G 2E9, Canada

Soil Biology, Volume 9
Microbial Root Endophytes
B. Schulz, C. Boyle, T. N. Sieber (Eds.)

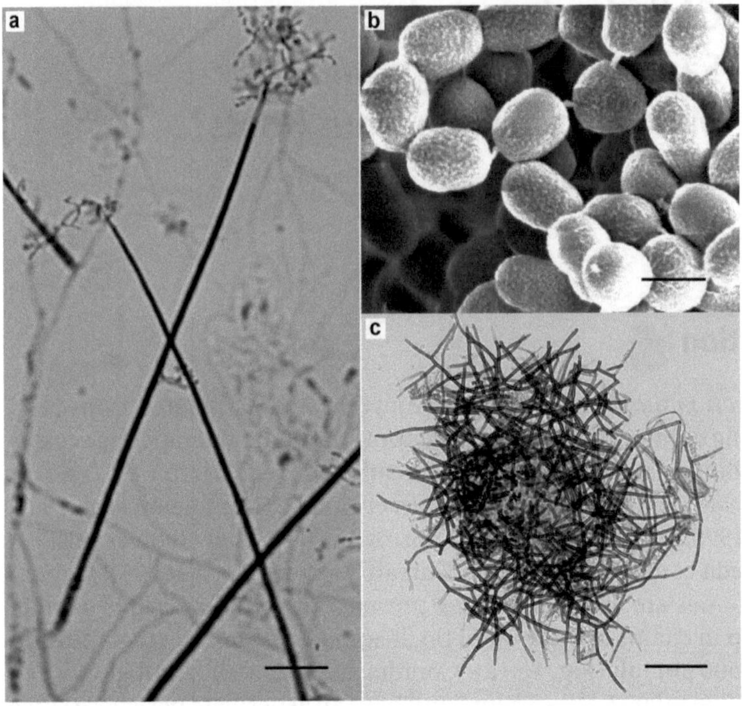

Fig. 13.1. a–c Morphology of *Oidiodendron maius* in axenic culture. **a** Tall, erect, melanized conidiophores and small, hyaline arthroconidia of *O. maius* (UAMH 9749) viewed under light microscopy. **b** Chains of arthroconidia of *O. maius* (UAMH 8920), showing asperulate ornamentation visible under scanning electron microscopy. **c** Sterile peridial elements produced by *O. maius* (UAMH 9749) on thalli of *Cladonia mitis*. The cage-like peridial elements resemble the cleistothecia of species of *Myxotrichum*. *Bars* **a** 40 μm, **b** 5 μm, **c** 80 μm

based on a predilection in both plants and fungus for acidic, nutrient-poor, organic soils; it is also possible that *O. maius* has some degree of dependency on a mycorrhizal or at least root-endophytic relationship with ericaceous plants.

Thus, two hypotheses can be advanced to explain the distribution of *O. maius*. The first hypothesis, that it is a competitive and effective saprobe on acidic, organic soils, is supported by its optimal growth in culture on acidic growth media (Rice and Currah 2001, 2005), its ability to grow and sporulate readily on *Sphagnum* plants (Rice and Currah 2002), and its ability to produce enzymes that degrade the cell walls of *Sphagnum* leaves in vitro (Tsuneda et al. 2001; Rice et al. 2006). The second hypothesis, that *O. maius* is a mycorrhizal endophyte of ericaceous shrubs, is supported by its frequent isolation from ericaceous roots (e.g. Douglas et al. 1989; Perotto et al. 1995, 1996; Hambleton and Currah 1997; Currah et al. 1999; Monreal et al. 1999; Chambers et al. 2000; Johansson 2001; Usuki et al. 2003) and by observations that it can form typical ericoid mycorrhizal infection units when reinoculated on Ericaceae grown in culture (e.g. Douglas et al. 1989; Xiao and Berch 1995, 1999; Monreal et al. 1999).

Since Douglas et al. (1989) described the ericoid mycorrhizas formed by *O. maius* in *Rhododendron,* most reports of *O. maius* have been from ericaceous plants. Its role in these associations has been considered one in which the plant derives some nutritional benefit (e.g. Perotto et al. 1995, 1996; Hambleton and Currah 1997, 2000; Johanson 2001). Oddly, the benefits accruing to the fungus in these relationships are rarely considered, possibly because they are considered secondary to the needs of the plant but perhaps also because they are difficult to determine (Douglas and Smith 1989). Unlike arbuscular mycorrhizal fungi and many nutritionally fastidious ectomycorrhizal basidiomycetes that are difficult to grow in culture or which require the addition of complex compounds in artificial media, *O. maius* does not display stringent demands for host-derived sugars and or complex growth factors, and grows readily in culture on many types of natural materials, including lichen thalli, *Sphagnum* gametophytes, context tissues of polypores, and on artificial growth media. In the absence of nutritional dependency, the benefits to the fungus in a mycorrhizal or root endophytic relationship are usually speculative and a number of possibilities could be considered. For example, the relationship may provide the fungus with a carbon source, growth factors, habitat, a preemptive position as a consumer of senescent tissue, or a competitive advantage over other saprobes. Alternatively, the fungus may not benefit from the association, with host plants exploiting their fungal partners, as in orchid mycorrhizas (Rasmussen 1995). In summary, there are two possible explanations for the distribution of *O. maius*: i.e. it may be a saprobe adapted to acidic conditions and enzymatically equipped to digest the intractable materials that accumulate in these areas, or it may be a type of root endophyte, possibly a mycorrhizal one, that requires its host plants to thrive in its habitats. These two suggested ecological roles are not necessarily mutually exclusive, i.e. the species could occupy both mycorrhizal and saprobic niches within suitable ecosystems.

One type of ecosystem that may be supporting *O. maius* both as a saprobe and a mycorrhizal associate is acidic *Sphagnum* peatlands, found throughout the circumboreal region. These peatlands include bogs and fens with an understory of ericaceous shrubs, including *Rhododendron, Andromeda,* and *Vaccinium* species, and a thick ground layer of *Sphagnum* spp. (e.g. Svensson 1995; Vitt et al. 1996; Hoosbeek et al. 2001). In Canada, many peatlands have a canopy of coniferous trees rooted in the *Sphagnum* (Vitt et al. 1996; Piercey et al. 2002). Bogs have a dense canopy of black spruce (*Picea mariana*) while poor fens are dominated by black spruce and larch (*Larix* spp.) (Vitt et al. 1996). European peatlands tend to have open canopies with few trees but, as in Canada, the common tree species in European peatlands are spruce (*Picea abies*) and larch (e.g. Peteet et al. 1998). Ericoid mycorrhizal fungi, including *O. maius*, have been isolated from ectomycorrhizas of conifers growing in such peatlands (Summerbell 1987; Schild et al. 1988; Perotto et al. 1995; Qian et al. 1998; Vrålstad et al. 2000); in some cases, *O. maius* was the most abundant sporulating species isolated from the roots (e.g. roots of sitka spruce sampled from blanket bogs; Schild et al. 1988). It has been proposed that these fungi may form associations with the roots of conifers and ericaceous shrubs in peatlands, and degrade the *Sphagnum* matrix (Piercey et al. 2002).

In this chapter, we first review the evidence suggesting that *O. maius* is a saprobe and then the evidence that it is an ericoid mycorrhizal fungus. Finally, we discuss why isolation records of this Helotialean anamorph point towards its simultaneous occupation of two apparently distinct niches.

13.2 *Oidiodendron maius* as a Saprobe

From 1962 to 1989, *Oidiodendron maius* was known only from scattered records from soils and other decaying organic debris (Barron 1962; Nordgren et al. 1985), where it was presumed to occupy a saprobic niche, and from the ectomycorrhizal root tips of sitka spruce (Schild et al. 1988). Although Schild et al. (1988) considered *O. maius* a cortical parasite of spruce, evidence of parasitism was not presented; instead, the evidence suggested that *O. maius* inhibited root pathogens, including *Phytophthora cinnamomi* and *Heterobasidium annosum*. The inhibitory activity of *O. maius* towards root pathogens was also suggested by Qian et al. (1998), who found that *O. maius* was a dominant inhabitant of the ectomycorrhizal root tips of Norway spruce under acidified conditions.

The first report of *O. maius* from the roots of ericaceous shrubs appeared in 1989, when it was isolated from ericoid mycorrhizal roots of *Rhododendron* (Douglas et al. 1989). Since then most records of *O. maius* have been based on isolates from presumably healthy ericaceous roots (e.g. Hambleton and Currah 1997; Currah et al. 1999; Monreal et al. 1999; Chambers et al. 2000; Usuki et al. 2003), but records from other substrates continue to appear (Nilsson et al. 1992; Qian et al. 1998; Lumley et al. 2001; Thormann et al. 2001, 2004; Rice and Currah 2002; Rice et al. 2006). The fungus is relatively easy to isolate from roots because it grows rapidly on artificial media, showing increases in colony radius of up to 1 mm/day on cornmeal agar (CMA) at pH 3 during periods of maximal growth (Rice and Currah 2001). The fungus has been shown to degrade a variety of carbon and nitrogen sources including tannic acid (a soluble phenolic compound), cellulose, starch (Rice and Currah 2001, 2005; Thormann et al. 2002), chitin, pectin (Rice and Currah 2001, 2005), and TWEEN 20, a lipid-based detergent (Rice and Currah 2005).

The presence of chitinases suggests that *O. maius* may obtain nutrients (e.g. nitrogen) from polymeric glucosamines found in insects and fungi. *Oidiodendron maius* grows luxuriantly on lichen (Rice and Currah 2002) and on the context tissue of the basidiocarps (polypores) of larger wood decay fungi (e.g. *Fomitopsis pinicola*). *Oidiodendron maius* also grows and sporulates abundantly on a lipid-rich growth medium with TWEEN 20. There are no data suggesting that *O. maius* is a fungal or arthropod parasite, but it could be a saprobe on materials rich in lipids and chitin, such as the remains of fungi and microfauna.

This suite of cultural characteristics is more indicative of a saprobic lifestyle rather than a mycorrhizal one that relies on biotrophically derived photosynthates (Hutchison 1990, 1991). Alternatively, because ecto- and ericoid-mycorrhizal fungi have also been shown to degrade a variety of organic substrates (e.g. Bajwa and Read 1985; Northup et al. 1995; Bending and Read 1996, 1997; Aerts 2002; Leake et al. 2002; Olsson et al. 2002; Simard et al. 2002), these abilities may enable the absorption and transfer of organic or non-mineralised nutrients directly from the substrate to the cytoplasm of a host plant, effectively short-circuiting the mineralisation steps in nutrient cycling ('organicization') (Northup et al. 1995; Aerts 2002; Leake et al. 2002). In this instance, it is the host that derives benefit, and reciprocity for the fungus is not evident.

The ability to sporulate in pure culture is much less common for mycorrhizal fungi than saprobes, with all arbuscular and most ectomycorrhizal fungi unable to reproduce in the absence of their hosts (Read 2002). Endorhizal fungi allied to the Helotiales are notable exceptions (Addy et al. 2005). For example, *Rhizoscyphus ericae*, an ericoid mycorrhizal fungus that also forms ectomycorrhizas (Vrålstad et al. 2000, 2002), produces chains of arthroconidia in culture, and in rare instances has formed discocarps (Hambleton et al. 1999). Unlike *O. maius*, *R. ericae* is unknown from non-mycorrhizal sources, although this may be due to its variable cultural morphology, and the concomitant difficulties in making definitive identification of this fungus (Hambleton and Currah 1997; Hambleton and Sigler 2005).

As with other microfungi, assumptions about distribution and habitat preferences are biased by isolation protocols, expertise in identification, and by the nature and objective of the reports in which taxa are listed. Nevertheless, peat is a likely substrate on which to find *O. maius* because it is acidic and rich in many of the organic compounds, including tannic acid, cellulose, pectin, and chitin, that *O. maius* is able to degrade. The first record of *O. maius* was from "peat soil" (Barron 1962) and subsequent reports of this species from peat (e.g. Nilsson et al. 1992; Thormann et al. 2001, 2004; Rice and Currah 2002, Rice et al. 2006) remain more common than reports from other organic debris, such as wood (Lumley et al. 2001). *O. maius* was more abundant in ectomycorrhizal root tips of spruce in blanket bogs than in mineral woodland soils (Schild et al. 1988) and in acidified rather than limed soils (Qian et al. 1998), further supporting an apparent preference for acidic substrates.

The scant isolation data from other materials is possibly the result of the biases mentioned above. For example, when Rice and Currah (2002) compared the isolation frequency of this taxon using agar media and moist chambering, *O. maius* was the most abundant sporulating species appearing directly on *Sphagnum* peat, but it was only rarely encountered when the same peat was placed on agar media (Rice and Currah 2002). *O. maius* was not isolated from ectomycorrhizal root tips when benomyl was added to the isolation medium (Schild et al. 1988). *O. maius* is capable of growing on benomyl-amended media (Rice and Currah 2002) but may not have been able to compete with basidiomycetes and other fungi favoured by the addition of this selective antifungal agent. Growth and sporulation of *O. maius* is restricted on rich artificial media, including the potato dextrose and malt extract agars that are commonly used to isolate fungi from substrates, further biasing against its recovery.

The isolation history of *O. maius* coupled with in vitro studies of its behaviour on *Sphagnum* peat strongly suggest that *O. maius* is abundant in this material and may degrade large quantities of the substrate under natural conditions (Thormann 2001; Piercey et al. 2002; Rice and Currah 2002, Rice et al. 2006). Three studies have shown that *O. maius* can cause significant mass losses of *Sphagnum* in vitro, ranging from 2–3% (Thormann 2001) to 10–12% (Piercey et al. 2002) and from negligible to almost 50% (Rice et al. 2006). These mass loss data may differ because intact *Sphagnum* was used by Thormann (2001) and ground *Sphagnum* by Piercey et al. (2002). Differences may also be due to the strains of *O. maius* used, as suggested by the wide range observed by Rice et al. (2006) for three different strains. Piercey et al. (2002) compared mass losses caused by two isolates of *O. maius* (UAMH 8919, 8920) with other ericoid mycorrhizal fungi (*R. ericae* and a non sporulating white-to-grey fungus designated "VWT") from the roots of peatland and heathland Ericaceae and found that the strains of *O. maius* caused the greatest losses. Thormann (2001) found that the mass losses caused by *O. maius* (UAMH 9749) were intermediate among five saprobic hyphomycetes [*O. maius*, *Acremonium* cf. *curvulum* (identified later as *Pochonia bulbillosa*, Thormann et al. 2004), *Penicillium thomii*, *O. scytaloides* (= *O. chlamydosporicum* sensu, Rice and Currah 2005), and *Trichoderma viride*] and greater than an unidentified basidiomycete. Tsuneda et al. (2001) used scanning electron microscopy to compare the ultrastructural patterns of decay of *Sphagnum* caused by *O. maius* and *P. bulbillosa*. *Sphagnum* cell walls are analogous to wood,

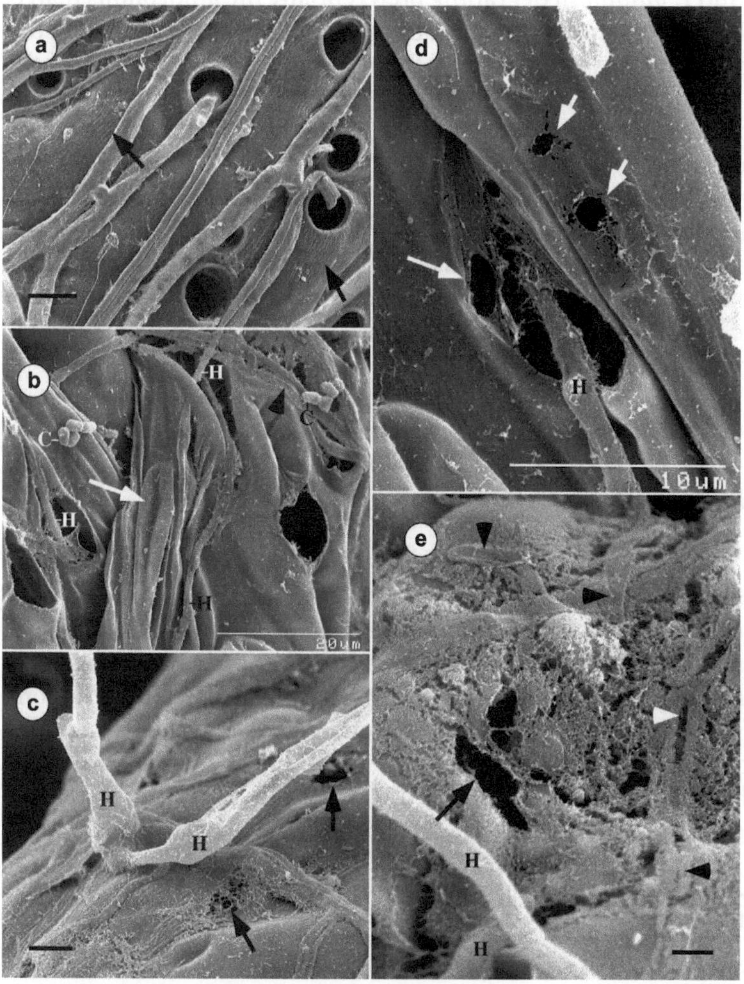

Fig. 13.2. a–e Degradation of *Sphagnum fuscum* leaf cell walls by *Oidiodendron maius* (UAMH 9749; Tsuneda et al. 2001). **a** Affected cell wall showing finely wavy deformations (*arrows*). **b** Severely distorted leaf cell wall (*arrow*). Note autolysing hypha (*arrowhead*) and degraded leaf cell wall in the immediate vicinity. *H* Hypha, *C* conidia. **c** Localized voids (*arrows*) and hyphae (*H*) emerging through the leaf cell wall. **d** More or less simultaneous degradation of the leaf cell wall by a hypha (*H*). *Arrows* Localized voids. **e** Enlarged view of an area showing the simultaneous degradation (*arrow*). *Arrowheads* Autolysing hyphae. *H* Sound, turgid hyphae. *Bars* **a** 3 μm, **b** 20 μm, **c** 5 μm, **d** 10 μm, **e** 2 μm. Reproduced with permission from Tsuneda et al. 2001

because both consist of cellulose microfibrils embedded in an amorphous matrix of phenolic polymers and polysaccharides (Tsuneda et al. 2001). Both species degraded *Sphagnum* leaves but decay patterns differed, with *O. maius* (UAMH 9749) eroding all cell wall components simultaneously (Fig. 13.2) and *P. bulbillosa* degrading preferentially the amorphous matrix material (Tsuneda et al. 2001). This pattern, analogous to the simultaneous white rot of wood, was confirmed in *O. maius* and other members of the Myxotrichaceae isolated from peat (Rice et al. 2006).

These in vitro enzymatic studies, mass loss experiments, and scanning electron microscopic examinations indicate that *O. maius* has the potential to degrade *Sphagnum* peat in nature. The abundance of *O. maius* conidia and conidiophores on peat (Rice and Currah 2002), and the relatively frequent isolation of *O. maius* from this material (Barron 1962; Nilsson et al. 1992; Thormann et al. 2001, 2004; Rice and Currah 2002; Rice et al. 2006), support the hypothesis that *O. maius* is an active component of the saprobic microfungal community in peatlands.

13.3 Ericoid Mycorrhizas

Cronquist (1988) recognised eight families within the globally distributed order Ericales that are integral components of many acidic, nutrient-poor ecosystems with organic soils. Four families, the Ericaceae, Empetraceae, Monotropaceae, and Pyrolaceae, are found in the northern hemisphere (Cronquist 1988). Molecular evidence suggests that these families, along with the Epacridaceae in the southern hemisphere, should be included together in the Ericaceae (Kron 1996; Kron et al. 2002). Ericoid and ectendomycorrhizas are common within the Ericaceae but there are also reports of ectomycorrhizas (Largent et al. 1980; Smith et al. 1995; Horton et al. 1999) and arbuscular mycorrhizas (Koske et al. 1990).

Most Ericaceae are dwarf shrubs adapted to harsh ecosystems including bogs, heaths, alpine and arctic regions, and boreal forests (Hambleton 1998). These woody plants have leathery, perennial leaves that minimise nutrient loss. Ericoid mycorrhizal associations may enhance the success of host plants in nutrient-poor, acidic, phenol-rich, and heavy metal-contaminated soils (Perotto et al. 1995; Hambleton 1998; Hambleton and Currah 2000). The below-ground network consists of well-developed mats of rhizomes and "hair roots" that form in the surface layers of organic soil (Read 1991). "Hair roots" have a narrow stele surrounded by an endodermis and one to two layers of cells, representing the cortex and/or epidermis (Read 1991; Smith and Read 1997). Hyphae penetrate the cell walls of the outer cell layers, form an interface with cell membranes (Read 1991; Smith

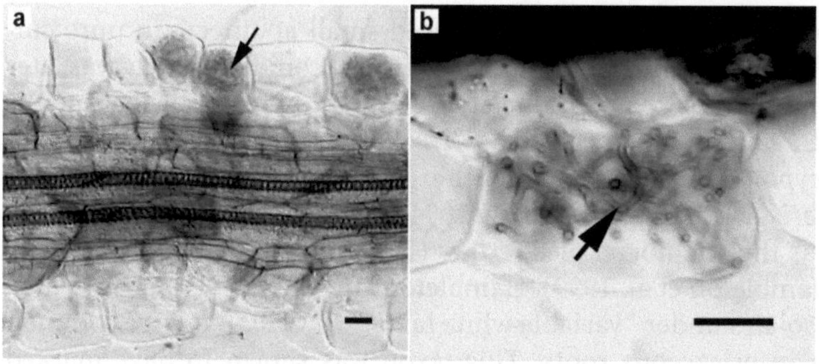

Fig. 13.3. a,b *Oidiodendron maius* (S. Hambleton personal collection, S-272a) colonising the roots of *Vaccinium vitis-idaea* in resynthesis studies (S. Hambleton, unpublished). **a** Longitudinal section of mycorrhizal root showing hyphal complexes formed in the outer layer of root cells (*arrow*). **b** Close up of hyphal complex (*arrow*) formed in the root cell. *Bars* 10 μm. Images provided by Sarah Hambleton

and Read 1997), and develop into complexes made up of densely intertwined, thin, lightly pigmented hyphae (Fig. 13.3). Hyphae also extend out of the root and absorb nutrients by decomposing organic matter; some of these nutrients, at least, are then supplied to the plant (Northup et al. 1995; Smith and Read 1997). Nutrient and carbon exchange is believed to occur across the interfaces between plant cell membranes and hyphal complexes for about 5 weeks until both the plant cell cytoplasm and the fungal hyphae within the cell degenerate (Read 1991; Smith and Read 1997). Ericoid mycorrhizal fungi have also been shown to break down phenolic compounds and sequester heavy metal ions, detoxifying the soil for their plant partner (Read 1991; Perotto et al. 1995; Smith and Read 1997; Yang and Goulart 2000). While the benefits to the host plant are readily demonstrated in resynthesis studies, the benefits to the mycobiont (fungal partner) are more difficult to measure; but it is assumed that the mycobiont receives photosynthates from the host plant (Smith and Read 1997).

Identification of mycorrhizal fungal symbionts requires isolation and identification of the fungi in culture, followed by resynthesis of the association (Smith and Read 1997; Hambleton 1998). Symbiont identification has relied upon morphological and cultural characteristics of the fungi; these sources of characters work well in conjunction with DNA fingerprinting and sequence techniques (e.g. Gardes et al. 1991; Simon et al. 1992; Egger 1995; Clapp et al. 2002; Erland and Taylor 2002; Allen et al. 2003). The first fungi confirmed, through resynthesis experiments, to form ericoid mycorrhizas did not sporulate and hence were unidentifiable (Doak 1928; Bain 1937; Gordon 1937; McNabb 1961). In 1973, Pearson and Read reported that some of their ericoid mycorrhizal isolates produced zigzag chains of arthroconidia in culture and one produced small apothecia in pure culture and in pots containing *Calluna vulgaris*. The conidial fungus was later named *Scytalidium vaccinii* (Dalpé et al. 1989) and the apothecial fungus was named *Pezizella ericae* (Read 1974). This species was transferred to *Hymnenoscyphus* (Kernan and Finocchio 1983) and recently to *Rhizoscyphus*, as *R. ericae* (Zhang and Zhuang 2004). *Scytalidium vaccinii* has been confirmed as the anamorph of *R. ericae* (Egger and Sigler 1993; Hambleton 1998; Hambleton et al. 1999). Hambleton and Currah (1997) described a series of isolates under "variable white taxon" (VWT) that were common endophytes in ericaceous roots. This taxon was shown to have marked affinities to *R. ericae* but produced neither a teleomorph nor conidia in culture. Recently, three phylogenetically distinct species in the VWT complex have been recognised in a new anamorphic genus (Hambleton and Sigler 2005). Comparison of fungi detected in the roots of *Gaultheria shallon* using culturing and molecular (DNA) methods revealed that an abundance of hyphae within hair roots belonged to an unculturable species of *Sebacina* (Allen et al. 2003). Fungi that were culturable included *R. ericae* and an isolate tentatively identified as a species of *Capronia* (Allen et al. 2003). As detection techniques and identification protocols improve, it is expected that additional fungal taxa will be described as ericoid mycorrhizal endophytes.

Species of *Oidiodendron* other than *O. maius* have also been reported from ericoid mycorrhizas (Pearson and Read 1973; Couture et al. 1983; Dalpé 1986, 1989, 1991, Douglas et al. 1989; Xiao and Berch 1992; Currah et al. 1993, 1999; Johansson 1994, 2001; Perotto et al. 1995, 1996; Hambleton and Currah 1997; Monreal et al. 1999; Chambers et al. 2000; Usuki et al. 2003). While many of the early reports implicated *O. griseum* in the associations, DNA analyses indicate that most, if not all, of these reports were based on misidentified strains of *O. maius* (Hambleton and Currah 1997; Hambleton et al. 1998).

Mycorrhizal resyntheses between host plants and fungi isolated from their ericoid mycorrhizas are generally assumed to be the definitive indicator that a fungus is mycorrhizal, but assumptions based on these data are tenuous at best. The Ericaceae is particularly problematic in this regard because, in axenic culture situations at least, the family appears to permit a wide range of fungi into the peripheral cells of hair roots where they form the typical coiled "infection units". For example, Dalpé (1986, 1989, 1991) found that blueberries (*Vaccinium angustifolium*) would form ericoid mycorrhizas with *Myxotrichum setosum*, *O. cerealis*, *O. chlamydosporicum*, *O. citrinum*, *O. flavum*, *O. griseum*, *O. periconioides*, *O. rhodogenum*, *O. scytaloides*, *Pseudogymnoascus roseus* (all members of the Myxotrichaceae, Leotiomycetes) as well as the unrelated species *Gymnascella dankaliense*s (a member of the Gymnoascaceae, Eurotiomycetes).

Salal (*Gaultheria shallon*) has formed in vitro mycorrhizal associations with *Acremonium strictum*, a fungus that was able to supply organic nitrogen and enhance host plant growth (Xiao and Berch 1999). Ericaceous plants may "prefer" some fungi over others in the field but when constrained, as in a culture situation, may form mycorrhizal relationships with, or exploit, a range of different fungal species.

13.4 *Oidiodendron maius* as an Ericoid Mycorrhizal Fungus

While the widespread isolation of *O. maius* from the roots of ericaceous roots supports the hypothesis that it may be mycorrhizal and as such, either a mutualist or a commensalist, it does not confirm the nature of the association. Many resynthesis studies have attempted to assess the morphological and functional aspects of the relationship; these studies have used a range of ericaceous shrubs and have reported positive, neutral, and negative effects on host plant growth (Douglas et al. 1989; Xiao and Berch 1995, 1999; Yang et al. 1998; Monreal et al. 1999; Bergero et al. 2000; Yang and Goulart 2000; Johansson 2001; Starrett et al. 2001; Piercey et al. 2002; Yang et al. 2002) using a range of *Oidiodendron* species (Couture et al. 1983; Dalpé 1986, 1989, 1991; Currah et al. 1993; Xiao and Berch 1995; Monreal et al. 1999). Characteristic hyphal complexes have been observed in roots of various ericaceous shrubs, including *Vaccinium vitis-idaea*, colonized by *Oidiodendron* species (Fig. 13.3) (Dalpé 1986, 1989, 1991; Douglas et al. 1989; Xiao and Berch 1995; Johansson 2001; S. Hambleton, personal communication). While the plants in these studies appeared healthy, the functional nature of the relationship between the fungi and the hosts was not determined.

Physiological evidence to support the mycorrhizal nature of the association between *O. maius* and ericaceous plants has been obtained from resynthesis studies. Yang et al. (1998) found that *O. maius* (UAMH 9263) did not affect the growth of blueberries (*Vaccinium corymbosum*), but later studies (Yang and Goulart 2000; Yang et al. 2002) on the same isolate of *O. maius* and the same plant species found positive effects of the fungus on plant growth, indicating that the nature of the relationship between *O. maius* and ericaceous shrubs may vary within fungal strains. Inoculation of salal (*Gaultheria shallon*) with four isolates of *O. maius* increased plant biomass regardless of the nitrogen source supplied (Xiao and Berch 1999). Inoculation with *O. maius* increased blueberry root and shoot dry mass as well as plant access to organic nitrogen (Yang et al. 2000). *O. maius* has also been shown to reduce aluminum uptake and increase the cation exchange capacity of blueberry (Yang and Goulart 2000).

In contrast, several resynthesis studies using *O. maius* (Dalpé 1991; Bergero et al. 2000; Piercey et al. 2002) have not resulted in the formation of distinctive infection units. These results may be due to the strong saprobic abilities of *O. maius*, strain specific effects, or to the sources of nutrients available to the plants. In these axenic culture systems, *O. maius* might have acquired sufficient carbon from the substrate (*Sphagnum* peat in the study by Piercey et al. 2002) and did not need the host plant for nutrition (Piercey et al. 2002). Other resynthesis studies have shown either neutral (Yang et al. 1998) or negative effects on the plants (Starrett et al. 2001). Starrett et al. (2001) inoculated microshoots of mountain andromeda (*Pieris floribunda*) with the "ericoid mycorrhizal fungi" *R. ericae*, *O. maius* (ATCC 66504), *O. griseum* and an unidentified species of *Oidiodendron*, and found that inoculation with all of these species caused shoot necrosis, but that this effect could be reduced by providing an alternative carbon source. Mitigation of shoot necrosis varied with carbon source and fungal isolate. Adding sucrose to the medium prevented *R. ericae*, but not the *Oidiodendron* species, from causing shoot necrosis while adding a peat-vermiculite mixture reduced the shoot necrosis caused by the *Oidiodendron* species. Additionally, the *Oidiodendron* species did not induce root formation by the microshoots to the same extent as *R. ericae*, leading Starrett et al. (2001) to conclude that the *Oidiodendron* species were not mycorrhizal symbionts.

None of the preceding studies investigated possible benefits to the mycobiont, so the mutualistic nature of the association has never been demonstrated. It is possible that the relationship is physiologically similar to the dynamics in orchid mycorrhizas in which the orchid "exploits" its saprobic or ectomycorrhizal mycobiont (e.g. Rasmussen 1995; McKendrick et al. 2000) or to the epiparasitic relationship of monotropes (non-photosynthetic Ericaceae) to ectomycorrhizal trees via their shared ectomycorrhizal fungi (Bidartondo et al. 2000). Many Ericaceae, similar to the Orchidaceae and the monotropes, are microspermous (e.g. *Rhododendron, Menziesia*). Perhaps the adoption of microspermy is a consequence of the host plants' ability to exploit fungi as a source of nutrients, and ericoid mycorrhizal associations may be a part of a mycoheterotrophic evolutionary trajectory. The discovery of *Sebacina*, a basidiomycete genus known to form orchid mycorrhizas (Currah et al. 1990; McKendrick et al. 2002), in ericoid mycorrhizal roots of Salal (Allen et al. 2003) is another similarity between ericoid and orchid mycorrhizal systems.

The nature of the relationship between *O. maius* and members of the Ericaceae is clearly complex and varies from one report to the next. Future research involving physiological assessment in laboratory, greenhouse, and field conditions coupled with morphological and molecular assessment in roots (Hambleton and Currah 2000) is required to identify and quantify possible benefits to both partners and to elucidate the factors determining the functional aspects of the relationship. For example, tracing the movement of radiolabeled carbon and nutrients between the partners could help determine whether *O. maius* obtains host photosynthate and if the plant receives any fungal-derived carbon or other nutrients. Additionally, in vitro studies could determine other potential benefits to either partner, such as competitive and growth advantages for *O. maius* and pathogen resistance for the plant. The potential role of ericoid mycorrhizal fungi in symbiotic seed germination should also be examined.

13.5 Significance and Relevance

The occupation of multiple niches within a given environment could confer survival and competitive advantages on *O. maius* by providing different refuges to the fungus. During periods of host plant dormancy, *O. maius* could thrive as a saprobe in the peat matrix, while host plant roots may serve as a refuge from competition with other saprobes and as a source of inoculum for colonisation of senescing surface peat. On the other hand, the prevalence of *O. maius* as a saprobe within the peat could ensure rapid colonisation of new ericaceous roots, perhaps at the expense of other species that are less able to degrade the surrounding substrate. While there is currently no experimental data to support either hypothesis (i.e. whether roots or peat serve as primary refugia for *O. maius*), testing both could provide information key to understanding of the ecological role of this species.

Ericoid and ectomycorrhizal fungi can degrade a variety of complex organic substrates (Bajwa and Read 1985; Read 1991; Northup et al. 1995; Bending and Read 1996, 1997; Smith and Read 1997; Aerts 2002; Leake et al. 2002; Olsson et al. 2002; Simard et al. 2002), with the abilities of ericoid mycorrhizal fungi possibly exceeding those of ectomycorrhizal fungi (Bajwa and Read 1985; Read 1991; Bending and Read 1996, 1997; Smith and Read 1997). These abilities are thought to aid in host plant nutrition by allowing the plant direct access to organic nutrient sources (Bajwa and Read 1985; Xiao and Berch 1999; Yang et al. 2002). Inorganic sources of nitrogen are scarce and organic sources relatively abundant when decomposition is slow, as in peatlands and heathlands, and when leaching is common (Perotto et al. 1995). Ericoid mycorrhizal fungi often produce phosphatase, enabling them to access and transfer organic phosphorus to their hosts (Aerts 2002). Ericaceous shrubs in these environments may rely on their mycorrhizal partners to supply them with sufficient carbon (Yang et al. 2002) and nutrients obtained from the organic sources (Northup et al. 1995; Xiao and Berch 1999; Yang et al. 2002). The abilities of ericoid mycorrhizal fungi, including

O. maius, to access carbon and nutrients from organic debris could reduce their reliance on host plant photosynthates for carbon (Piercey et al. 2002), while still supplying the host plant with nutrients.

Transfer of nutrients from organic matter to ericaceous shrubs via ericoid mycorrhizal fungi, such as *O. maius*, has important implications for nutrient cycling in ecosystems such as peatlands, where decomposition is slow and carbon and nutrients are sequestered in organic debris (Northup et al. 1995). *Sphagnum* decomposes slowly with relatively few fungi having the ability to cause significant mass losses (Thormann 2001). Decomposition of *Sphagnum* by *O. maius* may release a significant amount of carbon and nutrients from the peat and, instead of releasing carbon into the atmosphere and nitrogen and phosphorus into the pool of plant-available nutrients, these organic forms of nitrogen and phosphorus can be supplied directly to the ericaceous shrubs, giving them a competitive advantage over their neighboring plants (Aerts 2002; Leake et al. 2002). Short-circuiting nutrient cycles, by reducing the amount of decomposition required before nutrient absorption, results in increased supplies of nitrogen and phosphorus to mycorrhizal host plants and a resultant decrease in nitrogen and phosphorus available to saprobes and non-mycorrhizal plants (Leake et al. 2002).

Other *Oidiodendron* species are able to form ericoid mycorrhizal associations in vitro (Dalpé 1986, 1989, 1991; Currah et al. 1993). Species of *Oidiodendron* are the asexual states of myxotrichoid ascomycetes, and some of these, e.g. *Pseudogymnoascus roseus*, and other anamorphs, including species of *Geomyces*, also produce ericoid mycorrhizal associations in vitro (Dalpé 1989). However, only two species of *Oidiodendron* have been reported from ericaceous roots in situ. Currah et al. (1993) isolated *O. periconioides* from *Rhododendron brachycarpum* grown in pot cultures containing peat. The remaining species are known only as saprobes but since they share morphological characters, including dendritic arthroconidia and cage-like cleistothecial ascomata, and ecological characters, including the ability to degrade a variety of plant-based polymers and a predilection for cool, acidic conditions (Rice and Currah 2005), it is possible that these taxa could play biologically similar and significant roles to *O. maius* in cool, acidic soils. Additional surveys of ericoid mycorrhizal endophytes should employ a broad range of isolation and detection protocols to maximise the recovery of as wide a variety of fungi as possible.

The occupation of multiple niches by *O. maius* may parallel the situation observed for *Phialocephala fortinii*. Usually considered a root endophyte, *P. fortinii* is isolated most frequently from healthy roots of woody plants [e.g. see Chaps. 7 (Sieber and Grünig) and 15 (Schulz); Stoyke and Currah 1991; Menkis et al. 2004; Piercey et al. 2004] and, until recently, was unknown as a saprobe. However, Menkis et al. (2004) isolated *P. fortinii* from healthy wood in pine stems suggesting a possible role as a systemic endophyte, and in birch snags and birch and pine stumps, where it is presumably occupying a saprobic niche (Menkis et al. 2004), perhaps as an agent of soft rot (Sieber 2002; Menkis et al. 2004). Given the scattered and somewhat incidental reports of *O. maius* from wood and soil, a concerted and wider search for *O. maius* may reveal habitats in addition to ericaceous roots, where this species is abundant.

13.6 Conclusions

Oidiodendron maius forms associations with the roots of ericaceous shrubs, though the nature of the relationship remains uncertain. Is it a mutualistic mycorrhizal association, a preemptively colonised refugium for the fungus, a case of parasitism of the fungus by the plant, or some combination of the three? In vitro studies indicate that *O. maius* can improve host plant growth both by aiding plant nutrition and detoxifying the soil environment, although the benefits to *O. maius* are unclear. It remains necessary to investigate the benefits to both partners and demonstrate what environmental conditions determine the functional nature of the relationship. *O. maius* has the potential to degrade complex organic polymers within the soil, thus it is unlikely that it would rely on host photosynthate for survival. However, it is possible that *O. maius* receives some photosynthate, which could supplement saprobically derived carbon, potentially giving *O. maius* a competitive advantage over other soil fungi. The tendency towards microspermy in the Ericaceae and the saprobic abilities of ericoid endophytes suggests that ericoid mycorrhizal associations may represent another example of controlled parasitism of a fungal partner by the host plant, similar to the type that occurs with orchids. Entrapment of *O. maius* could confer a competitive advantage on the host plants by increasing the supply of organically bound nutrients-unavailable to plants that lack ericoid mycorrhizas, and by supplementing host plant photosynthesis with fungal-derived carbon. Given the wide taxonomic tolerance that ericaceous plants have for root endophytic fungi in vitro, the obvious need for more detailed studies of endophytic diversity of fungi growing in plants in situ, the enigmatic ecological roles of related Helotialean fungi (e.g. *P. fortinii*, *Geomyces*, etc.), much more exploratory and empirical research is needed before we will be able to answer the question posed at the outset of this chapter.

Acknowledgements. The authors thank S. Hambleton and A. Tsuneda for providing images. Comments on previous versions of this manuscript by H.D. Addy and the continuing support of the Natural Sciences and Engineering Research Council of Canada are gratefully acknowledged.

References

Addy HD, Piercey MM, Currah RS (2005) Microfungal endophytes in roots. Can J Bot 83:1–13
Aerts R (2002) The role of various types of mycorrhizal fungi in nutrient cycling and plant competition. In: van der Heijden MGA, Sanders IR (eds) Ecological studies, vol 157 Mycorrhizal ecology Springer, Berlin Heidelberg New York, pp 117–133
Allen TR, Millar T, Berch SM, Berbee ML (2003) Culturing and direct DNA extraction find different fungi from the same ericoid mycorrhizal roots. New Phytol 160:255–272
Bain HF (1937) Production of synthetic mycorrhiza in the cultivated cranberry. J Agric Res 55:811–835
Bajwa R, Read DJ (1985) The biology of mycorrhiza in the Ericaceae. IX. Peptides as nitrogen sources for the ericoid endophyte and for mycorrhizal and non-mycorrhizal plants. New Phytol 101:459–467
Barron GL (1962) New species and new records of *Oidiodendron*. Can J Bot 40:589–607
Bending GD, Read DJ (1996) Nitrogen mobilization from tannin-protein complexes by ericoid and ectomycorrhizal fungi. Soil Biol Biochem 28:1603–1612
Bending GD, Read DJ (1997) Lignin and soluble phenolic degradation by ectomycorrhizal and ericoid mycorrhizal fungi. Mycol Res 101:1348–1354
Bergero R, Perotto S, Girlanda MM, Vidano G, Luppi AM (2000) Ericoid mycorrhizal fungi are common root associates of a Mediterranean ectomycorrhizal plant (*Quercus ilex*). Mol Ecol 9:1639–1649
Bidartondo MI, Kretzer AM, Pine EM, Bruns TD (2000) High root concentrations and uneven ectomycorrhizal diversity near *Sarcodes sanguinea* (Ericaceae): A cheater that stimulates its victims? Am J Bot 87:1783–1788
Chambers SM, Liu G, Cairney WG (2000) ITS rDNA sequence comparison of ericoid mycorrhizal endophytes from *Woollsia pungens*. Mycol Res 104:168–174
Clapp JP, Helgason T, Daniell TJ, Young JPW (2002) Genetic studies of the structure and diversity of arbuscular mycorrhizal fungal communities. In: van der Heijden MGA, Sanders IR (eds) Ecological studies, vol 157. Mycorrhizal ecology Springer, Berlin Heidelberg New York, pp 201–224
Couture M, Fortin JA, Dalpé Y (1983) *Oidiodendron griseum* Robak: an endophyte of ericoid mycorrhiza in *Vaccinium* spp. New Phytol 95:375–380
Cronquist A (1988) The evolution and classification of flowering plants, 2nd edn. New York Botanic Garden, New York
Currah RS, Smreciu ES, Hambleton S (1990) Mycorrhizae and mycorrhizal fungi of boreal species of *Platanthera* and *Coeloglossum* (Orchidaceae). Can J Bot 68:1171–1181
Currah RS, Tsuneda A, Murakami S (1993) Conidiogenesis in *Oidiodendron periconioides* and ultrastructure of ericoid mycorrhizas formed with *Rhododendron brachycarpum*. Can J Bot 71:1481–1485
Currah RS, Niemi M, Huhtinen S (1999) *Oidiodendron maius* and *Scytalidium vaccinii* from the mycorrhizas of Ericaceae in northern Finland. Karstenia 39:65–68
Dalpé Y (1986) Axenic synthesis of ericoid mycorrhiza in *Vaccinium angustifolium* Ait. by *Oidiodendron* species. New Phytol 103:391–396

Dalpé Y (1989) Ericoid mycorrhizal fungi in the Myxotrichaceae and Gymnoascaceae. New Phytol 113:523–527
Dalpé Y (1991) Statut endomycorhizien du genre *Oidiodendron*. Can J Bot 69:1712–1714
Dalpé Y, Litten W, Sigler L (1989) *Scytalidium vaccinii* sp. nov., an ericoid endophyte of *Vaccinium angustifolium* roots. Mycotaxon 35:371–377
Doak KD (1928) The mycorrhizal fungus of *Vaccinium*. Phytopathology 18:148
Douglas AE, Smith DC (1989) Are endosymbioses mutualistic. Tree 4:350–352
Douglas GC, Heslin MC, Reid C (1989) Isolation of *Oidiodendron maius* from *Rhododendron* and ultrastructural characterization of synthesized mycorrhizas. Can J Bot 67:2206–2212
Egger KN (1995) Molecular analysis of ecto-mycorrhizal fungal communities. Can J Bot 73:S1415–S1422
Egger KN, Sigler L (1993) Relatedness of the ericoid endophytes *Scytalidium vaccinii* and *Hymenoscyphus ericae* inferred from analysis of ribosomal DNA. Mycologia 85:219–230
Erland S, Taylor AFS (2002) Diversity of ecto-mycorrhizal fungal communities in relation to the abiotic environment. In: van der Heijden MGA, Sanders IR (eds) Ecological studies, vol 157. Mycorrhizal ecology Springer, Berlin Heidelberg New York, pp 163–200
Gardes M, White TJ, Fortin JA, Bruns TD, Taylor JW (1991) Identification of indigenous and introduced symbiotic fungi in ecto-mycorrhizas by amplification of nuclear and mitochondrial ribosomal DNA. Can J Bot 69:180–190
Gordon HD (1937) Mycorrhiza in Rhododendron. Ann Bot 1:593–613
Hambleton S (1998) Mycorrhizas of the Ericaceae: Diversity and systematics of the mycobionts. PhD Dissertation, University of Alberta, Edmonton
Hambleton S, Currah RS (1997) Fungal endophytes from the roots of alpine and boreal *Ericaceae*. Can J Bot 75:1570–1581
Hambleton S, Currah RS (2000) Molecular characterization of the mycorrhizas of woody plants. In: Jain SM, Minocha SC (eds) Molecular biology of woody plants, vol 2. Kluwer, Dordrecht, pp 351–373
Hambleton S, Sigler L (2005) *Melinia*, a new anamorph genus for root-associated fungi with phylogenic affinities to *Rhizoscyphus ericae* (≡ *Hymenoscyphus ericae*), *Leotiomycetes*. Stud Mycol 53:1–27
Hambleton S, Egger KN, Currah RS (1998) The genus *Oidiodendron*: species delimitation and phylogenetic relationships based on nuclear ribosomal DNA analyses. Mycologia 90:854–869
Hambleton S, Huhtinen S, Currah RS (1999) *Hymenoscyphus ericae*: a new record from western Canada. Mycol Res 103:1391–1397
Hoosbeek M, van Breemen N, Berendse F, Grosvernier P, Vasander H, Wallén B (2001) Limited effect of increased atmospheric CO_2 concentration on ombrotrophic bog vegetation. New Phytol 150:459–463
Horton TR, Bruns TD, Parker VT (1999) Ectomycorrhizal fungi associated with *Arctostaphylos* contribute to *Pseudotsuga menziesii* establishment. Can J Bot 77:93–102
Hutchison LJ (1990) Studies on the systematics of ectomycorrhizal fungi in axenic culture. II. The enzymatic degradation of selected carbon and nitrogen compounds. Can J Bot 68:1522–1530
Hutchison LJ (1991) Description and identification of cultures of ectomycorrhizal fungi found in North America. Mycotaxon 42:387–504
Johansson M (1994) Quantification of mycorrhizal infection units in roots of *Calluna vulgaris* (L.) Hull from Danish heathland. Soil Biol Biochem 26:557–566
Johansson M (2001) Fungal associations of Danish *Calluna vulgaris* roots with special reference to ericoid mycorrhiza. Plant Soil 231:225–232
Kernan MJ, Finocchio AF (1983) A new discomycete associated with the roots of *Monotropa uniflora* (Ericaceae). Mycologia 75:916–920

Koske RE, Gemma JN, Englander L (1990) Vesicular-arbuscular mycorrhizae in Hawaiian Ericales. Am J Bot 77:64–68
Kron KA (1996) Phylogenetic relationships of Empetraceae, Epacridaceae, Ericaceae, Monotropaceae, and Pyrolaceae: evidence from nuclear ribosomal 18S sequence data. Ann Bot 77:293–303
Kron KA, Judd WS, Stevens PF, Crayne DM, Anderberg AA, Gadek PA, Quinn CJ, Luteyn JL (2002) Phylogenetic classification of Ericaceae: molecular and morphological evidence. Bot Rev 68:335–423
Largent DL, Sugihara N, Wishner C (1980) Occurrence of mycorrhizae on ericaceous and pyrolaceous plants in northern California. Can J Bot 58:2274–2279
Leake JR, Donnelly DP, Boddy L (2002) Interactions between ecto-mycorrhizal and saprotrophic fungi. In: van der Heijden MGA, Sanders IR (eds) Ecological studies, vol 157. Mycorrhizal Ecology. Springer, Berlin Heidelberg New York, pp 345–372
Lumley TC, Gignac LD, Currah RS (2001) Microfungus communities of white spruce and trembling aspen logs at different stages of decay in disturbed and undisturbed sites in the boreal mixedwood region of Alberta. Can J Bot 79:76–92
McKendrick SL, Leake JR, Read DJ (2000) Symbiotic germination and development of mycoheterotrophic plants in nature: transfer of carbon from ectomycorrhizal *Salix repens* and *Betula pendula* to the orchid *Corallorhiza trifida* through shared hyphal connections. New Phytol 145:539–548
McKendrick SL, Leake JR, Taylor DL, Read DJ (2002) Symbiotic germination and development of the myco-heterotrophic orchid *Neottia nidus-avis* in nature and its requirement for locally distributed *Sebacina*. New Phytol 154:233–247
McNabb RFR (1961) Mycorrhiza in the New Zealand Ericales. Aust J Bot 9:57–61
Menkis A, Allmer J, Vasiliauskas R, Lygis V, Stenlid J, Finlay R (2004) Ecology and molecular characterization of dark septate fungi in roots, living stems, coarse and fine woody debris. Mycol Res 108:965–973
Monreal M, Berch SM, Berbee M (1999) Molecular diversity of ericoid mycorrhizal fungi. Can J Bot 77:1580–1594
Nilsson M, Bååth E, Söderström B (1992) The microfungal communities of a mixed mire in Northern Sweden. Can J Bot 70:272–276
Nordgren A, Bååth E, Söderström B (1985) Soil microfungi in an area polluted by heavy metals. Can J Bot 63:448–455
Northup RR, Zengshou Y, Dahlgren RA, Vogt KA (1995) Polyphenol control of nitrogen release from pine litter. Nature 377:227–229
Olsson PA, Jakobsen I, Wallander H (2002) Foraging and resource allocation strategies of mycorrhizal fungi in a patchy environment. In: van der Heijden MGA, Sanders IR (eds) Ecological studies, vol 157. Mycorrhizal ecology Springer, Berlin Heidelberg New York, pp 93–115
Pearson V, Read DJ (1973) The biology of mycorrhiza in Ericaceae. I. The isolation of the endophyte and synthesis of mycorrhizas in aseptic culture. New Phytol 72:371–379
Perotto S, Perotto R, Faccio A, Schubert A, Varma A, Bonfante P (1995) Ericoid mycorrhizal fungi: cellular and molecular bases of their interactions with the host plant. Can J Bot 73:S557–S568
Perotto S, Actis-Perino E, Perugini J, Bonfante P (1996) Molecular diversity of fungi from ericoid mycorrhizal roots. Mol Ecol 5:123–131
Peteet D, Andreev A, Bardeen W, Mistretta F (1998) Long-term arctic peatland dynamics, vegetation and climate history of the Pur-Taz region, western Siberia. Boreas 27:115–126
Piercey MM, Thormann MN, Currah RS (2002) Saprobic characteristics of three fungal taxa from ericalean roots and their association with the roots of *Rhododendron groenlandicum* and *Picea mariana* in culture. Mycorrhiza 12:175–180

Piercey MM, Graham SW, Currah RS (2004) Patterns of genetic variation in *Phialocephala fortinii* across a broad latitudinal transect in Canada. Mycol Res 108:955–964

Qian XM, El-Ashker A, Kottke I, Oberwinkler F (1998) Studies of pathogenic and antagonistic microfungal populations and their potential interactions in the mycorrhizoplane of Norway spruce (*Picea abies* (L.) Karst.) and beech (*Fagus sylvatica* L.) on acidified and limed plots. Plant Soil 199:111–116

Rasmussen H (1995) Terrestrial orchids: From seed to mycotrophic plant. Cambridge University Press, Cambridge

Read DJ (1974) *Pezizella ericae* sp. nov., the perfect state of a typical mycorrhizal endophyte of Ericaceae. Trans Br Mycol Soc 63:381–383

Read DJ (1991) Mycorrhizas in ecosystems. Experientia 47:376–391

Read DJ (2002) Towards ecological relevance – progress and pitfalls in the path towards an understanding of mycorrhizal functions in nature. In: van der Heijden MGA, Sanders IR (eds) Ecological studies, vol 157. Mycorrhizal ecology. Springer, Berlin Heidelberg New York, pp 1–29

Rice AV, Currah RS (2001) Physiological and morphological variation in *Oidiodendron maius*. Mycotaxon 79:383–396

Rice AV, Currah RS (2002) New perspectives on the niche and holomorph of the myxotrichoid hyphomycete, *Oidiodendron maius*. Mycol Res 106:1463–1467

Rice AV, Currah RS (2005) *Oidiodendron*: a survey of the named species and unnamed anamorphs of *Myxotrichum*. Stud Mycol 53:83–120

Rice AV, Tsuneda A, Currah RS (2006) In vitro decomposition of *Sphagnum* by some microfungi resembles white rot of wood. FEMS Microbiol Ecol 56:372–382

Schild DE, Kennedy A, Stuart MR (1988) Isolation of symbiont and associated fungi from ectomycorrhizas of Sitka spruce. Eur J Forest Pathol 18:51–61

Seiber TN (2002) Fungal root endophytes. In Wasel Y, Eshel A, Kafkafi U (eds) Plant roots: the hidden half. Dekker, New York, pp 887–917

Simard SW, Jones MD, Durrall DM (2002) Carbon and nutrient fluxes within and between mycorrhizal plants. In: van der Heijden MGA, Sanders IR (eds) Ecological studies, vol 157. Mycorrhizal ecology Springer, Berlin Heidleberg New York, pp 33–74

Simon L, Lalonde M, Bruns TD (1992) Specific amplification of the 18S fungal ribosomal genes from VA endomycorrhizal fungi colonizing roots. Appl Environ Microbiol 58:291–295

Smith JE, Molina R, Perry D (1995) Occurrence of ectomycorrhizas on ericaceous and coniferous seedlings grown in soils from the Oregon Coast Range. New Phytol 129:73–81

Smith SE, Read DJ (1997) Mycorrhizal symbioses, 2nd edn. Academic Press, San Diego

Starrett MC, Blazich FA, Shafer SR, Grand LF (2001) In vitro colonization of micropropagated *Pieris floribunda* by ericoid mycorrhizae. I. Establishment of mycorrhizae on microshoots. HortScience 36:353–356

Stoyke G, Currah RS (1991) Endophytic fungi from the mycorrhizae of alpine ericoid plants. Can J Bot 69:347–352

Summerbell RC (1987) Microfungi associated with the mycorrhizal mantle and adjacent microhabitats within the rhizosphere of black spruce. Can J Bot 65:1085–1095

Svensson BM (1995) Competition between *Sphagnum fuscum* and *Drosera rotundifolia*: a case of ecosystem engineering. Oikos 74:205–212

Thormann MN (2001) The fungal communities of decomposing plants in southern boreal peatlands of Alberta, Canada. PhD Dissertation, University of Alberta, Edmonton

Thormann MN, Currah RS, Bayley SE (2001) Microfungi isolated from *Sphagnum fuscum* from a southern boreal bog in Alberta, Canada. Bryologist 104:548–559

Thormann MN, Currah RS, Bayley SE (2002) The relative ability of fungi from *Sphagnum fuscum* to decompose selected carbon substrates. Can J Microbiol 48:204–211

Thormann MN, Currah RS, Bayley SE (2004) Patterns of distribution of microfungi in decomposing bog and fen plants. Can J Bot 82:710–720

Tsuneda A, Currah RS (2004) Ascomatal morphogenesis in *Myxotrichum arcticum* supports the derivation of the Myxotrichaceae from a discomycetous ancestor. Mycologia 96:627–635

Tsuneda A, Thormann MN, Currah RS (2001) Modes of cell wall degradation of *Sphagnum fuscum* by *Acremonium cf. curvulum* and *Oidiodendron maius*. Can J Bot 79:93–100

Usuki F, Abe JP, Kakishima M (2003) Diversity of ericoid mycorrhizal fungi isolated from hair roots of *Rhododendron obtusum* var. *kaempferi* in a Japanese red pine forest. Mycoscience 44:97–102

Vitt DH, Halsey LA, Thormann MN, Martin T (1996) Peatland inventory of Alberta Phase 1: overview of peatland resources in the natural regions and subregions of the province. Alberta Peatland Resource Centre, Edmonton, Publication 96-1

Vrålstad T, Fossheim T, Schumacher T (2000) *Piceirhiza bicolorata* – the ectomycorrhizal expression of the *Hymenoscyphus ericae* aggregate? New Phytol 145:549–563

Vrålstad T, Schumacher T, Taylor AFS (2002) Mycorrhizal synthesis between fungal strains of the *Hymenoscyphus ericae* aggregate and potential ectomycorrhizal and ericoid hosts. New Phytol 153:143–152

Xiao G, Berch S (1992) Ericoid mycorrhizal fungi of *Gaultheria shallon*. Mycologia 84:470–471

Xiao G, Berch S (1995) The ability of known ericoid mycorrhizal fungi to form mycorrhizae with *Gaultheria shallon* on forest clearcuts. Mycologia 87:467–470

Xiao G, Berch S (1999) Organic nitrogen use by salal ericoid mycorrhizal fungi from northern Vancouver Island and impacts on growth in vitro of *Gaultheria shallon*. Mycorrhiza 9:145–149

Yang WQ, Goulart BL (2000) Mycorrhizal infection reduces short-term aluminum uptake and increases root cation exchange capacity of highbush blueberry plants. HortScience 35:1083–1086

Yang WQ, Goulart BL, Demchak K (1998) Mycorrhizal infection and plant growth of highbush blueberry in fumigated soil following soil amendment and inoculation with mycorrhizal fungi. HortScience 33:1136–1137

Yang WQ, Goulart BL, Demchak K, Li Y (2002) Interactive effects of mycorrhizal inoculation and organic soil amendments on nitrogen acquisition and growth of highbush blueberry. J Am Soc Hortic Sci 127:742–748

Zhang Y-H, Zhuang W-Y (2004) Phylogenetic relationships of some members in the genus *Hymenoscyphus* (Ascomycetes, Helotiales). Nova Hedwigia 78:475–484

14 Mycorrhizal and Endophytic Fungi of Epacrids (Ericaceae)

John W.G. Cairney

14.1 Introduction

Epacrids are a group of over 450 species of woody plants that were conventionally classified as the family Epacridaceae (Powell et al. 1996). Detailed phylogenetic analyses based on morphological and molecular data indicate that epacrids represent a lineage within Ericaceae, with a recent classification regarding epacrids as Styphelioideae, one of eight Ericaceae subfamilies (Kron et al. 2002). Although epacrids occur in several southern hemisphere locations, including New Zealand, south east Asia, Pacific Ocean Islands and Patagonia, they are primarily an Australian group (Copeland 1954; Powell et al. 1996). Species richness is greatest in Western Australia (WA), with some 181 named species occurring in this state, 98% of which are endemic to WA (Keighery 1996). Seven epacrid tribes are currently recognised (Kron et al. 2002), most of which comprise small-to-medium sized shrubby heath-like plants. Some Richeeae taxa, however, resemble large arborescent monocots (Allaway 1996). Epacrids occupy a range of habitats that includes dry sandy heathlands and sclerophyll forests, along with wet alpine bogs and Magellanic tundra (Read 1996). They generally occur in acidic or neutral soils, but are occasionally found in more basic soils (Keighery 1996). Despite being geographically and hydrologically disparate, these habitats characteristically encompass soils in which availability of mineral nutrients is relatively poor (Read 1996).

In common with many other Ericaceae taxa, epacrids produce extremely fine lateral roots known as hair roots. Up to three orders of hair roots may be present, and their structure is broadly similar in all taxa: a stele, surrounded by two layers of suberised cortical cells and an epidermis. Hair roots lack root hairs, but the surface is generally covered by a mucilage layer (Cairney and Ashford 2002). When collected from the field, a proportion of the epidermal cells of epacrid hair roots invariably contains hyphal structures that are morphologically similar to those produced by ericoid mycorrhizal

John W.G. Cairney: Centre for Plant and Food Sciences, Parramatta Campus, University of Western Sydney, Locked Bag 1797, Penrith South DC, NSW 1797, Australia, E-mail: j.cairney@usw.edu.au

Soil Biology, Volume 9
Microbial Root Endophytes
B. Schulz, C. Boyle, T. N. Sieber (Eds.)

fungi in roots of other Ericaceae subfamilies (Reed 1989, 1996). Up to 90% of the hair root length can display this type of colonisation (Davies et al. 2003), but in most cases considerably less root length is colonised and there is evidence that colonisation varies on a seasonal basis (Reed 1996; Hutton et al. 1994; Davies et al. 2003; Kemp et al. 2003).

Detailed analysis of mycorrhiza-like infections of epacrid hair roots generally reveals longitudinally oriented surface hyphae from which perpendicular branches arise. Penetration of individual epidermal cells is generally by a single hypha, sometimes involving an appressorium, and is followed by development of a hyphal coil in the periplasmic space (Cairney and Ashford 2002). These features, together with the fact that the fungi are generally ascomycetes, as evidenced by the presence of simple septa and Woronin bodies (Allen et al. 1989; Steinke et al. 1996; Briggs and Ashford 2001), are shared with ericoid mycorrhizal infection of other Ericaeae, and are generally taken to indicate mycorrhizal infection. Aside from putatively ericoid mycorrhizal ascomycetes, arbuscular mycorrhiza-like infections, suggesting the presence of Glomales taxa, have also been observed as apparent endophytes of field-collected epacrid hair roots (Khan 1978; McGee 1986; Bellgard 1991; McLean and Lawrie 1996; Reed 1996; Davies et al. 2003), and there is a single report of basidiomycete hyphae in epidermal cells of an epacrid (Allen et al. 1989). As emphasised by Reed (1996), however, the presence of infection does not necessarily indicate a mutualistic association. In the case of apparent arbuscular mycorrhizal infection at least, it seems most plausible that this reflects opportunistic infection by arbuscular mycorrhizal fungal hyphae that grew from neighbouring non-Ericaceae hosts (Reed 1996).

14.2 Endophytes of Epacrid Roots

To date, only some of the endophytes obtained from Australian epacrid roots have been tested for their abilities to form ericoid mycorrhizas with host plants and, where this has been conducted, mycorrhizal status has been inferred simply from production of mycorrhiza-like coils in epidermal cells. For this reason these endophytes are referred to as putative ericoid mycorrhizal fungi throughout this chapter.

Many endophytes have been isolated from surface-sterilised epacrid roots into axenic agar culture. These are typically isolated as sterile mycelia, rendering their classification on the basis of morphology difficult. Nonetheless, studies that grouped isolates on the basis of gross morphological characteristics of cultured mycelia and/or by pectic zymogram patterns established that multiple ascomycete taxa are likely to form ericoid mycorrhizal associations with epacrids in their native habitats (Reed 1989; Hutton et al. 1994, 1996b; Steinke et al. 1996). Subsequent molecular analyses of the isolated endophytes suggest that they are broadly similar to those obtained from other Ericaceae taxa in northern hemisphere habitats.

To date, molecular analysis of endophytes from epacrid roots has concentrated largely on the rDNA internal transcribed spacer (ITS) region. Assemblages of isolated endophytes have thus been grouped on the basis of gross morphological characteristics of cultures (McLean et al. 1999; Chambers et al. 2000) or as ITS-restriction fragment length polymorphism (RFLP) groups (Midgley et al. 2002, 2004a) and ITS sequences obtained for representative isolates of each morphological or RFLP group. Comparison of the ITS sequences with those available in the GenBank nucleotide database has confirmed that most endophytes isolated from epacrid roots are ascomycetes or their anamorphs. Indeed, there is only a single published account of isolation of a fungus from epacrid hair roots that appears, on the basis of ITS sequence identity, to be a basidiomycete (Midgley et al. 2004a). This isolate did not, however, form ericoid mycorrhiza in gnotobiotic culture with *Woollsia pungens* (see below) and seems most likely to represent a facultative root inhabitant. Recent analysis of epacrid endophytes by direct DNA extraction from hair roots has revealed the presence of other putative basidiomycetes. These, however, appear to be relatively uncommon and their mycorrhizal status is unclear (D. Bougoure and J.W.G. Cairney, unpublished data).

ITS sequence data suggest that many of the endophytes are Helotiales ascomycetes, several of which have affinities with the *Hymenoscyphus ericae* aggregate (McLean et al. 1999; Chambers et al. 2000; Sharples et al. 2000; Cairney and Ashford 2002). This aggregate encompasses *H. ericae*, *Phialophora finlandia* and related taxa, many of which form ericoid mycorrhiza with northern hemisphere Ericaceae (Vrålstad et al. 2002). These isolates have been obtained from epacrids at alpine, dry sclerophyll forest, sand mine and coastal heathland sites, indicating that they have a widespread distribution in a variety of Australian habitats (McLean et al. 1998, 1999; Midgley et al. 2002). Many of the other endophytes are part of a poorly defined Helotiales ascomycete group that probably incorporates a number of taxa and includes many ericoid mycorrhizal and other root-associated fungi from northern hemisphere Ericaceae (McLean et al. 1999; Chambers et al. 2000; Sharples et al. 2000; Cairney and Ashford 2002; Berch et al. 2002; Midgley et al. 2002, 2004a). Other isolates had closest ITS sequence similarity to *Capronia* (Chaetothyriales) and *Thielavia* (Sordariales) (Midgley et al. 2002, 2004a), and, while some of these are also endophytes of non-ericaceae hosts, similar isolates are known to form ericoid mycorrhiza with *Gaultheria shallon* in Canada (Berch et al. 2002). *Oidiodendron* spp. are widespread ericoid mycorrhizal fungi of northern hemisphere Ericaceae

(Perotto et al. 2002; see Chap. 13 by Rice and Currah). Endophytes with strong ITS sequence similarity to *O. maius* have recently been isolated, and may represent common endophytes of certain epacrids (Chambers et al. 2000; D. Bougoure and J.W.G. Cairney, unpublished data).

In addition to the putative ericoid mycorrhizal fungi, many fungi isolated from surface-sterilised epacrid roots show closest ITS sequence matches to fungi that are not known to be mycorrhizal fungi. These include isolates that have closest sequence matches to taxa that are regarded as saprotrophic or pathogenic, but also with dark septate endophytes (DSE) (Midgley et al. 2002, 2004a; Davies et al. 2003; D. Bougoure and J.W.G. Cairney, unpublished data). Isolates with close sequence similarity to *Phialocephala fortinii* and other DSE taxa, although occasionally obtained from lower elevation habitats, have been obtained primarily from epacrids in Australian sub-alpine and alpine habitats (Davies et al. 2003; Midgley et al. 2004a). Furthermore, Davies et al. (2003) found that infection of some epacrid roots by DSE at alpine sites can be more prevalent than ericoid mycorrhizal infection. Since only limited sampling has so far been undertaken in alpine habitats, and systematic comparisons of endophyte abundance in different habitats have yet to be conducted, the ecological significance of these observations is difficult to assess.

These observations are, however, based on fungi that have been isolated from epacrid hair roots and assume that all endophytic fungi are isolated with equal efficiency, or indeed are isolated at all. It is possible, for example, that fast-growing endophytes may mask slower growing endophytes that are present in the same root piece, resulting in only the faster-growing taxa being isolated (Hambleton and Currah 1997). Endophytes that are difficult or impossible to culture may also be present in epacrid hair roots. This may be the case in the North American Ericaeae taxon *Gaultheria shallon*, for which direct DNA extraction and subsequent cloning of ITS sequences revealed the presence of a *Sebacina*-like basidiomycete that was not present in cultured assemblages of fungi from the same roots (Berch et al. 2002; Allen et al. 2003). Furthermore, these sequences were obtained from root segments that appeared to have mycorrhizal infection in epidermal cells, yet did not yield culturable endophytes. In contrast, where culturable endophyte assemblages from the *Epacris pulchella* hair roots were compared to fungal sequences obtained following direct DNA extraction from roots, the most-commonly isolated fungi were also commonly represented in the directly extracted DNA (D. Bougoure and J.W.G. Cairney, unpublished data). This clearly suggests that in the case of these *E. pulchella* plants the most common endophytes were culturable and that the observations from *G. shallon* do not serve as a paradigm for Ericaceae in general.

14.3 Diversity and Spatial Distribution of Endophyte Taxa in Epacrid Root Systems

Where multiple isolations have been made from root systems of individual epacrids in the field, an assemblage of endophyte taxa has invariably been recovered (McLean et al. 1999; Chambers et al. 2000). Midgley et al. (2002, 2004a) undertook intensive isolations from 5.0 mm long root pieces from throughout the hair root systems of two *Woollsia pungens* plants and a *Leucopogon parviflorus* plant at a sclerophyll forest site in temperate eastern Australia. Between 99 and 227 isolates were obtained from each plant and ITS-RFLP analysis suggested that the assemblage from each plant comprised 5 to 17 taxa. Most isolates (76–85%) in each assemblage from a single plant, however, were of the same ITS-RFLP-type, with a single RFLP-type found to dominate the root system of the *L. parviflorus* and a neighbouring *W. pungens* plant. Mapping the distribution of the RFLP types according to their position in the root systems from which they were isolated, demonstrated that the isolates that dominated the assemblages were widespread throughout the hair root systems (Midgley et al. 2002, 2004a). These dominant RFLP-types were shown to form typical ericoid mycorrhiza-like coil structures in gnotobiotic mycorrhizal infection experiments with *W. pungens* as host (Midgley 2003; Midgley et al. 2004a), suggesting that each root system was dominated by a single putative ericoid mycorrhizal fungal taxon. Only a few other RFLP-types from each root system formed mycorrhiza-like coils in *W. pungens* hair roots, with the remainder failing to form typical mycorrhizal structures in epidermal cells.

Genetic diversity in populations of the dominant putative ericoid mycorrhizal fungal taxa isolated from epacrid roots has been investigated using inter-simple sequence repeat (ISSR) PCR (Midgley et al. 2002, 2004a). These investigations revealed that, in the case of *W. pungens* and *L. parviflorus* in a dry sclerophyll forest habitat, the population of the dominant taxon within a root system comprised three to six genotypes. The population from each root system was, however, dominated (81–96% of isolates) by a single, spatially widespread genotype of the most abundant putative mycorrhizal taxon, with the remaining genotypes being relatively uncommon. In the neighbouring *W. pungens* and *L. parviflorus* plants dominated by the same putative ericoid mycorrhizal fungal taxon, each root system was dominated by a different genotype of that taxon (Midgey et al. 2004a). The existence of genotypes that are widespread within root systems indicates that there is a high probability that isolates from individual root segments in the same root system will be from the same mycelial individual. This will complicate investigations of the community ecology of ericoid mycorrhizal fungi, particularly where comparisons of relative abundance of taxa between sites is concerned, and may ultimately inform the spatial scale at which future sampling should be conducted.

As proposed for populations of the DSE *Phialocephala fortinii* associated with roots in a conifer stand, domination of the root system by a single genotype might result from early establishment and/or greater competitiveness of the dominant genotype (Grünig et al. 2002). In either case, local micro-scale edaphic conditions are likely to play a strong selective role in structuring the endophyte communities and populations. Putative mycorrhizal fungi from epacrids vary in their responses to water stress (Hutton et al. 1996b; Chen et al. 2003). Furthermore, the extent of ericoid mycorrhiza-like hair root colonisation varies seasonally for *W. pungens* in eastern Australian sclerophyll forests (Kemp et al. 2003). Indeed, in epacrids that are subject to a Mediterranean-type climate in south-western Australia, few hair roots survive the summer and apparent ericoid mycorrhizal infection disappears during the driest months. Infection then increases progressively with decreasing temperature and increasing rainfall (Hutton et al. 1994). Although no investigations have so far been conducted, it is thus possible that the relative abundance of different taxa and/or genotypes within root systems will be found to show considerable seasonal variation.

The same genotype of a single putative ericoid mycorrhizal fungal taxon was found to be present in the root systems of the neighbouring *W. pungens* and *L. parviflorus* plants (Midgey et al. 2004a). Interestingly, Liu et al. (1998) isolated identical genotypes of what, based on the >99% ITS sequence similarity between the two (Midgley et al. 2004a), appears likely to be the same taxon identified by Midgley et al. (2004a) from hair roots of neighbouring *W. pungens* plants at a different sclerophyll forest site. This implies that the phenomenon is not uncommon for putative mycorrhizal fungi of these epacrids, but whether it reflects the presence of single genets that were continuous between the two plants, or ramets that were confined to one or other root system remains unresolved. In contrast to ecto- and arbuscular-mycorrhizal fungi, ericoid mycorrhizal fungi are generally regarded as producing only limited mycelial growth in soil (Smith and Read 1997). This tenet is, however, based primarily on observations of *H. ericae* in mor-humus heathland vegetation communities in the northern hemisphere. It is possible that some ericoid mycorrhizal fungi of epacrids are capable of growing further from a host root into soil. Even if this is not the case, it is possible that the root systems of neighbouring epacrids were juxtaposed and that hyphal bridges do in fact interconnect epacrid plants. Such interconnectedness might have significant physiological and ecological consequences for the symbiotic partners (Robinson and Fitter 1999), and further investigation in this area is clearly warranted.

14.4 Mycorrhizal Status of Endophytes from Epacrids

Due to problems of germinating seeds and maintaining epacrid seedlings under sterile conditions in the laboratory, some of these experiments have been conducted using northern hemisphere Ericaceae hosts such as *Vaccinium macrocarpon* or *V. corymbosum* (Liu et al. 1998; Davies et al. 2003). Recent success in sterile propagation of epacrids such as *Epacris impressa* by micropropagation and *Woollsia pungens* from seed, however, have facilitated the use of epacrids as hosts for mycorrhiza infection testing (Fig. 14.1a,b) (McLean et al. 1998; Midgley et al. 2004a). ITS sequence data indicate that those endophytes that have been shown to produce ericoid mycorrhiza-like infection have affinity with either the *Hymenoscyphus ericae* aggregate or two broad groups of Helotiales ascomycetes (equivalent to "Assemblage A" and "Unknown 2 & possible relatives" in Berch et al. 2002) (McLean et al. 1998; Chambers et al. 2000; Berch et al. 2002; Davies et al. 2003; Midgley et al. 2004a). Endophytes in the *H. ericae* aggregate and both Helotiales groups have also been isolated from, and shown to form ericoid mycorrhizas with, northern hemisphere Ericaceae (Berch et al. 2002), suggesting that these groups are important mycorrhizal fungi of Ericaceae worldwide.

While *O. maius* and some *Thielavia*-like and *Capronia*-like endophytes from northern hemisphere Ericaceae have been shown to be ericoid

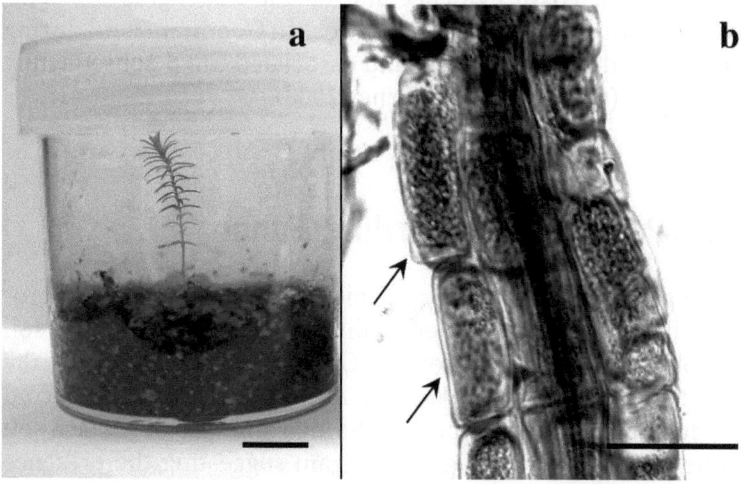

Fig. 14.1. a,b Testing for ericoid mycorrhiza formation by endophytes isolated from epacrid hair roots in gnotobiotic culture. **a** *Woollsia pungens* seedling in gnotobiotic culture with ericoid mycorrhizal fungi. **b** *W. pungens* hair root showing dense hyphal coils produced by an ericoid mycorrhizal fungus in epidermal cells (*arrows*) *Bars* **a** 1 cm, **b** 50 μm

mycorrhizal fungi (Berch et al. 2002), none of the isolates from epacrids that have close sequence identity to these fungi formed mycorrhiza-like structures in gnotobiotic infection experiments (Chambers et al. 2000; Midgley et al. 2004a). Certain *Thielavia*-like and *Capronia*-like isolates from northern hemisphere Ericaceae hosts also appear to be non-mycorrhizal, suggesting that further work is required to ascertain the extent of mycorrhiza-forming abilities of isolates in these groups. The only basidiomycete so far isolated from an epacrid hair root system failed to form ericoid mycorrhiza-like structures in *W. pungens* roots (Midgley et al. 2004a), suggesting that it, and the basidiomycete hyphae observed in *Dracophyllum secundum* epidermal cells by Allen et al. (1989), was probably a saprotroph. Although they can infect Ericaceae epidermal cells, the coils formed by isolates from epacrids that have close ITS sequence identity to *Phialocephala fortinii* are typical of those formed by DSE rather than ericoid mycorrhizal fungi (Davies et al. 2003; Midgley et al. 2004a). Interactions between DSE and their plant hosts remain poorly understood, and it appears that they may, under different circumstances, have mildly parasitic, neutral, or mildly beneficial effects on their hosts [Jumpponen 2001; see Chaps. 7 (Sieber and Grünig) and 15 (Schulz)]. The significance of infection of epacrids by DSE thus remains unclear.

The broad taxonomic congruence between the fungi that appear to form ericoid mycorrhizas with epacrids and those that form associations with Ericaceae from other continents is consistent with the hypothesis that ericoid mycorrhizal associations evolved in a common ancestral host plant group (Cullings 1996). Fossil evidence and extant biogeographical patterns suggest that the association probably arose in Gondwanaland during the Cretaceous period, and that hosts and associated fungi subsequently radiated northwards (Cullings 1996; Cairney 2000).

14.5 Saprotrophic Potential of Mycorrhizal Fungi

The presence of propagules of putative ericoid mycorrhizal fungi in sclerophyll forest soil has been demonstrated by direct DNA extraction from soil from which roots were removed (Chen and Cairney 2002). Similarly, in the seasonally drought-affected soils of Western Australian heathlands, ericoid mycorrhizal inoculum appears to persist through summer in surface soil above the zone of most hair root growth, again suggesting the presence of host-free fungal propagules (Hutton et al. 1996a). Further evidence of the abilities of ericoid mycorrhizal fungi to persist in the absence of their Ericaceae hosts has recently been obtained for the northern hemisphere taxon *Erica arborea*, which became infected by ericoid mycorrhizal fungi when planted in a mature *Quercus ilex* forest that lacked Ericaceae vegetation (Bergero et al. 2003). Unfortunately the techniques used in these studies cannot discriminate between actively growing mycelia and inactive propagules such as asexual spores. However, the activities of the fungi in axenic culture suggest that, to some extent at least, they are capable of saprotrophic growth in the absence of an Ericaceae host.

It is well established that, via production of an array of extracellular enzymes, *H. ericae* has considerable ability to exist as a free-living saprotroph (reviewed by Cairney and Burke 1998), and there is evidence that other mycorrhizal fungi of northern hemisphere Ericaceae have similar abilities (Piercey et al. 2002; Varma and Bonfante 1994). Relatively little is known regarding the saprotrophic potential of endophytes from epacrids. Midgley et al. (2004c), however, have shown that two frequently isolated putative ericoid mycorrhizal fungal taxa from *W. pungens* can utilise a range of compounds as sole carbon sources during growth in axenic culture. Thus, along with hexoses, these taxa can derive carbon from xylan and cellulose, suggesting production of 1-4-β-xylanase and β-D-xylosidase, along with a complete cellulase complex (Midgley et al. 2004c). While these enzyme activities are doubtless important in the penetration of host epidermal cell walls during the establishment of mycorrhizal symbiosis, they probably also facilitate a degree of saprotrophic growth and may be functionally important in the process of symbiotic nutrient acquisition.

14.6 Symbiotic Functioning of Mycorrhizal Fungi

The mutualistic nature of ericoid mycorrhizal infection of epacrids has not yet been confirmed by gnotobiotic culture nutrient transfer experiments (see McLean et al. 1998; Anthony et al. 2000; Cairney and Ashford 2002). Nonetheless, the putative ericoid mycorrhizal fungi can utilise a range of amino acids and simple proteins as sole nitrogen sources, along with inositol hexaphosphate and DNA as sole sources of phosphorus (Chen et al. 1999; Whittaker and Cairney 2001; Midgley et al. 2004b). Their abilities to access nitrogen and phosphorus thus appear to be on a par with *H. ericae* and other ericoid mycorrhizal fungi from northern hemisphere Ericaceae, which are known to acquire nitrogen and phosphorus from these substrates and effect transfer of the elements to plant hosts (see Smith and Read 1997; Xiao and Berch 1999). Given the relatedness of the putative ericoid mycorrhizal fungi of epacrids to these known mycorrhizal fungi, the similarities in the infections formed in host epidermal cells and their abilities to utilise organic nitrogen and phosphorus sources, however, it seems probable that they function in a manner similar to their northern hemisphere counterparts.

In addition to utilising organic nitrogen substrates, Midgley et al. (2004b) found that, as is the case for *H. ericae*, putative mycorrhizal fungi from epacrids are efficient users of nitrate. This contrasts with ectomycorrhizal fungi from the same eastern Australian sclerophyll forest habitats, which have limited abilities to utilise nitrate (Anderson et al. 1999; Sawyer et al. 2003). Although the soils in such forests are likely to be ammonifying (Connell et al. 1995), the ability to utilise nitrate may be advantageous to the mycorrhizal fungi and their epacrid hosts following fire, when nitrate may be relatively abundant (Stewart et al. 1993). Many epacrids are killed, but regenerate rapidly from seed, following fire, while others may survive fire by resprouting (Bell et al. 1996). There is also evidence that epacrid abundance increases following low intensity fires (Morrison 2002) and that putative ericoid mycorrhizal fungi can be present in some sclerophyll forest soils within a few days of a low intensity fire (Chen and Cairney 2002). Since non-mycorrhizal epacrids have only limited abilities to utilise NO_3^- for growth (Stewart et al. 1993), these observations suggest that the fungi may be of particular importance to the success of epacrids in fire-prone vegetation communities.

It has been postulated that different mycorrhizal fungal taxa or genotypes might differentially access nutrient sources in soil, and that this might be important in increasing overall nitrogen and/or phosphorus uptake by their plant hosts (Cairney et al. 2000; Koide 2000). Midgley et al. (2004b) investigated the relative abilities of six genotypes of an *H. ericae*-like putative mycorrhizal fungus from *W. pungens* and six isolates of an unknown Helotiales putative mycorrhizal fungal taxon from *W. pungens* and *L. parviflorus* to utilise a range of simple inorganic and organic nitrogen and phosphorus sources for growth. All isolates were found to utilise inorganic forms of the elements, acidic, neutral and basic amino acids, along with protein, phosphomonoesterase and phosphodiesterase. Although some intraspecific variation was observed on all substrates, one taxon produced significantly more biomass on most substrates than the other. For all substrates, however, the difference between the two taxa was considerably less than an order of magnitude, leading Midgley et al. (2004b) to conclude that both taxa would confer broadly similar nutritional benefits to their hosts. This, and the fact that screening of single isolates of a range of putative ericoid mycorrhizal fungal taxa from *W. pungens* suggests that all are broadly similar in their abilities to utilise organic nitrogen and phosphorus substrates (Chen et al. 1999; Whittaker and Cairney 2001), might appear to contradict the proposal that different isolates/genotypes vary in their abilities to enhance host nutrition. This, however, may not be the case. The efficiency with which the various fungi transfer nitrogen and/or phosphorus from the substrates to the host plant remains to be tested, and different taxa/genotypes may differ considerably in this respect.

14.7 Conclusions

Although research to date has been confined to a small number of epacrid taxa from a limited range of habitats, it appears that most endophytes that have been isolated from epacrid hair roots are probably ericoid mycorrhizal fungi. An array of mainly Helotiales ascomycetes forms putative ericoid mycorrhizal associations with epacrids, but root systems of individual plants in the field are dominated by a small number of taxa, and populations of these dominated by single genotypes. Studies of mycorrhizal fungal diversity in natural habitats are at an early stage and in the future will need to take into account the spatial distribution of mycelial individuals if meaningful comparisons of communities in different habitats are to be made. Functional aspects of the symbiotic relationships between putative ericoid mycorrhizal fungi and their epacrid hosts remain to be investigated, however it is likely that these will follow the established paradigm for ericoid mycorrhizal associations of northern hemisphere Ericaecae.

Acknowledgements. I thank D.J. Midgley for his permission to use Fig. 14.1b.

References

Allaway WG (1996) Biology of the Epacridaceae. Ann Bot 77:290–291

Allen TR, Millar T, Berch SM, Berbee ML (2003) Culturing and direct DNA extraction find different fungi from the same ericoid mycorrhizal roots. New Phytol 160:255–272

Allen WK, Allaway WG, Cox GC, Valder PG (1989) Ultrastructure of mycorrhizas of *Dracophyllum secundum* R. Br. (Ericales: Epacridaceae). Aust J Plant Physiol 16:147–153

Anderson IC, Chambers SM, Cairney JWG (1999) Intra- and interspecific variation in patterns of organic and inorganic nitrogen utilisation by three Australian *Pisolithus* species. Mycol Res 103:1579–1587

Anthony J, McLean CB, Lawrie AC (2000) In vitro propagation of *Epacris impressa* (Epacridaceae) and the effects of clonal material. Aust J Bot 48:215–221

Bell TL, Pate JS, Dixon KW (1996) Relationships between fire response, morphology, root anatomy and starch distribution in south-western Australian Epacridaceae. Ann Bot 77:357–364

Bellgard SE (1991) Mycorrhizal associations of plant species in Hawkesbury sandstone vegetation. Aust J Bot 39:357–364

Berch SM, Allen TR, Berbee ML (2002) Molecular detection, community structure and phylogeny of ericoid mycorrhizal fungi. Plant Soil 244:55–66

Bergero R, Girlanda M, Bello F, Luppi AM, Perotto S (2003) Soil persistence and biodiversity of ericoid mycorrhizal fungi in the absence of the host plant in a Mediterranean ecosystem. Mycorrhiza 13:69–75

Briggs CL, Ashford AE (2001) Structure and composition of the thick wall in hair root epidermal cells of *Woollsia pungens*. New Phytol 149:219–232

Cairney JWG (2000) Evolution of mycorrhiza systems. Naturwissenschaften 87:467–475

Cairney JWG, Ashford AE (2002) Biology of mycorrhizal associations of epacrids (Ericaceae). New Phytol 154:305–326

Cairney JWG, Burke RM (1998) Extracellular enzyme activities of the ericoid mycorrhizal endophyte *Hymenoscyphus ericae* (Read) Korf & Kernan: their likely roles in decomposition of dead plant material in soil. Plant Soil 205:181–192

Cairney JWG, Sawyer NA, Sharples JM, Meharg AA (2000) Intraspecific variation in nitrogen source utilisation by isolates of the ericoid mycorrhizal fungus *Hymenoscyphus ericae* (Read) Korf & Kernan. Soil Biol Biochem 32:1319–1322

Chambers SM, Liu G, Cairney JWG (2000) ITS rDNA sequence comparison of ericoid mycorrhizal endophytes from *Woollsia pungens*. Mycol Res 104:168–174

Chen A, Chambers SM, Cairney JWG (1999) Utilisation of organic nitrogen and phosphorus sources by mycorrhizal endophytes of *Woollsia pungens* (Cav.) F. Muell. (Epacridaceae). Mycorrhiza 8:181–187

Chen DM, Cairney JWG (2002) Investigation of the influence of prescribed burning on ITS profiles of ectomycorrhizal and other soil fungi at three Australian sclerophyll forest sites. Mycol Res 106:532–540

Chen DM, Khalili K, Cairney JWG (2003) Influence of water stress on biomass production by isolates of an ericoid mycorrhizal endophyte of *Woollsia pungens* and *Epacris microphylla* (Ericaceae). Mycorrhiza 13:173–176

Connell MJ, Raison RJ, Khanna PK (1995) Nitrogen mineralisation in relation to site history and soil properties for a range of Australian forest soils. Biol Fert Soil 20:213–220

Copeland HF (1954) Observations on certain Epacridaceae. Am J Bot 41:215–222

Cullings KW (1996) Single phylogenetic origin of ericoid mycorrhizae within the Ericaceae. Can J Bot 74:1896–1909

Davies PW, Mclean CB, Bell TL (2003) Root survey and isolation of fungi from alpine epacrids (Ericaceae). Aust Mycol 22:4–10

Grünig CR, Sieber TN, Rogers SO, Holdernrieder O (2002) Spatial distribution of dark septate endophytes in a confined forest plot. Mycol Res 106:832–840

Hambleton S, Currah RS (1997) Fungal endophytes from the roots of alpine and boreal Ericaceae. Can J Bot 75:1570–1581

Hutton BJ, Dixon KW, Sivasithamparam K (1994) Ericoid endophytes of Western Australian heaths (Epacridaceae). New Phytol 127:557–566

Hutton BJ, Dixon KW, Sivasithamparam K, Pate JS (1996a) Inoculum potential of ericoid endophytes of Western Australian heaths (Epacridaceae). New Phytol 134:665–672

Hutton BJ, Sivasithamparam K, Dixon KW, Pate JS (1996b) Pectic zymograms and water stress tolerance of endophytic fungi isolated from Western Australian heaths (Epacridaceae). Ann Bot 77:399–404

Jumpponen A (2001) Dark septate endophytes – are they mycorrhizal? Mycorrhiza 11:207–211

Keighery GJ (1996) Phytogeography, biology and conservation of Western Australian Epacridaceae. Ann Bot 77:347–355

Kemp E, Adam P, Ashford AE (2003) Seasonal changes in hair roots and mycorrhizal colonization in *Woollsia pungens* (Cav.) F. Muell. (Epacridaceae). Plant Soil 250:241–248

Khan AG (1978) Vesicular arbuscular mycorrhizas in plants colonising black wastes from bituminous coal mining in the Illawarra region of New South Wales. New Phytol 81:53–63

Koide RT (2000) Functional complementarity in the arbuscular mycorrhizal symbiosis. New Phytol 147:233–235

Kron KA, Judd WS, Stevens PF, Crayn DM, Anderberg AA, Gadek PA, Quinn CJ, Luteyn JL (2002) Phylogenetic classification of Ericaceae: molecular and morphological evidence. Bot Rev 68:335–423

Liu G, Chambers SM, Cairney JWG (1998) Molecular diversity of ericoid mycorrhizal endophytes isolated from *Woollsia pungens*. New Phytol 140:145–153

McGee P (1986) Mycorrhizal associations of plant species in a semiarid community. Aust J Bot 34:585–593

McLean C, Lawrie AC (1996) Patterns of root colonization in epacridaceous plants collected from different sites. Ann Bot 77:405–411

McLean CB, Anthony J, Collins RA, Steinke E, Lawrie AC (1998) First synthesis of ericoid mycorrhizas in the Epacridaceae under axenic conditions. New Phytol 139:589–593

McLean CB, Cunnington JH, Lawrie AC (1999) Molecular diversity within and between ericoid endophytes from the Ericaceae and Epacridaceae. New Phytol 144:351–358

Midgley DJ (2003) Diversity and distribution of mycorrhizal and root-associated fungal endophytes in *Leucopogon parviflorus* and *Woollsia pungens* (Ericaceae). PhD Thesis, University of Western Sydney, Australia

Midgley DJ, Chambers SM, Cairney JWG (2002) Spatial distribution of fungal endophyte genotypes in a *Woollsia pungens* (Ericaceae) root system. Aust J Bot 50:559–565

Midgley DJ, Chambers SM, Cairney JWG (2004a) Distribution of ericoid mycorrhizal endophytes and root-associated fungi in neighbouring Ericaceae plants in the field. Plant Soil 259:137–151

Midgley DJ, Chambers SM Cairney JWG (2004b) Inorganic and organic substrates as sources of nitrogen and phosphorus for multiple genotypes of two ericoid mycorrhizal fungal taxa from *Woollsia pungens* Cav. (Muell.) and *Leucopogon parviflorus* (Andr.) Lindl. (Ericaceae). Aust J Bot 52:63–71

Midgley DJ, Chambers SM, Cairney JWG (2004c) Utilisation of carbon substretae by multiple genotypes of ericoid mycorrhizal fungal endophytes from eastern Australian Ericaceae. Mycorrhiza 14:245–251

Morrison DA (2002) effects of fire intensity on plant species composition of sandstone communities in the Sydney region. Aust Ecol 27:433–441

Perotto S, Girlanda M, Martino E (2002) Ericoid mycorrhizal fungi: some new perspectives on old acquaintances. Plant Soil 244:41–53

Piercey MM, Thormann MN, Currah RS (2002) Saprobic characteristics of three fungal taxa from ericalean roots and their association with the roots of *Rhododendron groenlandicum* and *Picea mariana* in culture. Mycorrhiza 12:175–180

Powell JM, Crayn DM, Gadek PA, Quinn CJ, Morrison DA, Chapman AR (1996) A reassessment of relationships within Epacridaceae. Ann Bot 77:305–315

Read DJ (1996) The structure and function of the ericoid mycorrhizal root. Ann Bot 77:365–374

Reed ML (1989) Ericoid mycorrhizas of Styphelieae: intensity of infection and nutrition of the symbionts. Aust J Plant Physiol 16:155–160

Reed ML (1996) Diversity of mycorrhizal fungi in the roots of epacrids. In: Hopper SD, Chappill JA, Harvey MS, George AS (eds) Gondwanan heritage: past, present and future of the Western Australian biota. Surrey Beattie, Chipping Norton, Australia, pp 309–313

Robinson D, Fitter A (1999) The magnitude and control of carbon transfer between plants linked by a common network. J Exp Bot 50:9–13

Sawyer NA, Chambers SM, Cairney JWG (2003) Utilisation of inorganic and organic nitrogen sources by *Amanita* species native to temperate eastern Australia. Mycol Res 107:413–420

Sharples JM, Chambers SM, Meharg AA, Cairney JWG (2000) Genetic diversity of root-associated fungal endophytes from *Calluna vulgaris* at contrasting field sites. New Phytol 148:153–162

Smith SE, Read DJ (1997) Mycorrhizal symbiosis. Academic Press, London

Steinke E, Williams PG, Ashford AE (1996) The structure and fungal associates of mycorrhizas in *Leucopogon parviflorus* (Andr.) Lindl. Ann Bot 77:413–419

Stewart GR, Pate JS, Unkovich M (1993) Characteristics of inorganic nitrogen assimilation of plants in fire-prone Mediterranean type vegetation. Plant Cell Environ 16:351–363

Varma A, Bonfante P (1994) Utilisation of cell-wall related carbohydrates by ericoid mycorrhizal endophytes. Symbiosis 16:301–313

Vrålstad T, Myhre E, Schumacher T (2002) Molecular diversity and phylogenetic affinities of symbiotic root-associated ascomycetes of the Helotiales in burnt and metal polluted habitats. New Phytol 155:131–148

Whittaker SP, Cairney JWG (2001) Influence of amino acids on biomass production by ericoid mycorrhizal endophytes from *Woollsia pungens* (Epacridaceae). Mycol Res 105:105–111

Xiao G, Berch SM (1999) Organic nitrogen use by salal ericoid mycorrhizal fungi from northern Vancouver Island and impacts on growth in vitro of *Gaultheria shallon*. Mycorrhiza 9:145–149

15 Mutualistic Interactions with Fungal Root Endophytes

Barbara Schulz

15.1 Introduction

With few exceptions, colonisation of a plant host is beneficial for the fungus, assuring it a supply of nutrients and shelter from most abiotic stressors. Previously, within plant roots only the symbioses of mycorrhizal fungi were considered to be mutualistic. Recently, it has been recognised that many other fungi, and in particular endophytic fungi, can participate in mutualistic root symbioses (e.g. Sieber 2002; Brundrett 2002; Schulz and Boyle 2005).

There are various potential benefits for the host in mutualistic interactions with endophytic fungi, for example induction of defence metabolites potentially active against pathogens (Schulz et al. 1999; Mucciarelli et al. 2003; Arnold and Herre 2003), endophytic secretion of phytohormones (Holland 1997; Rey et al. 2001; Römmert et al. 2002; Tudzynski and Sharon 2002), mobilisation of nutrients for the host from the rhizosphere (Jumpponen et al. 1998; Caldwell et al. 2000; Usuki et al. 2002) and/or an alteration of host metabolism (Jallow et al. 2004). Colonisation by fungal root endophytes may lead to induced disease resistance (Picard et al. 2000; Benhamou and Garand 2001), improved growth of the host (Kimura et al. 1992; Jumpponen 2001; Mucciarelli et al. 2002; Ernst et al. 2003), abiotic stress tolerance (Barrow and Aaltonen 2001; Redman et al. 2002; Barrow 2003) or protection from pathogenic competitors and insect predators of the host through synthesis of antagonistic fungal secondary metabolites (Schulz et al. 1995; Hallmann and Sikora 1996; Schulz et al. 2002; Miller et al. 2002; Selosse et al. 2004; see Chap. 8 by Bacon and Yates). This chapter reviews results dealing with mutualistic interactions between non-mycorrhizal root-colonising endophytic fungi with their plant hosts. When reading this chapter it is important to bear in mind that within the realms of our present knowledge, far from all of the interactions with fungal root endophytes are beneficial for both partners.

Barbara Schulz: Technical University of Braunschweig, Institute of Microbiology, Spielmannstraße 7, 38106 Braunschweig, Germany, E-mail: b.schulz@tu-bs.de

Soil Biology, Volume 9
Microbial Root Endophytes
B. Schulz, C. Boyle, T. N. Sieber (Eds.)

15.2 Colonisation and Histology

To the extent that histological studies have been undertaken, growth of fungal endophytes in roots is usually extensive and can be inter- and/or intra-cellular (Boyle et al. 2001; Sieber 2002; Schulz and Boyle 2005). Most of the histological studies have dealt with endophytic colonisation by *Fusarium* and dark septate endophytes (DSE; Jumpponen and Trappe 1998; Sieber 2002).

Histological studies of colonisation by DSE have focussed on *Phialocephala fortinii*, but also on other DSE including *Phialophora* spp., other *Phialocephala* spp., *Chloridium paucisporum*, *Heteroconium chaetospira* and *Leptodontidium orchidicola*. The hyphae of DSE vary from thin and hyaline to melanised, sometimes forming black microsclerotia. The hyaline hyphae grow both intercellularly and within the vascular cylinder and, upon favourable nutrient status of the host, contain lipid vacuoles (Barrow and Aaltonen 2001; Barrow 2003). Barrow (2003) suggests that polymorphic fungal structures, which are found primarily within the sieve elements and accumulate massive quantities of lipids, are protoplasts. Not only DSE, but also other endophytes, e.g. *Cryptosporiopsis* sp., colonise the vascular cylinder (Schulz and Boyle 2005). The hyphae of root endophytes often develop intracellular coils, e.g. *H. chaetospira* and *Oidiodendron maius* (Usuki and Narisawa 2005), as does the basidiomycete, *Piriformospora indica* (Varma et al. 2000). These presumably increase the absorption of assimilates or the exchange of nutrients.

In axenic culture, the DSE *Phialocephala* colonised the surface and the inner cortex of the roots of seedlings of Norway spruce and Scotch pine (Sieber 2002) and *Larix decidua* (Schulz et al. 1999; A.K. Römmert unpublished) with mats of mycelia growing within the roots both inter- and intra-cellularly. From roots of Norway spruce and Scotch pine collected in the field, the density of mycelial and intracortical development was much lower (Sieber 2002), suggesting that, under natural conditions, plant defence limited the density of colonisation. However, in arid habitats, Barrow (2003) found a highly branched extraradical hyphal network of hyaline vacuolated hyphae in a mucilaginous gel covering roots of *Bouteloua*. The polysaccharide mucilage, which was apparently produced by the fungi, stores moisture under arid conditions (see Sect. 15.6).

Depending on the host being colonised, some DSE can develop mycorrhizal structures. For example, in the roots of at least 19 plant species, the dematiaceous hyphomycete *H. chaetospira* grew intra- or intercellularly within the cortical cells (Narisawa et al. 1998; Usuki et al. 2002; Ohki et al. 2002). However, within the roots of *Rhododendron obtusum* var. *kaempferi*, it developed typical ericoid mycorrhizas (Usuki and Narisawa

2005). *P. fortinii* has even been found to form a Hartig net and a thin patchy mantle, considered the anatomical hallmarks of ectomycorrhizae, both in axenic culture of seedlings of *Salix glauca* (Fernando and Currah 1996) and *Betula platyphylla* (Hashimoto and Hyakumachi 2001), with the roots of some nursery stocks of *Pinus banksiana*, *Pinus contorta* and *Pinus glauca* (Danielson and Visser 1990) and with the roots of *Populus tremula* x *Populus tremuloides* grown in an experimental field plot (Kaldorf et al. 2004). This is also the case for the DSE *Phialophora finlandia*, which formed ectendomycorrhizal associations with the roots of *Pinus resinosa* (Lobuglio and Wilcox 1988) and ectomycorrhizal associations with the roots of several trees (Vralstad et al. 2002).

The growth modus of avirulent strains of *Fusarium* spp. is variable, but often differs from that of virulent strains. Bacon and Hinton (1996) found that only pathogenic strains of *F. moniliforme* colonised maize both inter- and intracellularly; growth of the avirulent isolate was intercellular. However, other avirulent strains of *Fusarium* spp. colonised the roots of barley (Boyle et al. 2001) and pea (Benhamou and Garand 2001) both inter- and intra-cellularly. Whereas active growth of the avirulent strain of *Fusarium oxysporum* was restricted to the root surface, epidermis and outer cortex of the pea roots, the pathogenic strain of *F. oxysporum* rapidly colonised epidermis, cortex, endodermis and paratracheal parenchyma cells (Benhamou and Garand 2001).

The colonisation modus of *Stagonospora* spp. in *Phragmites australis* is exemplary for fungi that initially colonise the roots, but then apparently grow systemically within the entire plant following vertical transmission (Ernst et al. 2003). Similarly, Sieber et al. (1988) isolated *Stagonospora nodorum* as an endophyte from both roots and shoots of wheat. Another fungus that colonises both roots and shoots is a non-sporulating endophyte of *Mentha piperita*. The endophyte formed an external enveloping mycelial sheet around the root with only little penetration of the root cortex (Mucciarelli et al. 2003), but was detected in the parenchymatic cells of mature leaves, especially during senescence (Mucciarelli et al. 2002).

The histology of fungal root colonisations in mutualistic interactions is extremely variable, in part because any one species can exhibit extreme phenotypic plasticity. Colonisation can be inter- and/or intra-cellular, limited to the roots or apparently systemic within the entire plant. The morphologies vary from apparent protoplasts to very fine undifferentiated hyaline hyphae to highly differentiated ectomycorrhizas. So, it seems that neither fungal morphology nor mode of colonisation is decisive for determining whether or not an interaction is mutualistic. But perhaps the quantity of colonisation is a decisive factor, since extensive colonisation is common to all of the investigated mutualistic interactions.

15.3 Secondary Metabolites

A strikingly high proportion of endophytic fungi (80%) produce biologically active metabolites in vitro in tests for antibacterial, fungicidal and herbicidal activities (Schulz et al. 2002). Most of the investigations dealing with the synthesis of endophytic metabolites active against phytopathogens (Schulz et al. 1995; Tan and Zou 2001; Schulz et al. 2002; Fang-ting et al. 2004) and against predators (Azevedo et al. 2000) have not specified from which organ the endophytes were isolated.

Fungal secondary metabolites may play a role within the host, and/or have an ecological significance. As hypothesised by Demain (1980): "If a fungus can produce metabolites in vitro, they must also have a function in nature" Fungi would not retain the multienzyme reaction sequences required for the synthesis of secondary metabolites without some beneficial effect for survival. As virulence factors, we have hypothesised that secondary metabolites are involved in maintaining a balance of antagonisms in the interaction with the host (Schulz et al. 1999; Schulz and Boyle 2005; see Chap. 1 by Schulz and Boyle). However, they may also have antagonistic functions.

Some endophytes, for example *Fusarium* spp., that colonise both the shoots and roots of numerous hosts, synthesise a number of toxins including beauvericin, a cyclic hexadepsipeptide with insecticidal properties (see Chap. 8 by Bacon and Yates; Kuldau and Yates 2000; Miller 2001). Beauvericins are produced by at least 12 *Fusarium* species, and protect infected plants against herbivoric insects (see Chap. 8 by Bacon and Yates). Fungal root endophytes also synthesise metabolites toxic to the nematode *Meloidogyne incognita*. For example, both phomalactone, synthesised by *Verticillium chlamydosporium* (Khambay et al. 2000), and metabolites present in the culture filtrate of a non-pathogenic *F. oxysporum* reduced mobility of the nematode within 10 min of exposure (Hallmann and Sikora 1996).

Many root endophytic fungi produce antimicrobial metabolites (Schulz et al. 2002), e.g. the antibacterial and antifungal metabolite(s) produced in vitro by *Cryptosporiopsis* sp. from *Larix decidua* (Schulz et al. 1995). These included the antifungal metabolite mycorrhizin, which in situ could protect the host from phytopathogenic fungi (Schulz et al. 1995). Hallmann and Sikora (1996) found that the culture filtrate of *F. oxysporum* strain 162 was not only toxic to *M. incognita*, but was also antifungal, inhibiting soil-borne plant pathogens.

Bultman and Murphy (2000) suggested that endophytic *Neotyphodium* spp. are stimulated to increase production of mycotoxins in shoots of grasses after damage to the host has occurred, the adaptive significance being apparent. A similar effect could occur when roots colonised by endophytic fungi are injured. Fungal secondary metabolites may also play roles in signalling and growth enhancement (see Sect. 15.4).

In conclusion, since most of the fungal root isolates can produce biologically active secondary metabolites in vitro, it seems probable that they are also produced in planta. They can be antagonistic against plant predators and microbial antagonists, in both cases suppressing disease. But they may also play roles within the host, e.g. in maintaining a balance of antagonism between endophyte and host.

15.4 Growth Enhancement

The non-mycorrhizal fungal root colonisers that have been reported to improve growth of their hosts have various growth modi and belong to diverse genera, including *Chaetomium, Cladorrhinum, Cryptosporiopsis, Fusarium, Heteroconium, Oidiodendron, Phialocephala, Piriformospora* and *Stagonospora.*

Stagonospora spp., and a non-sporulating endophyte of Mentha piperita, are exemplary for endophytes that grow systemically in roots and shoots of their hosts (see Sect. 15.2). When isolates of three species of *Stagonospora* were reinoculated into axenic host seedlings of *Phragmites australis*, all increased growth significantly (Ernst et al. 2003). The authors note that this is only the third reported case, besides those dealing with *Neotyphodium*, in which a seed-transmitted fungus enhanced the biomass of its host. Mucciarelli et al. (2002, 2003) speculated that improved growth of *M. piperita* might be due to a better nutrient supply or, alternatively, to the synthesis of plant growth hormones by a non-sporulating endophytic fungus.

In contrast, colonisation in most of the studied interactions is limited to the roots, e.g. that by *Phoma* in *Vulpia ciliata* spp. *ambigua*, which increased shoot biomass, root lengths, and tiller numbers of the host (Newsham 1994), and that of barley by *Chaetomium* or *Chaetomium globosum*, which increased root fresh weight (Vilich et al. 1998). Gasoni and Stegman de Gurfinkel (1997) suggested that increased phosphorous uptake, as observed in cotton roots colonised by *Cladorrhinum foecumdissimum*, was responsible for promoting growth of the host. Similarly, results obtained by Jumpponen and Trappe (1998) led Jumpponen (1999) to the conclusion that growth enhancement by the DSE may be due to improved phosphorous and nitrogen uptake (Jumpponen et al. 1998) or, in a closed system, to increased availability of carbohydrates and/or CO_2, both resulting from fungal metabolism (Jumpponen and Trappe 1998).

It is also possible that growth enhancement is due to indirect acquisition of nutrients obtained saprophytically from the endophyte from the rhizosphere. Caldwell et al. (2000) suggested that due to their hydrolytic capabilities, DSE are able to grow both biotrophically and saprophytically. Of note: they found no evidence for lignolytic enzymes. They hypothesised that the fungi may access litter and detrital carbon, nitrogen and phosphorous, making this available to the host. On the basis of finding a continuous hyphal network extending from the rhizosphere to the vascular cylinder, Barrow (2003) also suggested that there is potential for bidirectional carbon transport. The situation could be similar to that hypothesised for *Oidiodendron maius*, which not only develops ericoid mycorrhizas within the host, but also can grow saprophytically in peat (Rice and Currah 2002, see Chap. 13 by Rice and Currah) and perhaps supplies its host with organic nutrients from the rhizosphere. Inter-plant connections may also be involved in growth enhancement, as suggested by Sen et al. (1999), who found evidence that a common population of *Ceratobasidium cornigerum* occupies roots of orchid and pine. DSE may even replace arbuscular mycorrhiza (AM) and ectomycorrhizal fungi at sites with extreme environmental conditions (Sieber 2002; see Chap. 7 by Sieber and Grünig). Co-inoculation of tomato with a non-pathogenic strain of *Fusarium oxysporum* and the AM fungus *Glomus coronatum* resulted in better mycorrhization, but did not improve growth more than mono-inoculation (Diedhiou et al. 2003).

Colonisation by the DSE *Phialocephala* does not always exert positive effects on the host: in axenic culture it significantly retarded growth of *Betula platyphylla* var. *japonica* seedlings. Nevertheless, *P. fortinii* formed typical ectomycorrhiza with *B. platyphylla*, underlining the plasticity of the interaction (Hashimoto and Hyakumachi 2001). Whether or not colonisation by DSE improves growth of the host depends on environmental conditions and on the metabolic status of the host. Usuki et al. (2002) found that variance of the growth medium affected the mode of colonisation by *Heteroconium chaetospira* in seedlings of Chinese cabbage. Colonisation and growth enhancement were best, and intracellular, in peat moss amended with 0.1% glucose. At higher glucose concentrations both frequency of colonisation and plant growth decreased, and fungal colonisation was restricted mostly to the intercellular regions of epidermal and cortical cells, leading to the formation of microslerotia.

Not only colonisation, but also culture extracts of the endophyte can improve growth. Colonisation of the roots of seedlings of *Larix decidua* by the endophytes *P. fortinii* and *Cryptosporiopsis* sp., both of which had been isolated from roots of the host, significantly improved lengths (Schulz et al. 2002) and dry weights (Römmert et al. 2002) of both roots and shoots (Fig. 15.1). Additionally, disease symptoms resulting from the stress of axenic culture decreased (Schulz et al. 1999; Römmert et al. 2002). Similarly,

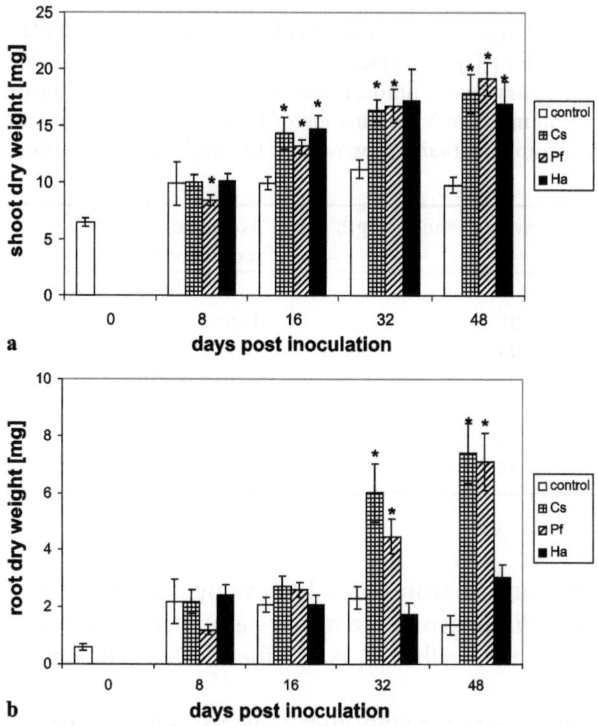

Fig. 15.1. Influence of colonisation on dry weights of shoots (**a**) and roots (**b**) of *Larix decidua*, cultivated for 3 months axenically in a synthetic medium in expanded clay and inoculated at day 0 with either an endophyte, *Cryptosporiopsis* sp. (Cs) or *Phialocephala fortinii* (Pf), or a pathogen, *Heterobasidion annosum* (Ha). $n = 29-55$, $*P < 0.05$, significant in comparison to the control according to Kruskal-Wallis ANOVA on ranks (Schulz et al. 2002)

Mucciarelli et al. (2003) found that colonisation by a non-sporulating endophyte increased the biomass of the roots of peppermint. Application of a methanolic culture extract of *P. fortinii* to the seedlings also increased host biomass (Table 15.1; Römmert et al. 2002), as Varma et al. (2000) had reported for shoots of maize not only when colonised by *Piriformospora indica*, but also following treatment with methanolic mycelial culture extracts.

It is possible that the enhancement of growth of *L. decidua* was, perhaps in part, due to synthesis of indole acetic acid (IAA) by the endophytes studied, since both the endophytes synthesised IAA in vitro (Römmert et al. 2002). This hypothesis is supported by the fact that colonisation by the pathogen *Heterobasidion annosum*, which also produced IAA in vitro, initially improved growth of shoots of *L. decidua* (Fig. 15.1; Römmert et al. 2002). The synthesis of phytohormones is not unusual for fungal pathogens

Table 15.1. Addition of methanolic culture extract of *Phialocephala fortinii* grown in MMNA liquid medium (Schulz et al. 1999) for 14 days to surface-sterilised seedlings of *Larix decidua* (approx. 3 months old) and *Triticum aestivum* (approx. 1 week old) on filter paper, previously germinated on biomalt agar medium (5% w/v). Evaluation of *L. decidua* after 28 days, of *T. aestivum* after 7 days of incubation. Methanolic extracts of the sterile culture medium were used as the controls

	Average shoot length (cm)	Average dry weight of root+shoot (mg)
L. decidua		
+ Culture extract ($n = 8$)	1.0*	44.0*
Control ($n = 10$)	0.08	16.9
T. aestivum L.		
+ Culture extract ($n = 20$)	0.84	1.6
Control ($n = 20$)	0.82	2.4

$*P < 0.05$

(Tudzynski 1997; Tudzynski and Sharon 2002). For example, the yield loss caused by infections with *Pythium* group F, a minor pathogen that induces symptomless infections in tomato cultures, is ascribed to IAA (Rey et al. 2001).

However, small molecular weight metabolites may also be responsible for growth promotion, as was found by Kimura et al. (1992), who identified the novel altechromones A and B in the culture broth of an *Alternaria* sp. When these substances were applied to the seedlings of lettuce, root growth was improved by ~ 50%. Similarly, fumonisin, which on the one hand is a potent mycotoxin of *Fusarium*, also improves development of the roots of maize (Yates et al. 1997).

Host adaptation may be involved in the growth-enhancing effects of some root endophytes. In contrast to colonisation of the roots of seedlings of *L. decidua* by *P. fortinii*, colonisation of *Triticum aestivum* L. (wheat) led to no significant enhancement of growth (data not shown). The culture extract was also ineffective when applied to this non-host (Table 15.1). Similarly, inoculation of the roots of aseptically grown seedlings of *Carex firma* and *C. curvula* with a DSE led to a significant increase in production of dry matter in *C. firma*, but not when the fungus was inoculated into *C. curvula* (Haselwandter and Read 1982).

In this context, Holland (1997) provided evidence for a very intriguing hypothesis, giving a clue as to one reason why plants may need microbial interactions. He suggested that the microbial synthesis of so-called phytohormones might well be microbial signals to the plant that active growth can be accelerated, that endophytic microbes are present to degrade and recycle plant waste products resulting from growing tissue. For example,

Methylobacterium spp. that synthesise cytokinin in vitro, are found endophytically in actively growing tissues of many or even all plants. They are present in the apoplast and could degrade metabolic wastes, producing e.g. ammonium ions, which could be metabolised by the host. Holland concluded that since endophyte-free plant tissue has not been conclusively shown to synthesise cytokinins, "microbial symbionts are not accidental visitors. They are co-evolved participants in plant physiology." Similar signals seem to be generated by fungal endophytes, which not only synthesise the metabolites reported above (Kimura et al. 1992; Yates et al. 1997; Rey et al. 2001; Römmert et al. 2002), but also cytokinins (Petrini 1991; Tudzynski 1997; Tudzynski and Sharon 2002).

The plasticity of fungal root endophytes is demonstrated by the multiple options they have for regulating plant growth: synthesis of secondary metabolites and fungal phytohormones; directly providing nutrients, i.e. nitrogen and phosphate from the rhizosphere; but also bidirectional carbon transport. These results suggest that fungal endophytes are not only involved in a balanced antagonism with their hosts, but also may be responsible for a balanced regulation of growth. Thus, in at least some interactions, fungal root endophyte and host seem to be highly adapted to one another, resulting in a balanced interaction, as has also developed during the co-evolution of fungus and algae in lichens.

15.5 Disease Suppression

The idea of endophyte mediated induced resistance is an extension of the defensive mutualism hypothesis, i.e. plants may acquire protection through constitutive and induced resistance (Bultman and Murphy 2000). As defined by Schönbeck et al. (1993), "induced resistance describes the improvement of natural resistance of a plant without alterations of the genome". It is a non-specific general response that precludes absence of toxic effects for the parasite of the inducing agent as well as the absence of a dosage-response correlation, and it necessitates a time interval between application and response. Colonisation with fungal root endophytes has been reported to negatively influence growth of nematodes, insects and microbial pathogens (Selosse et al. 2004). However, it is not always clear whether induced resistance, altered plant nutrition, metabolism and/or sink effects are responsible for the increased mortality of predators when systemic effects are observed and mycotoxins are not involved.

Systemic fungal root infections have been reported to lead to acquisition of induced resistance in both roots and/or shoots (Bargmann and Schönbeck 1992; Hallmann and Sikora 1994; Sieber 2002). For example, root inoculations of cabbage plants with soilborne endophytic *Acremonium alternatum* decreased damage due to larvae of the diamondback moth on the leaves (Raps and Vidal 1998). However, the authors suggested that competition for phytosterol, which both fungi and insects obtain from their host, rather than induced resistance could account for reduced larval growth on inoculated cabbage plants. They furthermore hypothesise that a disjunction of fungal colonisation and predation is responsible for the observed effect, the roots serving as a sink for the plant metabolite.

Non-pathogenic *Fusarium* spp. have been found to induce resistance or afford the host cross-protection against pathogenic fungi (Kuldau and Yates 2000; Sieber 2002). For example, colonisation of the roots of pea and tomato by non-pathogenic strains of *Fusarium oxysporum* reduced disease by fungal root pathogens (Benhamou and Garand 2001) and predation by nematodes (Diedhiou et al. 2003), respectively. Fungal growth within the roots of pea was restricted to the outermost cell layers. Benhamou and Garand (2001) and Narisawa et al. (2004) found that resistance was induced, at least in part, by a massive elaboration of cell wall appositions and the deposition of electron-opaque material surrounding hyphae in the roots of pea and Chinese cabbage, respectively. This might be due to the accelerated deposition of lignin, a mechanical barrier for invading organisms, as was observed in shoots of maize when the roots were colonised by an avirulent isolate of *F. moniliforme* (Yates et al. 1997). Such host defence responses are initially produced to limit colonisation of the fungal endophytic invader, presumably resulting in a balance of antagonisms between host and fungus and in an asymptomatic endophytic infection. However, the host mechanical defence response also limits colonisation by subsequent pathogens.

Only a few of the endophytic isolates from a given host are effective protectants when reinoculated into the host. Of 322 endophytic fungi isolated from Chinese cabbage (*Brassica campestris*), only 16 isolates, including *Heteroconium chaetospira*, *Mortierella elongata*, *Westerdykella* sp. and three non-sporulating isolates, almost completely suppressed disease caused by *Plasmodiophora brassicae* when reinoculated into the host in sterile soil (Narisawa et al. 1998; Usuki et al. 2002). The authors suggest that disease suppression by DSE may be strain specific, as also seems to be the case with growth enhancement (see Sect. 15.4). Other DSE have also been reported to protect roots against phytopathogenic fungi (Sieber 2002; Narisawa et al. 2002). For example, three endophytic fungi reduced disease caused by *Verticillium longisporum* in Chinese cabbage: two were *Phialocephala* and one an unidentified DSE (Narisawa et al. 2004). The latter DSE, which extensively colonised root cells of the cortex (in contrast to the *P. fortinii* isolates), was very effective in preventing disease even in field experiments. In addition, of 16 isolates of *H. chaetospira* that suppressed disease in axenically cultured seedlings, only 2 were also effective in non-sterile soil

(Narisawa et al. 2002). This could be due to the fact that in non-sterile soil the roots are naturally colonised by endophytes.

Root colonisation with fungal endophytes can activate not only mechanical defence, but also induce the synthesis of defence metabolites in the roots, i.e. induce resistance. For example, colonisation of the roots of seedlings of *Larix decidua* with either the endophyte *Cryptosporiopsis* sp. or *P. fortinii* increased concentrations of soluble proanthocyanidins, and colonisation of barley roots with *Fusarium* sp. led to higher concentrations of phenylpropanoids in the roots than in those of the controls (Schulz et al. 1999). Similarly, Mucciarelli et al. (2003) found that the concentration of total phenolics increased significantly when *Mentha piperita* was colonised by a non-sporulating endophyte, though they found no significant change in the concentrations of total terpenoids.

Picard et al. (2002) investigated what fungal factors were responsible for inducing systemic resistance when the mycoparasite *Pythium oligandrum* asymptomatically colonised the roots of tomato without inducing extensive cell damage. They demonstrated that colonisation led to cell wall appositions and to the deposition of phenolic metabolites (Benhamou et al. 1997). Picard et al. (2000) found that oligandrin, an elicitin-like protein produced by *P. oligandrum*, migrates within the vascular system of the host and induces systemic resistance.

Improving plant resistance without the use of pesticides is one goal of biocontrol in agriculture. Only a small proportion of the fungal root endophytes of any one host seem capable of inducing systemic resistance. Are these endophytes present under natural environmental conditions at a sufficient colonisation density to induce resistance? In most cases, too little is presently known about what factor(s) induce resistance in the field. This is a broad and important field for future investigations.

15.6 Stress Tolerance

Whereas mycorrhizal fungi can alleviate abiotic stresses, e.g. salt (Rabie 2005) and osmotic stress (Ruiz-Lozano 2003), there have been only a few investigations studying the influence of endophytic colonisation on abiotic stress tolerance. One example strikingly demonstrates induction of stress tolerance to abiotic stress: a novel endophytic *Curvularia* sp., which colonised both roots and shoots of the host, increased host tolerance to temperatures of up to 65°C (Redman et al. 2002). Another abiotic stress is transplantation to polluted sites or micropropagation; stresses that can be counteracted by hardening. This was achieved not only by mycorrhization, but also by colonisation by the non-obligate biotrophic endophyte

Piriformospora indica (Sahay and Varma 1999). An arid climate also poses an abiotic stress, which DSE may help alleviate. Barrow and Aaltonen (2001) suggest that DSE may play such a role, because they found that under xeric, in contrast to mesic, conditions, colonisation by DSE was more prevalent than that by aseptate hyphae (AM). It is also possible that lipid accumulation within hyaline hyphae of DSE serves as an energy-rich carbon reserve to sustain plants during extended drought (Barrow 2003).

The potential of fungal root endophytes to alleviate other abiotic and biotic stresses is a field that has not been adequately investigated. Their capacity to induce tolerance to other individual stressors, as well as to accumulated stress, should be tested both with individual and with a mixture of stressors at sublethal levels to determine the limits and potentials for induction of stress tolerance by fungal root endophytes, but also for future use in sustainable agriculture.

15.7 Factors Determining the Status of the Interaction

Establishment of any interaction is always a multifactorial process. One factor that may contribute to the interaction being mutualistic is the pure extent of colonisation, since in the reported mutualistic interactions with fungal root endophytes, growth has been extensive (see Sect. 15.2.; Stone et al. 2000; Sieber 2002; Schulz and Boyle 2005). However, other factors e.g. biotic and abiotic stressors, nutrient support, and ontogenetic status, may also be relevant. As hypothesised by Schulz and Boyle (2005), mutualistic interactions have more frequently developed between microorganisms and the roots, because (1) the roots are a natural carbon sink of the plants and can supply dual and multi-organism symbioses with nutrients, (2) infection is less limited by xeromorphic structures, and (3) roots are in close contact with an environment harbouring many different, mainly degradatively active, micro-organisms.

In spite of the examples of mutualistic interactions with fungal root endophytes presented in this chapter, it is important not to draw the conclusion that all interactions with fungal root endophytes are mutualistic. Only a small percentage of these endophytes seem to have the capability to interact mutualistically with their hosts (see Sects. 15.5, 15.6), perhaps because interactions with their hosts depend on the genetic predisposition and momentary metabolic status of the host, as well as on environmental factors. The same endophytes that under certain conditions interact mutualistically with their hosts may become pathogenic, for example when the host is stressed and the balance of the antagonism is tilted "in favour" of the fungus (Kuldau and Yates 2000; Jumpponen 2001; Sieber 2002; Schulz and Boyle 2005).

15.8 Conclusions

Fungal endophytic colonisation of the roots of plants is very variable (Table 15.2), reflecting the plasticity of individual fungal endophytes, but also of endophytes as a whole. The hyphae can be hyaline and very thin, melanised, or develop microsclerotia. Colonisation is often extensive, may be inter- and/or intra-cellular, and is sometimes limited to the epidermis or cortex. The hyphae may extensively colonise the vascular cylinder and in particular the sieve elements, but also grow on the root surface and into the rhizosphere. Some DSE have even been observed to develop ericoid, ectendo- and ectomycorrhizal-like structures.

Benefits for the fungal partner are a stable nutrient source and some protection from abiotic stresses. Factors that fungal root endophytes may contribute to a mutualistic interaction include secondary metabolites, phytohormones, nutrients, elicitins and colonisation (Fig. 15.2). These factors may have more than one benefit for the host. Potential advantages of the interactions for the host are summarised in Table 15.3 and include improved growth, induced resistance, stress tolerance, and protection from microbial and insect predators by mycotoxins.

Morphologically and physiologically, endophytic root colonisations mirror the variability and thus the plasticity of endophytic interactions, but also

Table 15.2. Endophytic plasticity

Average shoot length (cm)	Average dry weight of root+shoot (mg)
Attribute	In planta
Colonisation	Intercellular, intracellular, on/in: root surface, root hairs, epidermis, cortex, and/or vascular cylinder (xylem and phloem), systemic within roots, systemic within roots and shoots, local colonisation (?)
Fungal morphologies	Lipid-filled hyaline hyphae and protoplasts (?), melanised hyphae, microsclerotia, coiled hyphae, appressoria
Fungal-plant associations	No specialised structures Ericoid, ectendo-, and ectomycorrhizas
Secondary metabolites	Growth enhancement, growth inhibition, elicitation, balanced antagonism, microbial antagonism
Phytohormones	Modulate host growth
Nutrient source	Parasitically from host assimilates, saprophytically

Table 15.3. Endophytes/metabolites known to be involved in mutualistic interactions. *DSE*: Dark septate endophyte

Advantage for host	Endophyte / metabolite	Host	Reference
Growth enhancement	*Cryptosporiopsis* sp.	*Larix decidua*	Schulz et al. 2002
	Chaetomium, C. globosum	Barley	Vilich et al. 1998
	Cladorrhinum foecumdissimum	Cotton	Gasoni and Stegman de Gurfinkel 1997
	Fusarium	Maize	Yates et al. 1997
	Heteroconium chaetospira	Chinese cabbage	Usuki et al. 2002
	Oidiodendron maius	Ericaceous plants	Rice and Currah 2002
	Phialocephala fortinii	*Pinus contorta*	Jumpponen and Trappe 1998; Jumpponen et al. 1998
	P. fortinii	*Larix decidua*	Schulz et al. 2002
	P. fortinii	*Carex firma, C. curvula*	Haselwandter and Read 1982
	Phoma fimeti	*Vulpia ciliate*	Newsham 1994
	Stagonospora spp.	*Phragmites australis*	Ernst et al. 2003
	n.i.[a]	*Mentha piperita*	Mucciarelli et al. 2002, 2003
	Altechromones A and B	Lettuce	Kimura et al. 1992
	Fumonisin	Maize	Yates et al. 1997
Disease suppression/ induced resistance	*Acremonium*	Tomato	Raps and Vidal 1998
	Fusarium moniliforme	Maize	Yates et al. 1997
	Fusarium oxysporum	Pea	Benhamou and Garand 2001
	F. oxysporum	Tomato	Diedhiou et al. 2003
	Fusarium spp.	Various crops	Kuldau and Yates 2000
	H. chaetospira, Mortierella elongata, Westerdykella sp.	Chinese cabbage	Narisawa et al. 1998; Usuki et al. 2002
	P. fortinii	Chinese cabbage	Narisawa et al. 2004
	Pythium oligandrum	Tomato	Benhamou et al. 1997; Picard et al. 2002
	Oligandrin	Tomato	Picard et al. 2000
Stress tolerance	*Curvularia* sp.	*Dicanthelium lanuginosum*	Redman et al. 2002
	Piriformospora indica	Tobacco	Sahay and Varma 1999
	DSE	*Atriplex canescens*	Barrow and Aaltonen 2001

Table 15.3. (continued)

Advantage for host	Endophyte / metabolite	Host	Reference
Mycotoxins	*Cryptosporiopsis* sp.		Schulz et al. 1995
	Fusarium oxysporum		Khambay et al. 2000
	Fusarium spp		Chap. Bacon and Yates; Kuldau and Yates 2000; Miller 2001
	Verticillium chlamydosporium		Hallmann and Sikora 1996

[a]Not identified

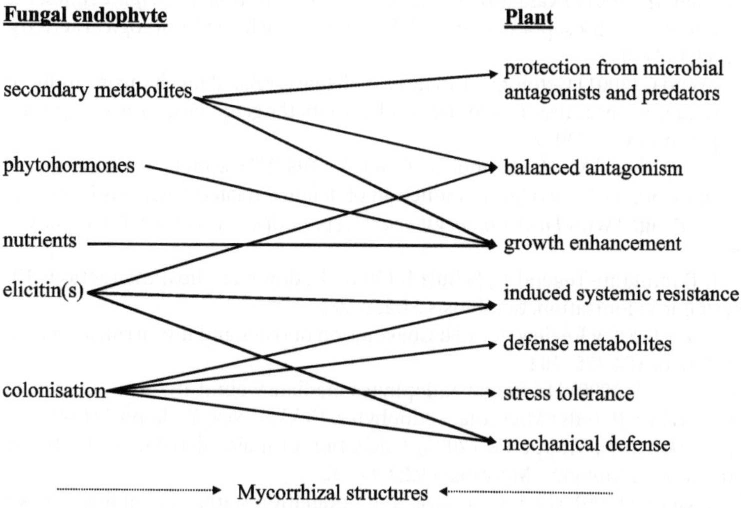

Fig. 15.2. Potential contributions of fungal root endophytes to mutualistic interactions, assuming extensive colonisation of the root. "Colonisation" indicates that it is unknown what aspect of fungal colonisation induces the response

the evolutionary developmental stages from ubiquitous endophyte to mutualistic endophyte to specialised mycorrhizal fungus. Not only the exploitive, but also the mutualistic life history strategy becomes more prevalent with increasing specialisation (Brundrett 2002; see Chap. 16 by Brundrett).

Acknowledgements. My thanks go to Anne-Kathrin Römmert and Miruna Oros-Sichler for permission to use unpublished results and to Christine Boyle for constructive discussions that helped develop the manuscript and ideas presented above.

References

Arnold AE, Herre EA (2003) Canopy cover and leaf age affect colonisation by tropical fungal endophytes: ecological pattern and process in *Theobroma cacao* (Malvaceae). Mycologia 95:388–398

Azevedo JL, Maccheroni W, Pereira JO, de Araujo WL (2000) Endophytic microorganisms: a review on insect control and recent advances on tropical plants. Electron J Biotechnol 3:1–36

Bacon CW, Hinton NS (1996) Symptomless endophytic colonisation of maize by *Fusarium*. Can J Bot 74:1195–1202

Bargmann C, Schoenbeck F (1992) *Acremonium kiliense* as inducer of resistance to wilt disease on tomatoes. Z Pflanzenkr Pflanzenschutz 99:266–272

Barrow JR (2003) Atypical morphology of dark septate fungal root endophytes of *Bouteloua* in arid southwestern USA rangelands. Mycorrhiza 13:239–247

Barrow JR, Aaltonen RE (2001) Evaluation of the internal colonisation of *Atriplex canescens* (Pursh) Nutt. roots by dark septate fungi and the influence of host physiological activity. Mycorrhiza 11:199–205

Benhamou N, Garand C (2001) Cytological analysis of defense-related mechanisms induced in pea root tissues in response to colonisation by nonpathogenic *Fusarium oxysporum* Fo47. Phytopathology 91:730–740

Benhamou N, Rey P, Chérif M, Hockenhull J, Tirilly Y (1997) Treatment with the mycoparasite *Pythium oligandrum* triggers induction of defense-related reactions in tomato roots when challenged with *Fusarium osysporum* f. sp. *radicis-lycopersici*. Phytopathology 87:108–122

Boyle C, Götz M, Dammann-Tugend U, Schulz B (2001) Endophyte - host interactions III. Local vs. systemic colonisation. Symbiosis 31:259–281

Brundrett MC (2002) Tansley Review no. 134: Coevolution of roots and mycorrhizas of land plants. New Phytol 154:275–304

Bultman TL, Murphy JC (2000) Do fungal endophytes mediate wound-induced resistance? In: Bacon CW, White JF (eds) Microbial endophytes. Dekker, New York, pp 421–455

Caldwell BA, Jumpponen A, Trappe JM (2000) Utilization of major detrital substrates by dark-septate, root endophytes. Mycologia 92:230–232

Danielson RM, Visser S (1990) The mycorrhizal and nodulation status of container-grown trees and shrubs reared in commercial nurseries. Can J For Res 20:609–614

Demain AL (1980) Do antibiotics function in nature? Search 11:148–151

Diedhiou PM, Hallmann J, Oerke EC (2003) Effects of arbuscular mycorrhizal fungi and a non-pathogenic *Fusarium oxysporum* on *Meloidogyne incognita* infestation of tomato. Mycorrhiza 13:199–204

Ernst M, Mendgen KW, Wirsel SG (2003) Endophytic fungal mutualists: seed-borne *Stagonospora* spp. enhance reed biomass production in axenic microcosms. Mol Plant-Microbe Interact 16:580–587

Fang-ting W, Dai-jie C, Xiu-ping Q (2004) Recent studies of bioactive substances produced by endophyte. Chin J Antibiot 29:184–192

Fernando AA, Currah RS (1996) A comparative study of the effects of the root endophytes *Leptodontidium orchidicola* and *Phialocephala fortinii* (Fungi imperfecti) on the growth of some subalpine plants in culture. Can J Bot 74:1071–1078

Gasoni L, Stegman De Gurfinkel B (1997) The endophyte *Cladorrhinum foecundissimum* in cotton roots: phosphorus uptake and host growth. Mycol Res 101:867–870

Hallmann J, Sikora RA (1994) Influence of *Fusarium oxysporum*, a mutualistic fungal endophyte on *Meloidogyne javanica* infection of tomato. J Plant Dis Prot 101:475–481

Hallmann J, Sikora RA (1996) Toxicity of fungal endophyte secondary metabolites to plant-parasitic nematodes and soil-borne plant-pathogenic fungi. Eur J Plant Pathol 102:155–162
Haselwandter K, Read DJ (1982) The significance of a root-fungus association in two *Carex* species of high-alpine plant communities. Oecologia 53:352–254
Hashimoto Y, Hyakumachi M (2001) Effects of isolates of ectomycorrhizal fungi and endophytic *Mycelium radicis atrovirens* that were dominant in soil from disturbed sites on growth of *Betula platyphylla* var. *japonica* seedlings. Ecol Res 16:117–125
Holland MA (1997) Occam's razor applied to hormonology. Are cytokinins produced by plants? Plant Physiol 115:865–868
Jallow JFA, Digassa-Gobena D, Vidal S (2004) Indirect interaction between an unspecialized endophytic fungus and a polyphagous moth. Basic Appl Ecol 5:183–191
Jumpponen A (1999) Spatial distribution of discrete RAPD phenotypes of a root endophytic fungus, *Phialocephala fortinii*, at a primary successional site on a glacier forefront. New Phytol 141:333–344
Jumpponen A (2001) Dark septate endophytes - are they mycorrhizal? Mycorrhiza 11:207–211
Jumpponen A, Trappe JM (1998) Performance of *Pinus contorta* inoculated with two strains of root endophytic fungus, *Phialocephala fortinii*: effects of synthesis system and glucose concentration. Can J Bot 76:1205–1213
Jumpponen A, Mattson KG, Trappe JM (1998) Mycorrhizal functioning of *Phialocephala fortinii* with *Pinus contorta* on glacier forefront soil: interactions with soil nitrogen and organic matter. Mycorrhiza 7:261–265
Kaldorf M, Renker C, Fladung M, Buscot F (2004) Characterization and spatial distribution of ectomycorrhizas colonising aspen clones released in an experimental field. Mycorrhiza 14:295–306
Khambay BPS, Bourne JM, Carmeron S, Kerry BR, Zaki MJ (2000) A nematicidal metabolite from *Verticillium chlamydosporium*. Pest Manage Sci 56:1098–1099
Kimura Y, Mizuno T, Nakajima H, Hamasaki T (1992) Altechromones A and B, new plant growth regulators produced by the fungus *Alternaria* sp. Biosci Biotechnol Biochem 56:1664–1665
Kuldau GA, Yates IE (2000) Evidence for *Fusarium* endophytes in cultivated and wild plants. In: Bacon CW, White JF (eds) Microbial endophytes. Dekker, New York, pp 85–120
Lubuglio KF, Wilcox HE (1988) Growth and survival of ectomycorrhizal and ectoendomycorrhizal seedlings of *Pinus resinosa* on iron tailings. Can J Bot 66:55–60
Miller JD (2001) Factors that affect the occurrence of fumonisin. Environ Health Perspect 109:321–324
Miller JD, Mackenzie S, Foto M, Adams GW, Findlay JA (2002) Needles of white spruce inoculated with rugulosin-producing endophytes contain rugulosin reducing spruce budworm growth rate. Mycol Res 106:471–479
Mucciarelli M, Scannerini S, Bertea C, Maffei M (2002) An ascomycetous endophyte isolated from *Mentha piperita* L.: biological features and molecular studies. Mycologia 94:28–39
Mucciarelli M, Scannerini S, Bertea C, Maffei M (2003) In vitro and in vivo peppermint (*Mentha piperita*) growth promotion by nonmycorrhizal fungal colonisation. New Phytol 158:579–591
Narisawa K, Tokumasu S, Hashiba T (1998) Suppression of clubroot formation in Chinese cabbage by the root endophytic fungus, *Heteroconium chaetospira*. Plant Pathol 47:206–210
Narisawa K, Kawamata H, Currah RS, Hashiba T (2002) Suppression of *Verticillium* wilt in eggplant by some fungal root endophytes. Eur J Plant Pathol 108:103–109

Narisawa K, Usuki F, Hashiba T (2004) Control of Verticillium yellows in Chinese cabbage by the dark septate endophytic fungus LtVB3. Phytopathology 94:412–418

Newsham KK (1994) First record of intracellular sporulation by a coelomycete fungus. Mycol Res 98:1390–1392

Ohki T, Masuya H, Yonezawa M, Usuki F, Narisawa, K, Hashiba T (2002) Colonisation process of the root endophytic fungus *Heteroconium chaetospira* in roots of Chinese cabbage. Mycoscience 43:191–194

Petrini O (1991) Fungal endophytes of tree leaves. In: Andrews J, Hirano S (eds) Microbial ecology of leaves. Springer, New York Berlin Heidelberg, pp 179–197

Picard K, Ponchet M, Blein J-P, Rey P, Tirilly Y, Benhamou N (2000) Oligandrin. A proteinaceous molecule produced by the mycoparasite *Pythium oligandrum* induces resistance to *Phytophthora parasitica* infection in tomato plants. Plant Physiol 124:379–395

Rabie GH (2005) Influence of arbuscular mycorrhizal fungi and kinetin on the response of mungbean plants to irrigation with seawater. Mycorrhiza 15:225–230

Raps A, Vidal S (1998) Indirect effects of an unspecialized endophytic fungus on specialized plant-herbivorous insect interactions. Oecologia 114:541–547

Redman RS, Sheehan KB, Stout RG, Rodriguez RJ, Henson JM (2002) Thermotolerance generated by plant/fungal symbiosis. Science 298:1581

Rey P, Leucart S, Désilets H, Bélanger RR, Larue JP, Tirilly Y (2001) Production of indole-3-acetic acid and tryptophol by *Pythium* group F: possible role in pathogenesis. Eur J Plant Pathol 107:895–904

Rice AV, Currah RS (2002) New perspectives on the niche and holomorph of the myxotrichoid hyphomycete, *Oidodendron maius*. Mycol Res 106:1463–1467

Römmert AK, Oros-Sichler M, Aust H-J, Lange T, Schulz B (2002) Growth promoting effects of endophytic colonisation of the roots of larch (*Larix decidua*) with *Cryptosporiopsis* sp. and *Phialophora* sp. 7th International Mycological Congress, Oslo, Norway, p 309

Ruiz-Lozano JM (2003) Arbuscular mycorrhizal symbiosis and alleviation of osmotic stress. New perspectives for molecular studies. Mycorrhiza 13:309–317

Sahay NS, Varma A (1999) *Piriformospora indica*: a new biological hardening tool for micropropagated plants. FEMS Microbiol Lett 181:297–302

Schönbeck F, Steiner U, Kraska T (1993) Induzierte Resistenz: Kriterien, Mechanismen, Anwendung und Bewertung. J Plant Dis Prot 100:541–557

Schulz B, Boyle C (2005) The endophytic continuum. Mycol Res 109:661–687

Schulz B, Sucker J, Aust H-J, Krohn K, Ludewig K, Jones PG, Döring D (1995) Biologically active secondary metabolites of endophytic *Pezicula* species. Mycol Res 99:1007–1015

Schulz B, Römmert A-K, Dammann U, Aust H-J, Strack D (1999) The endophyte-host interaction: a balanced antagonism. Mycol Res 103:1275–1283

Schulz B, Boyle C, Draeger S, Römmert A-K, Krohn K (2002) Endophytic fungi: a source of biologically active secondary metabolites. Mycol Res 106:996–1004

Selosse MA, Baudoin E, Vandenkoorhuyse P (2004) Symbiotic microorganisms, a key for ecological success and protection of plants. C R Biol 327:639–648

Sen R, Hietala AM, Zelmer CD (1999) Common anastomosis and internal transcribed spacer RFLP groupings in binucleate *Rhizoctonia* isolates representing root endophytes of *Pinus sylvestris*, *Ceratorhiza* spp. from orchid mycorrhizas and a phytopathogenic anastomosis group. New Phytol 44:331–341

Sieber TN (2002) Fungal root endophytes In: Waisel Y, Eshel A, Kafkafi U (eds) The hidden half. Dekker, New York, pp 887–917

Sieber TN, Riesen TK, Müller E, Fried PM (1988) Endophytic fungi in four winter wheat cultivars (*Triticum aestivum* L.) differing in resistance against *Stagonospora nodorum* (Berk.) Cast. & Germ. = *Septorianodorum* (Berk.) Berk. J Phytopathol 122:289–306

Stone JK, Bacon CW, White JF (2000) An overview of endophytic microbes: endophytism defined. In: Bacon CW, White JF (eds) Microbial endophytes. Dekker, New York, pp 3–30
Tan RX, Zou WX (2001) Endophytes: a rich source of functional metabolites. Nat Prod Rep 18:448–459
Tudzynski B (1997) Fungal phytohormones in pathogenic and mutualistic associations. In: Carroll GC, Tudzynski P (eds) The mycota V. Springer, Berlin Heidelberg New York, pp 167–184
Tudzynski B, Sharon A (2002) Biosynthesis, biological role and application of fungal phytohormones In: Osiewacz HD (ed) The mycota X. Industrial applications. Springer, Berlin Heidelberg New York, pp 183–212
Usuki F, Narisawa K (2005) Formation of structures resembling ericoid mycorrhizas by the root endophytic fungus *Heteroconium chaetospira* within roots of *Rhododendron obtusum* var. *kaempferi*. Mycorrhiza 15:61–64
Usuki F, Narisawa K, Yonezawa M, Kakishima M, Hashiba T (2002) An efficient method for colonisation of Chinese cabbage by the root endophytic fungus *Heteroconium chaetospira*. J Gen Plant Pathol 68:326–332
Varma A, Singh A, Sahay NS, Sharma J, Roy A, Kumari M, Raha D, Thakran S, Deka D, Bharti K, Hurek T, Blechert O, Rexer K-H, Kost G, Hahn A, Maier W, Walter M, Strack D, Kranner I (2000) *Piriformospora indica*: an axenically culturable mycorrhiza-like endosymbiotic fungus In: Hock B (ed) The mycota, vol IX Fungal associations. Springer, Berlin Heidelberg New York, pp 125–150
Vilich V, Dolfen M, Sikora RA (1998) *Chaetomium* spp. colonisation of barley following seed treatment and its effect on plant growth and *Erysiphe graminis* f. sp. *hordei* disease severity. Z Pflanzenkr Pflanzenschutz 105:130–139
Vrålstad T, Schumacher T, Taylor AFS (2002) Mycorrhizal synthesis between fungal strains of the Hymenoscyphus ericae aggregate and potential ectomycorrhizal and ericoid hosts. New Phytol 153:143–152
Yates IE, Bacon CW, Hinton DM (1997) Effects of endophytic infection by *Fusarium moniliforme* on corn growth and cellular morphology. Plant Dis 81:723–728

16 Understanding the Roles of Multifunctional Mycorrhizal and Endophytic Fungi

Mark C. Brundrett

16.1 How Mycorrhizal Fungi Differ from Endophytes

16.1.1 Definitions

The most appropriate definition of endophytism is that of symptomless associations by organisms that grow within living plant tissues (see other chapters in this book). Many fungi colonise the cortex of living roots without causing disease, including pathogenic or necrotrophic fungi with latent phases as well as beneficial fungi that offer protection against pathogens, but it is difficult to precisely categorise these fungi [Saikkonen et al. 1998; Sivasithamparam 1998; see Chaps. 1 (Schulz and Boyle) and 8 (Bacon and Yates)]. A new definition of mycorrhizal associations was required to adequately separate them from other root-fungus associations (Brundrett 2004). According to this definition, endophytic associations differ from mycorrhizas primarily by the absence of a localised interface of specialised hyphae, the absence of synchronised plant-fungus development, and the lack of plant benefits from nutrient transfer – the three key defining features of mycorrhizas. However, plants may benefit indirectly from endophytes by increased resistance to herbivores, pathogens or stress, or by other unknown mechanisms, as described elsewhere in this book (see in particular Chap. 15 by Schulz).

There are a number of distinct categories of mycorrhizal fungi, all of which differ from other fungi primarily because they are dual soil-plant inhabitants that are efficient at growth and nutrient uptake in both soils and plants (Brundrett 2002). Conversely, endophytes and pathogens are plant inhabitants, which do not require efficient means of acquiring nutrients from soils to supply plants (Table 16.1). The examples in this chapter primarily concern root-fungus associations, but mycorrhizal fungi and

Mark C. Brundrett: School of Plant Biology, Faculty of Natural and Agricultural Sciences, The University of Western Australia, Crawley, WA 6009, Australia, E-mail: mark.brundrett@environmnent.wa.gov.au

Soil Biology, Volume 9
Microbial Root Endophytes
B. Schulz, C. Boyle, T. N. Sieber (Eds.)

Table 16.1. Comparison of mycorrhizal, parasitic and endophytic root associations (Brundrett 2004). *VAM* Vesicular arbuscular mycorrhizas, *ECM* ectomycorrhizas

Criteria	Mycorrhizal	Parasitic	Endophytic
Morphology	Specialised hyphae in specialised plant organ	Specialised hyphae	Relatively unspecialised hyphae
Development	Synchronised	Often synchronised	Not synchronised
Impact on fungus	Obligate requirement for plant supplied nutrients (VAM and ECM)	Fungus obligately or facultatively dependent on plant	Fungus moderately or weakly dependent
Impact on plant	Strong or weak benefit	Strong or weak harm	Weak harm or benefit
Nutrient transfer	Synchronised transfer, fungus a strong sink	Active or passive transfer, fungus a strong or weak sink	Passive transfer, fungus not a strong sink

endophytes also occupy other substrate-contacting organs (e.g. rhizomes, stems, scale leaves, etc.).

16.1.2 Roles of Endophytes and Mycorrhizal Fungi

It is harder to separate mycorrhizal fungi from other functional groups of fungi than it is to separate mycorrhizal associations from other plant-fungus relationships, as mycorrhizas are defined by morphological criteria (Brundrett 2004). Differences between endophytic associations and mycorrhizal associations are summarised in Table 16.1. The most important of these differences are the absence of substantial fungus-to-plant nutrient transfer in endophytic associations and the lack of synchronised development between endophytes and plants. Endophytic associations lack the specialised interface, formed by complex hyphal growth synchronised with substantial cytoplasm synthesis by host cells, that is diagnostic of mycorrhizas. The complex host-fungus interface in mycorrhizal associations allows rapid bi-directional transfer of nutrients over a relatively short time (a few weeks). However, nutrient transfer to fungi in endophytes is still likely to be important to the fungus if it continues over much longer time-spans (months or years) than the active phase of mycorrhizal associations. We would also not expect plant-fungus coevolution in endophytic associations if plants do not receive substantial direct benefits from these organisms. However, the results of plant evolution to become more efficient at forming mycorrhizas (e.g. roots that evolved to house fungi, Brundrett 2002) would ultimately have also benefitted endophytes.

While we can safely say that endophytes are not mycorrhizal (as defined above), it seems likely that mycorrhizal fungi have endophytic phases involving long-term coexistence with plants without active growth or nutrient transfer. For example, old roots are an important source of inoculum for Glomeromycete fungi, which can survive as endophytes in living roots for up to 10 years after arbuscules collapse (Brundrett and Kendrick 1988). These fungi can also have a necrotrophic phase in dying roots, providing the fungus with first access to nutrients that can be transferred to hyphae within other plants (Eason et al. 1991). Examples of endophytic activity by mycorrhizal fungi are summarised in the next section.

16.2 Endophytic Activity by Mycorrhizal Fungi

Endophytic growth of mycorrhizal fungi in plants is common and differs from mycorrhizal associations primarily by the absence of defining morphological features, as discussed above. It is not difficult to explain the endophytic competence of mycorrhizal fungi in non-host plants, as we would expect that they would normally have the highest inoculum potentials of soil organisms and these fungi require the capacity to rapidly and efficiently colonise plant roots. It has been suggested that plants must employ effective defences if they are to remain free of mycorrhizal fungi (Brundrett 1991).

Another difference between mycorrhizal and endophytic fungi is that the nutritional resources provided by endophytic activity (access to endophytic exudates) do not seem to be sufficient to sustain mycorrhizal fungi, which generally cannot persist in soils without forming mycorrhizas with host plants (Brundrett 1991). It has been suggested that mycorrhizal fungi may endophytically colonise roots and other soil organisms for shelter rather than food, to avoid the many soil organisms that prey on them (Koske 1984). Such use of roots as shelters by mycorrhizal fungi likely would be of limited immediate significance to the plant in which this occurs, but should ultimately benefit other plants in the same ecosystem, by maintaining a reservoir of inoculum. Examples of suggested endophytic associations by mycorrhizal fungi and other fungi that frequent mycorrhizas are listed in sections below, with fungi classified by their primary roles.

16.2.1 Glomeromycotan (Vesicular-Arbuscular Mycorrhizal) Fungi

Glomeromycotan fungi, forming vesicular arbuscular mycorrhizas [VAM, or arbuscular mycorrhizas (AM)], are ubiquitous soil organisms that proliferate within patches of soil organic material (St John et al. 1983; Joner and Jakobsen 1995; Azcón-Aguilar et al. 1999). As shown in Fig. 16.1A, B, these fungi commonly grow in non-host plants (Brundrett and Kendrick 1988; Imhof 2001). They also occupy plant organs other than roots, dead soil animals and spores of other VAM fungi, presumably to acquire nutri-

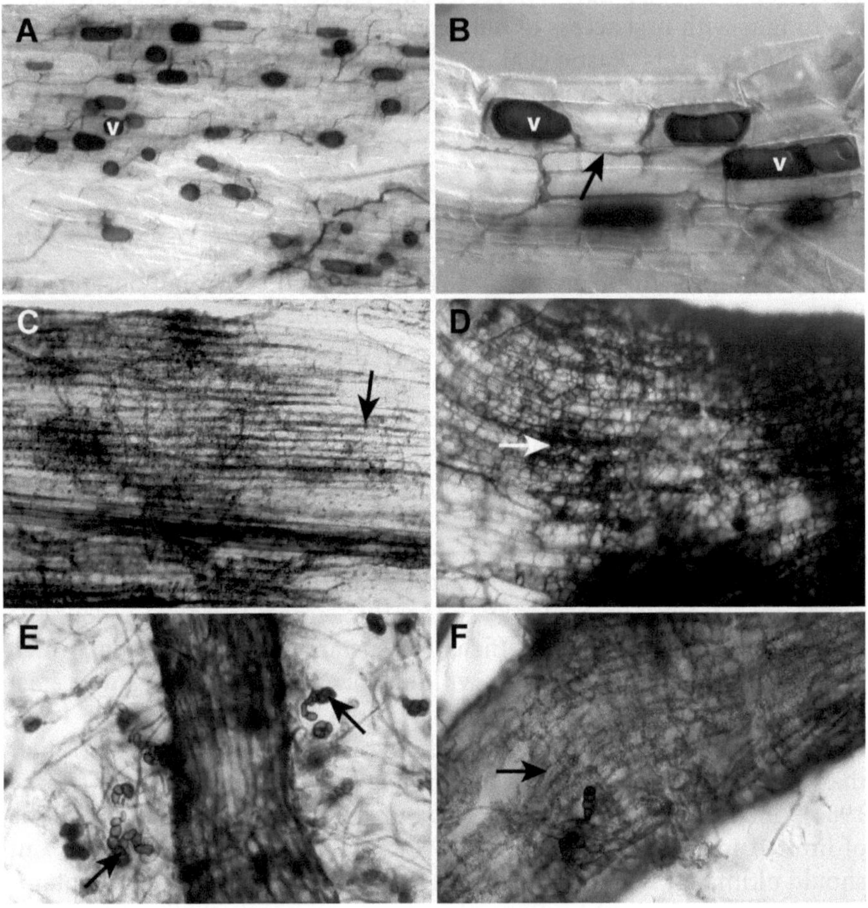

Fig. 16.1. A–D Roots cleared in potassium hydroxide and stained with Chlorazol Black E in lactoglycerol. **A, B** Endophytic growth of a Glomeromycotan fungus in the nonmycorrhizal plant *Hydrophyllum virginianum* consisting of vesicles (*v*) and hyphae (*arrow*) in a rhizome scale **A** and nonmycorrhizal root **B. C, D** Dense growth by unknown fungi (*arrows*) on the epidermis of roots of *Scaevola crassifolia* **C** and *Acer saccharum* **D.** These plants also have vesicular-arbuscular mycorrhizas. **E, F** Unstained roots of *Eucalyptus globulus* grown in pasteurised soil colonised by an unidentified opportunistic conidial fungus. **E** Hyphae and conidia (*arrows*) formed in soil. **F** Hyphae on the surface of long roots forming a patchy Hartig-net-like structure (*arrow*)

ents or avoid predation (Rabatin and Rhodes 1982; Warner 1984; St John et al. 1983; Koske 1984; Brundrett and Kendrick 1988). Mycorrhizal colonisation of some mutants of normally mycorrhizal species also resembles endophytic activity by these fungi, as they form hyphae and vesicles but not arbuscles (Demchenko et al. 2004). This suggests that VAM fungi will switch between endophytic and mycorrhizal activity in plants in response to signals provided by the plant.

Nonmycorrhizal plants are defined as those that normally exclude mycorrhizal fungi from their healthy young roots (Brundrett 1991). As implied by this definition, endophytic colonisation of nonmycorrhizal plant roots by VAM fungi is common in older roots, but is considered to be of limited functional significance because it does not result in plant growth responses (Ocampo 1986; Muthukumar et al. 1997; Giovannetti and Sbrana 1998). However, there may be a fine line between nonmycorrhizal and facultatively mycorrhizal species in which VAM associations are present or absent as a result of soil conditions (see Brundrett 1991). These issues are best illustrated by the debate about the role of mycorrhizal associations in sedges (Cyperaceae and allied families) that has been ongoing for decades (Powell 1975; Tester et al. 1987; Brundrett 1991; Muthukumar et al. 2004). The statement by Muthukumar et al. (2004) that the Cyperaceae is a mycorrhizal family, is in disagreement with the opinion of earlier reviewers (e.g. Powell 1975; Tester et al. 1987; Brundrett 1991) and is not as clear-cut as it seems. Half of sedges examined in studies summarised by Muthukumar et al. (2004) contained mycorrhizal fungi. However, they could not clearly distinguish endophytic from mycorrhizal associations using the information provided in many studies, and they suspected that these associations were often non-functional. Detailed studies of sedges such as *Carex* spp. have found them to be nonmycorrhizal throughout the year with endophytic VAM hyphae in older roots (Brundrett and Kendrick 1988; Cornwell et al. 2001). There may be a continuum from mycorrhizal to nonmycorrhizal species in plant families such as the Cyperaceae. In facultatively mycorrhizal species, VAM fungi may sometimes function more as endophytes than as mycorrhizal associates, as they can contribute to disease suppression in the absence of growth responses (Newsham et al. 1995; Cordier et al. 1998). It is often assumed that sparse or inconsistent mycorrhizal formation is of limited benefit to plants, but this has rarely been tested in experiments using appropriate soil conditions.

A second common example of cases where the hyphal growth of Glomeromycotan fungi is difficult to interpret is their colonisation of roots of ectomycorrhizal (ECM) plants in species with dual mycorrhizal associations. In plants with both ECM and VAM, their relative importance can vary due to the age of plants and the habitats in which they grow (Chen et al. 2000; van der Heijden 2001). Most trees with dual ECM/VAM associations are considered to benefit primarily from ECM, but large growth responses to VAM occur in experiments, especially when plants are young (Brundrett et al. 1996; Chen et al. 2000). A further complication is the fact that Glomeromycete fungi are commonly present as endophytes in roots of ECM plants (e.g. Harley and Harley 1987; Cázares and Trappe 1993; Smith, et al. 1998). This endophytic activity is distinguished from functional dual ECM/VAM associations by the absence of arbuscules. Glomeromycotan fungi can also grow endophytically in orchids (Hall 1976).

16.2.2 Ectomycorrhizal Fungi

Ectomycorrhizal associations are defined by a key morphological criterion: labyrinthine Hartig net hyphae in an interface between cells of the primary root cortex or epidermis. However, designation of ECM without a well-defined Hartig net is not always clear cut, as "superficial" ECM roots with a thin or patchy Hartig net occur in synthesis experiments using host-fungus combinations that are not fully compatible (Burgess et al. 1994; Peterson and Massicotte 2004). These also occur in nature, where they are believed to benefit plants (Malajczuk et al. 1987). When initiated by spores or other limited sources of inoculum in experiments, ECM fungi weakly colonise root surfaces before they have sufficient energy to form typical mycorrhizas (Chilvers and Gust 1982). This establishment phase may equate to a transition from endophyte-like to mutualistic activity. Some fungi form associations with a mantle but no Hartig net on non-host roots, such as those of *Morchella* sp. on Pinaceae (Dahlstrom et al. 2000), *Cortinarius* sp. on *Carex* (Harrington and Mitchell 2002) and *Tricholoma* sp. on *Pinus* (Gill et al. 1999). The opportunistic growth of fungal hyphae on roots is common in nature and could be considered a form of endophytism in which fungi feed on root exudates without penetrating cells (Fig. 16.1C, D).

Other examples of fungi that seem to occupy intermediate positions on an ECM-endophyte continuum are opportunistic fungi that weakly colonise seedling roots in the nursery (see Fig. 16.1E, F). These nursery fungi fill an empty niche in potting mixes that lack mycorrhizal fungus propagules or that are not conducive to growth by mycorrhizal fungi. These nursery fungi are probably of limited functional significance, as suggested by similar fungal growth on both long and short roots, a patchy Hartig net and the absence of morphological responses by short roots (root swelling). Ectendomycorrhizas, a variant of ECM, also tend to occur in substrates lacking other fungi and where nutrient levels are high, and so are considered to be of limited benefit to plants (Yu et al. 2001).

16.2.3 Fungi in Orchids

Most fungi that form orchid mycorrhizas are basidiomycetes belonging to a diverse assemblage of fungi assigned by morphological features to the asexual form genus *Rhizoctonia* (see also Chap. 9 by Bayman and Otero). These fungi are also more precisely defined using anastomosis reactions, enzyme assays and DNA-based methods (Currah et al. 1997; Taylor et al. 2002; McKormick et al. 2004). Mycorrhizal associates of orchids occur in all three clades of the polyphyletic *Rhizoctonia* complex (Sebacinaceae, Tulasnellales, Ceratobasidiales), but are rare in sister clades of basidiomycetes (Taylor et al. 2002). As shown in Table 16.2, *Sebacina* members include ECM fungi, orchid mycorrhizal associates, and bryophyte fungi, and also occur in ericoid plants, but there is some doubt about how much roles overlap between separate clades within this genus (Warcup 1981; Selosse et al. 2002a, 2002b). Another multifunctional orchid fungus is *Thanatephorus gardneri*, the mycorrhizal associate of the underground orchid (*Rhizanthella gardneri*), which also forms ECM with *Melaleuca uncinata* and is capable of endophytic activity (Table 16.2). This multifunctional fungus is the only known member of the *Rhizoctonia* alliance outside of *Sebacina* that forms ECM.

Attempts to isolate mycorrhizal fungi from orchids often produce cultures of bacteria, actinomycetes and common endophytes such as *Fusarium* and *Verticillium*, as well as ECM fungi and ericoid fungi (Richardson and Currah 1995; Currah et al. 1997; Kristiansen et al. 2001; Bayman et al. 2002; Otero et al. 2002, Bidartondo et al. 2004). Most of these other fungi seem to be endophytes, but *Fusarium* isolates are also capable of forming orchid mycorrhizas (Vujanovic et al. 2000; Abdul Karim 2005). We have observed that the frequency of isolation of non-*Rhizoctonia* fungi decreases substantially when fungi are isolated from single pelotons rather than from surface-sterilised tissue blocks, providing further evidence that many fungi isolated by the latter method are endophytes. It is recommended that only isolates that germinate orchids to an advanced seedling stage with a green leaf be designated as orchid mycorrhizal fungi (Batty et al. 2002).

Kottke et al. (2003) found members of the *Rhizoctonia* complex (*Tulasnella* and *Sebacina*) closely related to orchid fungi that formed mycorrhiza-like associations with hepatics. These bryophytes also contained ascomycetes, and the nature of these associations requires further investigation. Members of the *Rhizoctonia* complex include major pathogens of seedlings of crop and horticultural species (Sivasithamparam 1998), but the degree of overlap between pathogens and orchid associates is unknown (Batty et al. 2002). In general, fungi associating with orchids differ substantially from other mycorrhizal fungi, because they do not belong to discrete

Table 16.2. Examples of fungi with multifunctional roles. Fungi within a row have been shown to be closely related by molecular identification

Fungus	Endophyte	Pathogen	Ectomycorrhizal	Ericoid mycorrhizas	Orchid mycorrhizas	Other role
Tomentella spp. (Thelephoraceae)			Primary role (Köljalg et al. 2000)		Myco-heterotrophic orchids (McCormick et al. 2004; Taylor et al. 2002)	Decomposition of wood?
Rhizoctonia complex S.L.	Common: e.g. *Pinus sylvestris* (Sen et al. 1999)	Major role: (see text)	See below		Many orchid associates (see text)	Mycorrhizas in hepatics (Kottke et al. 2003)[a]. Saprophytic
Sebacina spp. (in *Rhizoctonia* complex)	Ericaceae? (Allen et al. 2003)		(Glen et al. 2002; Selosse et al. 2002a; Urban et al. 2003)[a]	Suspected role? (Allen et al. 2003)	Some orchids (Warcup 1981; Y. Bonnardeaux, personal communication)	Myco-heterotrophic orchids (Selosse et al. 2002b; Taylor et al. 2003)[a]. Saprophytic?
Thanatephorus gardneri (in *Rhizoctonia* complex)	Some endophytic activity (Mursidawati 2003)		*Melaleuca uncinata* and other plants (Warcup 1985; Mursidawati 2003)		Myco-heterotrophic orchid *Rhizanthella gardneri* (Warcup 1985; Mursidawati 2003)	Saprophytic?
Phialocephala spp. (DSE fungi)	Widespread (see text)		ECM of trees (Harney et al. 1997)			
Fusarium spp.	Common (Kuldau and Yates 2000; Redman et al. 2001)	e.g. Pamphile and Azevedo 2002			Tropical orchids (Bayman et al. 2002; Abdul Karim 2005)	Saprophytic phases

Table 16.2. (continued)

Fungus	Endophyte	Pathogen	Ectomycorrhizal	Ericoid mycorrhizas	Orchid mycorrhizas	Other role
Hymenoscyphus ericae	*Picea mariana* roots (Piercey et al. 2002), Bryophyte (Chambers et al. 1999)		Six tree spp. (Vrålstad et al. 2000)[a]	Primary role (see text)		Mycorrhizas of bryophytes (Duckett and Read 1995)[a]
Oidiodendron spp.	ECM roots of *Quercus ilex* (Bergero et al. 2000)			Primary role? (see text)		Some isolates are strong saprophytes (Piercey et al. 2002)
Mycelium radicis atrovirens	In ECM of *Betula platyphylla* (Hashimoto and Hyakumachi 2001)			Trees (Sakakibara et al. 2002)[a]		

[a]Role not fully-established

evolutionary or taxonomic groups, and orchid mycorrhizal associations are not their primary ecological role (Brundrett 2002). These fungi are efficient plant colonisers where each has multiple roles as endophytes, parasites, ECM fungi and saprophytes. We require knowledge about the biology of these fungi in natural ecosystems to help us understand and control their plant parasitic activities, as well as their beneficial mycorrhizal associations with orchids.

16.2.4 Ericoid Mycorrhizal Fungi

Mycorrhizal fungi that associate with members of the Ericaceae (including the Epacridaceae) include several discrete groups of ascomycetes (McLean et al. 1999; Monreal et al. 1999; Sharples et al. 2000). Ericoid fungi with dual roles include *Hymenoscyphus ericae*, which forms ECM and can also associate with bryophytes (Table 16.2). However, ericoid fungi also endophytically colonise roots of ECM hosts without forming mycorrhizas (Bergero et al. 2000; Piercey et al. 2002). Some strains of the ericoid fungus *Oidiodendron maius*, which do not form mycorrhizal associations (Piercey et al. 2002), appear to be efficient saprophytes (see Chap. 13 by Rice and Currah). The primary role of ericoid fungi is not clear, as their occurrence in soils is independent of their host plants [Sharples et al. 2000; Bergero et al. 2003; see Chaps. 12 (Girlanda et al.) and 14 (Cairney)] and they also have a high degree of endophytic, or saprophytic competence, or have ECM associations with trees (Table 16.2).

16.2.5 Endophytic Fungi in Mycorrhizal Roots

Dark septate fungi (called DSF or DSE fungi) are common root endophytes in many ecosystems – in some cases with dual roles as ECM fungi (Stoyke and Currah 1991; O'Dell and Trappe 1992; Ahlich and Sieber 1996; see Chap. 7 by Sieber and Grünig). The DSE root endophytes may provide benefits to plants (Jumpponen and Trappe 1998). They include *Phialocephala* spp. with close relatives that are ECM or ericoid fungi (Vrålstad et al. 2002). However, DSE fungi do not form mycorrhizal associations as defined by morphological criteria. *Phialocephala fortinii*, an ECM fungus that may not benefit host plants, is closely related to other dematiaceous fungi, which are common root endophytes (Harney et al. 1997). Root endophytes can act as antagonists of ECM fungi, apparently by competing for space in roots (Hashimoto and Hyakumachi 2000, 2001). The role of fungi such as DSE and MRA [*Mycelium radicis atrovirens*; see Chaps. 7 (Sieber and Grünig) and 12 (Girlanda et al.)], which commonly share roots with mycorrhizal fungi, has not been well established.

16.3 Issues with the Identification and Categorisation of Fungi in Roots

Distinguishing endophytic activity from mycorrhizal associations caused by the same fungi is often problematic, especially for multifunctional fungi that are both endophytes and mycorrhizas. The examples discussed above illustrate how it is essential to use consistent definitions of mycorrhizal associations based on the structure and development of associations. A key defining feature of mycorrhizal associations is that they develop in young roots (Brundrett 2004). Consequently, mycorrhizal formation requires root growth while endophytic activity is not confined to young healthy roots.

Despite the common occurrence of endophytic activity by mycorrhizal fungi, we have very little knowledge of its ecological significance. Damage to roots of non-hosts plants caused by attempted colonisation by VAM and ECM fungi can lead to substantial growth reduction (Allen et al. 1989; Plattner and Hall 1995; Muthukumar et al. 1997). However, cases of antagonistic interactions caused by the endophytic activity by mycorrhizal fungi in non-host plants seem to be rare.

The common occurrence of endophytic fungi in roots may cause confusion with mycorrhizal fungi, especially when they are identified from DNA. In most cases, DNA extraction will detect several different fungi within a root and it may not be clear which are most abundant. Examples of cases where roles of fungi are uncertain include the study by Allen et al. (2003), where *Sebacina* spp. dominated DNA sequences from *Gaultheria shallon* (Ericaceae) roots, but did not form ericoid mycorrhizas. Another study by Bidartondo et al. (2004) found ECM fungi in addition to endophytes and orchid associates in five orchid species. They concluded that the ECM fungi were mycorrhizal associates of these orchids, but did not confirm this by mycorrhizal synthesis. There are numerous other examples in the recent literature of fungi identified in roots that have been designated as mycorrhizal without providing sufficient evidence.

Molecular methods used to detect fungi in roots or soils are still relatively new and more research is required to test the relative efficiency with which they detect different categories of fungi. In contrast, identification of mycorrhizal associations by morphology allows a much higher degree of replication and can be more accurate than DNA-based studies, which sometimes fail to detect the most important fungi in roots. The interpretation of fungal presence data provided by molecular tools that allow miniscule traces of fungi to be detected from almost any substrate will require further testing of key questions. These questions include: (1) Are the most common fungi in roots amplified most often? (2) Which of the detected fungi occur on the surface of roots or are endophytes? (3) How often are traces of DNA that contaminate samples detected? These questions can be answered by increasing the replication of sampling strategies, or combining isolation attempts with DNA extractions (e.g. Allen et al. 2003), but it may be even better to develop integrated approaches combining molecular and microscopic techniques. One such study by Sakakibara et al. (2002) found that microscopic identification of fungi by morphotypes and molecular methods (PCR-RFLP) were in close agreement, but multiple fungi were obtained from many mycorrhiza types.

Mycorrhizologists often use the term "endophytes" to refer to both mycorrhizal and non-mycorrhizal fungi in roots. However, due to the increasing recognition of the importance of "endophytes" as a specific category of specialised plant-inhabiting fungi, it is no longer appropriate to call mycorrhizal fungi endophytes. Types and categories of mycorrhizas are defined by morphological criteria (see Brundrett 2004), so mycorrhizal fungi need to be linked to a properly identified mycorrhizal structure. We should designate fungi as mycorrhizal only if they are isolated from a mycorrhiza by reliable means and belong to known groups of mycorrhizal associates. It will be necessary to confirm the mycorrhizal status of fungi by resynthesis, if it is not clear if they are endophytic or mycorrhizal.

16.4 Evolution of Root-Fungus Associations

An understanding of the capacity for mycorrhizal fungi to grow endophytically has led to a new theory of how these associations evolved (Brundrett 2002). This theory suggests that the switch to a new mycorrhizal association type starts with endophytic occupation of roots by fungi, and culminates in fully functional associations with coordinated development and synchronised nutrient transfer. The high degree of endophytic competence of Glomeromycotan fungi apparently has resulted in reacquisition of VAM by some plant lineages such as the Ericaceae in Hawaii (Koske et al. 1990). However, the reverse trend, where plant families become nonmycorrhizal is more common (Brundrett 2002). Other examples of new mycorrhizal associations, which probably started by endophytic fungal activity, include plants with dual ECM and VAM associations. These are common in some plant families that have ancestors with VAM (Brundrett 2002). There also seem to be intermediate types of ECM associations that involve new lineages of fungi that are not fully functional, as described above.

The Orchidaceae contain the most examples of plants acquiring new lineages of symbiotic fungi, in both myco-heterotrophic and green species (Taylor et al. 2002; Bidartondo et al. 2004). The fungal associates of these orchids are amazingly diverse, and the only theme that unifies them is that most are known to be efficient plant colonists as pathogens, endophytes or ectomycorrhizal associates of other species (Brundrett 2002). This suggests that most of these fungi were endophytes within orchids before the orchid evolved means to exploit them for its own purposes. All orchids seem to require is a fungus that can efficiently invade their roots or stems, but they seem to have much more success controlling fungi in the *Rhizoctonia* alliance than other fungi with similar endophytic competence for reasons we do not understand.

16.5 Conclusions

As demonstrated by the examples described above, fungi often have several roles and the primary roles of multifunctional fungi may not be certain. However, the number of roles fungi have does seem to be limited, presumably because they cannot be proficient at all of these roles. Consequently, we would expect multifunctional fungi to lose out due to competition from more efficient fungi that are more highly specialised. There also seem to be advantages to versatility in fungi. For example, multifunctional fungi seem to colonise new habitats or substrates more rapidly than specialised fungi, as shown by studies of mycorrhizal fungal succession after disturbance and the associations of nursery-grown plants (Jumpponen and Trappe 1998; Yu et al. 2001). Mycorrhizal fungi take part in a continuum of association types starting with endophytic associations and concluding with mycorrhizal associations. However, the endophytic and intermediate roles of these fungi seem to be less common than mycorrhizal associations, suggesting that there are clear advantages to fungi from their primary roles with plants. Recent advances in molecular biology have revealed a much more diverse and complex picture of the multifunctional nature of mycorrhizal fungi. Further research is required to determine the relative importance of fungi detected in roots, as it is not always clear which are endophytes and which are mycorrhizal associates, or how these roles change with time.

References

Abdul Karim N (2005) Molecular and enzymic groupings of fungi from tropical orchids of Western Australia and their patterns of tissue colonisation. PhD Thesis. The University of Western Australia

Ahlich K, Sieber TN (1996) The profusion of dark septate endophytic fungi in non-ectomycorrhizal fine roots of forest trees and shrubs. New Phytol 132:259–270

Allen MF, Allen EB, Friese CF (1989) Responses of the non-mycotrophic plant *Salsola kali* to invasion by vesicular-arbuscular mycorrhizal fungi. New Phytol 111:45–49

Allen TR, Millar T, Berch SM, Berbee ML (2003) Culturing and direct DNA extraction find different fungi from the same ericoid mycorrhizal roots. New Phytol 160:255–272

Azcón-Aguilar C, Bago B, Barea JM (1999) Saprophytic growth of arbuscular mycorrhizal fungi. In: Varma A, Hock B (eds) Mycorrhiza, structure, function, molecular biology and biotechnology. Springer, Berlin Heidelberg New York, pp 391–408

Batty AJ, Dixon KW, Brundrett MC, Sivasithamparam K (2002) Orchid conservation and mycorrhizal associations. In: Sivasithamparam K, Dixon KW, Barrett RL (eds) Microorganisms in plant conservation and biodiversity. Kluwer, The Netherlands, pp 195–226

Bayman P, Gonzalez EJ, Fumero, JJ, Tremblay RL (2002) Are fungi necessary? How fungicides affect growth and survival of the orchid *Lepanthes rupestris* in the field. J Ecol 90:1002–1008

Bergero R, Perotto S, Girlanda M, Vidano G, Luppi AM (2000) Ericoid mycorrhizal fungi are common root associates of a Mediterranean ectomycorrhizal plant (*Quercus ilex*). Mol Ecol 9:1639–1649

Bergero R, Girlanda M, Bello F, Luppi AM, Perotto S (2003) Soil Persistence and biodiversity of ericoid mycorrhizal fungi in the absence of the host plant in a Mediterranean ecosystem. Mycorrhiza 13:69–75

Bidartondo MI, Burghardt B, Gebauer G, Bruns TD, Read DJ (2004) Changing partners in the dark: isotopic and molecular evidence of ectomycorrhizal liaisons between forest orchids and trees. Proc R Soc London B 271:1799–1806

Brundrett MC (1991) Mycorrhizas in natural ecosystems. Adv Ecol Res 21:171–313

Brundrett MC (2002) Coevolution of roots and mycorrhizas of land plants. New Phytol 154:275–304

Brundrett MC (2004) Diversity and classification of mycorrhizal associations. Biol Rev 79:473–495

Brundrett MC, Kendrick WB (1988) The mycorrhizal status, root anatomy, and phenology of plants in a sugar maple forest. Can J Bot 66:1153–1173

Brundrett M, Bougher N, Dell B, Grove T, Malajczuk N (1996) Working with mycorrhizas in forestry and agriculture. Australian Centre for International Agricultural Research, Canberra

Burgess T, Dell B, Malajczuk N (1994) Variations in mycorrhizal development and growth stimulation by 20 *Pisolithus* isolates inoculated on to *Eucalyptus grandis* W. Hill ex Maiden. New Phytol 127:731–739

Cázares E, Trappe JM (1993) Vesicular endophytes in roots of the Pinaceae. Mycorrhiza 2:153–56

Chambers SM, Williams PG, Seppelt RD, Cairney JWG (1999) Molecular identification of *Hymenoscyphus* sp. from rhizoids of the leafy liverwort *Cephaloziella exiliflora* in Australia and Antarctica. Mycol Res 103:286–288

Chen YL, Dell B, Brundrett MC (2000) Effects of ectomycorrhizas and vesicular-arbuscular mycorrhizas, alone or in competition, on root colonization and growth of *Eucalyptus globulus* and *E. urophylla*. New Phytol 146:545–556

Chilvers GA, Gust LW (1982) Comparison between the growth rates of mycorrhizas, uninfected roots and a mycorrhizal fungus of *Eucalyptus st-johnii* R. T. Bak. New Phytol 91:453–66

Cordier C, Pozo MJ, Barea JM, Gianinazzi S, Gianinazzi-Pearson V (1998) Cell defence responses associated with localized and systemic resistance to *Phytophthora parasitica* induced in tomato by an arbuscular mycorrhizal fungus. Mol Plant-Microb Interact 11:1017–1028

Cornwell WK, Bedford BL, Chapin CT (2001) Occurrence of arbuscular mycorrhizal fungi in a phosphorus-poor wetland and mycorrhizal responses to phosphorus fertilization. Am J Bot 88:1824–1829

Currah RS, Zelmer CD, Hambleton S, Richardson KA (1997) Fungi from orchid mycorrhizas. In: Arditti J, Pridgeon AM (eds) Orchid biology: reviews and perspectives, VII. Kluwer, Dordrecht, pp 117–170

Dahlstrom JL, Smith JE, Weber NS (2000) Mycorrhiza-like interaction by *Morchella* with species of the Pinaceae in pure culture synthesis. Mycorrhiza 9:279–285

Demchenko K, Winzer T, Stougaard J, Parniske M, Pawlowski K (2004) Distinct roles of *Lotus japonicus* SYMRK and SYM15 in root colonisation and arbuscular formation. New Phytol 163:381–392

Duckett JG, Read DJ (1995) Ericoid mycorrhizas and rhizoid-ascomycete associations in liverworts share the same mycobiont: isolation of the partners and resynthesis of the associations in vitro. New Phytol 129:439–447

Eason WR, Newman EI, Chuba PN (1991) Specificity of interplant cycling of phosphorus: the role of mycorrhizas. Plant Soil 137:267–274

Gill WM, Lapeyrie F, Gomi T, Suzuki K (1999) *Tricholoma matsutake* – an assessment of in situ and in vitro infection by observing cleared and stained roots. Mycorrhiza 9:227–231

Giovannetti M, Sbrana C (1998) Meeting a non-host: the behaviour of AM fungi. Mycorrhiza 8:123–130

Glen M, Tommerup IC, Bougher NL, O'Brien PA (2002) Are Sebacinaceae common and widespread ectomycorrhizal associates of *Eucalyptus* species in Australian forests. Mycorrhiza 12:243–247

Hall IR (1976) Vesicular mycorrhizas in the orchid *Corybas macranthus*. Trans Br Mycol Soc 66:160

Harley JL, Harley EL (1987) A check-list of mycorrhiza in the British flora. New Phytol 105[Suppl 2]:1–102

Harney SK, Rogers SO, Wang CJK (1997) Molecular characterization of dematiaceous root endophytes. Mycol Res 101:1397–1404

Harrington TJ, Mitchell DT (2002) Colonization of root systems of *Carex flacca* and *C. pilulifera* by *Cortinarius* (*Dermocybe*) *cinnamomeus*. Mycol Res 106:452–459

Hashimoto Y, Hyakumachi M (2000) Quantities and types of ectomycorrhizal and endophytic fungi associated with *Betula platyphylla* var. *japonica* seedlings during the initial stage of establishment of vegetation after disturbance. Ecol Res 15:21–31

Hashimoto Y, Hyakumachi M (2001) Effects of isolates of ectomycorrhizal fungi and endophytic *Mycelium radicis atrovirens* that were dominant in soil from disturbed sites on growth of *Betula platyphylla* var. japonica seedlings. Ecol Res 16:117–125

Imhof S (2001) Subterranean structures and mycotrophy of the achlorophyllous *Dictyostega orobanchoides* (Burmanniaceae). Rev Biol Trop 49:239–247

Joner EJ, Jakobsen I (1995) Growth and extracellular phosphate activity of arbuscular mycorrhizal hyphae as influenced by soil organic matter. Soil Biol Biochem 9:1153–1159

Jumpponen A, Trappe JM (1998) Dark septate endophytes: a review of facultative biotrophic root-colonizing fungi. New Phytol 140:295–310

Kõljalg U, Dahlberg A, Taylor AFS, Larsson E, Hallenberg N, Stenlid J, Larsson K-H, Fransson PM, Kårén O, Jonsson L (2000) Diversity and abundance of resupinate thelephoroid fungi as ectomycorrhizal symbionts in Swedish boreal forests. Mol Ecol 9:1985–1996

Koske RE (1984) Spores of VAM fungi inside spores of VAM fungi. Mycologia 76:853–862

Koske RE, Gemma JN, Englander L (1990) Vesicular-arbuscular mycorrhizae in Hawaiian Ericales. Am J Bot 77:64–68

Kottke I, Beiter A, Weiss M, Haug I, Oberwinkler F, Nebel M (2003) Heterobasidiomycetes form symbiotic associations with hepatics: Jungermanniales have sebacinoid mycobionts while *Aneura pinguis* (Metzgeriales) is associated with a *Tulasnella* species. Mycol Res 107:957–968

Kristiansen KA, Taylor DL, Kjøller R, Rasmussen HN, Rosendahl S (2001) Identification of mycorrhizal fungi from single pelotons of *Dactylorhiza majalis* (Orchidaceae) using single-strand conformation polymorphism and mitochondrial ribosomal large subunit DNA sequences. Mol Ecol 10:2089–2093

Kuldau GA, Yates IE (2000) Evidence for *Fusarium* endophytes in cultivated and wild plants. In: Bacon CW, White JF Jr (eds) Microbial endophytes. Decker, New York, pp 85–117

Malajczuk N, Dell B, Bougher NL (1987) Ectomycorrhiza formation in Eucalyptus III. Superficial ectomycorrhizas initiated by *Hysterangium* and *Cortinarius* species. New Phytol 105:421–428

McKormick MK, Whigham DF, O'Neill (2004) Mycorrhizal diversity of photosynthetic terrestrial orchids. New Phytol 163:425–438

McLean CB, Cunnington JH, Lawrie AC (1999) Molecular diversity within and between ericoid endophytes from the Ericaceae and Epacridaceae. New Phytol 144:351–358

Monreal M, Berch SM, Berbee M (1999) Molecular diversity of ericoid mycorrhizal fungi. Can J Bot 77:1580–1594

Muthukumar T, Udaiyan K, Karthikeyan A, Manian S (1997) Influence of native endomycorrhiza, soil flooding and nurse plant on mycorrhizal status and growth of purple nutsedge (*Cyperus rotundus* L.). Agric Ecosyst Environ 61:51–58

Muthukumar T, Udaiyan K, Shanmughavel P (2004) Mycorrhiza in sedges - an overview. Mycorrhiza 14:65–77

Mursidawati S (2003) Mycorrhizal association, propagation and conservation of the mycoheterotrophic orchid *Rhizanthella gardneri.* MSc. Thesis, The University of Western Australia

Newsham KK, Fitter AH, Watkinson AR (1995) Arbuscular mycorrhiza protect an annual grass from root pathogenic fungi in the field. J Ecol 83:991–1000

Ocampo JA (1986) Vesicular-arbuscular mycorrhizal infection of "host" and "non-host" plants: effect on the growth responses of the plants and competition between them. Soil Biol Biochem 18:607–610

O'Dell T, Trappe JM (1992) Root endophytes of lupin and some other legumes of Northwestern U.S.A. New Phytol 122:479–485

Otero JT, Ackerman JD, Bayman P (2002) Diversity and host specificity of endophytic *Rhizoctonia*-like fungi from tropical orchids. Am J Bot 89:1852–1858

Pamphile JA, Azevedo JL (2002) Molecular characterization of endophytic strains of *Fusarium verticillioides* (= *Fusarium moniliforme*) from maize (*Zea mays* L). World J Microbiol Biotechnol 18:391–396

Peterson RL, Massicotte HB (2004) Exploring structural definitions of mycorrhizas, with emphasis on nutrient-exchange interfaces. Can J Bot 82:1074–1088

Piercey MM, Thormann MN, Currah RS (2002) Saprobic characteristics of three fungal taxa from ericalean roots and their associations with the roots of *Rhododendron groenlandicum* and *Picea mariana* in culture. Mycorrhiza 12:175–180

Plattner I, Hall IR (1995) Parasitism of non-host plants by the mycorrhizal fungus *Tuber melanosporum.* Mycol Res 99:1367–1370

Powell CL (1975) Rushes and sedges are non-mycotrophic. Plant Soil 42:481–484

Rabatin SC, Rhodes LH (1982) *Acaulospora bireticulata* inside orbatid mites. Mycologia 74:859–861

Redman RS, Dunigan DD, Rodriguez RJ (2001) Fungal symbiosis from mutualism to parasitism: who controls the outcome, host or invader? New Phytol 151:705–716

Richardson KA, Currah RS (1995) The fungal community associated with the roots of some rainforest epiphytes of Costa Rica. Selbyana 16:49–73

Saikkonen K, Faeth SH, Helander M, Sullivan TJ (1998) Fungal endophytes: a continuum of interactions with plants. Annu Rev Ecol Syst 29:319–343

Sakakibara SM, Jones MD, Gillespie M, Hagerman SM, Forrest ME, Simard SW, Durall DM (2002) A comparison of ectomycorrhiza identification based on morphotyping and PCR-RFLP analysis. Mycol Res 106:868–878

Selosse MA, Bauer R, Moyersoen B (2002a) Basal hymenomycetes belonging to the Sebacinaceae are ectomycorrhizal on temperate deciduous trees. New Phytol 155:183–195

Selosse M-A, Weiss M, Jany J-L, Tillier A (2002b) Communities and populations of sebacinoid basidiomycetes associated with the achlorophyllous orchid *Neottia nidus-avis* (L.) L.C.M. Rich. and neighbouring tree ectomycorrhizae. Mol Ecol 11:1831–1844

Sen R, Hietala AM, Zelmer CD (1999) Common anastomosis and internal transcribed spacer RFLP groupings in binucleate *Rhizoctonia* isolates representing root endophytes of *Pinus sylvestris*, *Ceratorhiza* spp. from orchid mycorrhizas and a phytopathogenic anastomosis group. New Phytol 144:331–341

Sharples JM, Chambers SM, Meharg AA, Cairney JWG (2000) Genetic diversity of root-associated fungal endophytes from *Calluna vulgaris* at contrasting field sites. New Phytol 148:153–162

Sivasithamparam K (1998) Root cortex – the final frontier for biocontrol of root-rot with fungal antagonists: a case study on a sterile red fungus. Annu Rev Phytopathol 36:439–452

Smith JE, Johnson KA, Cázares E (1998) Vesicular mycorrhizal colonisation of seedlings of Pinaceae and Betulaceae after spore inoculation with *Glomus intraradices*. Mycorrhiza 7:279–285

St John TV, Coleman DC, Reid CPP (1983) Growth and spatial distribution of nutrient-absorbing organs: selective exploitation of soil heterogeneity. Plant Soil 71:487–493

Stoyke G, Currah RS (1991) Endophytic fungi from the mycorrhizae of alpine ericoid plants. Can J Bot 69:347–352

Taylor DL, Bruns TD, Leake JR, Read DJ (2002) Mycorrhizal specificity and function in myco-heterotrophic plants. In: Van der Heijden MGA, Sanders IR (eds) The ecology of mycorrhizas. Springer, Berlin Heidelberg New York, pp 375–414

Tester M, Smith SE, Smith FA (1987) The phenomenon of "nonmycorrhizal" plants. Can J Bot 65:419–431

Urban A, Weiss M, Bauer R (2003) Ectomycorrhizas involving sebacinoid mycobionts. Mycol Res 107:3–14

Van der Heijden EW (2001) Differential benefits of arbuscular mycorrhizal and ectomycorrhizal infection of *Salix repens*. Mycorrhiza 10:185–193

Vrålstad T, Fossheim T, Schumacher T (2000) *Piceirhiza bicolorata* – the ectomycorrhizal expression of the *Hymenoscyphus ericae* aggregate? New Phytol 145:549–563

Vrålstad T, Schumacher T, Taylor FS (2002) Mycorrhizal synthesis between fungal strains of the *Hymenoscyphus ericae* aggregate and potential ectomycorrhizal and ericoid hosts. New Phytol 153:143–152

Vujanovic V, St-Arnaud M, Barabé D, Thibeaults G (2000) Viability testing of orchid seed and the promotion of colouration and germination. Ann Bot 86:79–86

Warcup JH (1981) The mycorrhizal relationships of Australian orchids. New Phytol 87:371–38

Warcup JH (1985) *Rhizanthella gardneri* (Orchidaceae), its *Rhizoctonia* endophyte and close association with *Melaleuca uncinata* (Myrtaceae) in Western Australia. New Phytol 99:273–280

Warner A (1984) Colonisation of organic matter by mycorrhizal fungi. Trans Br Mycol Soc 82:352–354

Yu TEJ-C, Egger KN, Peterson RL (2001) Ectendomycorrhizal associations – characteristics and functions. Mycorrhiza 11:167–177

17 Isolation Procedures for Endophytic Microorganisms

Johannes Hallmann, Gabriele Berg, Barbara Schulz

17.1 Introduction

There are good reasons for isolating endophytic microorganisms, e.g. for their characterisation, for studying population dynamics and diversity, use of microbial inoculants to improve plant growth and plant health, and as sources of novel biologically active secondary metabolites (Schulz et al. 2002; Strobel 2003; Schulz and Boyle 2005). The isolation procedure is a critical and important step in working with endophytic bacteria and fungi. It should be sensitive enough to recover endophytic microorganisms, but at the same time be strong enough to eliminate epiphytes from the root surface. In practice, this is often difficult, because microorganisms attach to plant cells/surfaces or hide in intercellular niches, thus evading the isolation procedure. Besides, there is no sharp delineation between the external and internal plant environment. Some bacteria constantly move between these two microhabitats, e.g. fungal mycelia grow from the root surface into the roots. Therefore, in order to be effective, the isolation procedure must be adapted to the respective tissues and microorganisms. Some isolation procedures are suitable only for certain plant tissues, whereas others favour local or systemic colonisers.

The most commonly used isolation procedures combine surface sterilisation of the root tissue with either maceration of the plant tissue and streaking onto nutrient agar, or plating small sterilised segments onto nutrient agar. Methods that avoid surface sterilisation are also available, e.g. vacuum or pressure extraction. In the past, isolation procedures focused primarily on culturable microorganisms. However, increasing interest in nonculturable endophytic microorganisms has recently led to the application

Johannes Hallmann: Federal Biological Research Centre for Agriculture and Forestry, Institute for Nematology and Vertebrate Research, Toppheideweg 88, 48161 Münster, Germany, E-mail: j.hallmann@bba.de

Gabriele Berg: Technical University of Graz, Department of Environmental Biotechnology, Petersgasse 12, 8010 Graz, Austria

Barbara Schulz: Institute of Microbiology, Technical University of Braunschweig, Spielmannstraße 7, 38106 Braunschweig, Germany

Soil Biology, Volume 9
Microbial Root Endophytes
B. Schulz, C. Boyle, T. N. Sieber (Eds.)

of molecular methods for their identification. This chapter summarises isolation procedures for culturable and detection methods for non-culturable endophytic bacteria and fungi of plant roots and discusses the strengths and weaknesses of each procedure. Examples are given on how the choice of the procedure can affect the microbial spectrum being isolated. The information provided here should enable the researcher to specifically select the method most suitable to meet their research objectives. The reader is also referred to the review by Sieber (2002), who summarised the available sterilisation and isolation procedures for fungi in plant roots.

17.2 Surface Sterilisation

Theoretically, the sterilising agent should kill any microbe on the plant surface without affecting the host tissue and the endophytic microorganisms. However, this is difficult to achieve because in time the agent may penetrate the root tissue. Conditions required to kill the last microbe on the surface may already be lethal for some endophytic microorganisms. In general, surface sterilisation of root tissue consists of the following steps: (1) thorough washing of the root tissue under tap water to remove adhering soil particles and the majority of microbial surface epiphytes and incidentals, (2) pre-treatment (optional) to eliminate hydrophobic substances on the plant surface and to provide better access for the sterilising agent to the root surface, (3) surface sterilisation to eliminate remaining microbial colonisers from the root surface, (4) several rinses under aseptic conditions in a sterile washing solution and (5) sterility check to confirm complete sterilisation of the root surface. It is very important that sterility is guaranteed for all tools and during all steps of this procedure, and that the researcher optimises the procedure for each plant tissue, since sensitivity varies with species, age and surface properties.

All steps in the sterilisation procedure should be conducted with sterilised tools, e.g. pincers, and preferably under a laminar flow hood. Between steps the tissue should be blotted on filter paper to avoid dilution of the sterilising agents.

17.2.1 Pre-treatment

Roots usually do not require special pre-treatment, since their surfaces do not contain hydrophobic substances like, e.g., the waxes of leaves. Washing with tap water and/or soft brushing is usually adequate. However, sonification may be used to dislodge soil and organic matter from the roots prior to surface sterilisation (Holdenrieder and Sieber 1992; Coombs and Franco 2003a).

17.2.2 Sterilising Agents

Commonly used sterilising agents are sodium hypochlorite (Gardner et al. 1982; Quadt-Hallmann et al. 1997; Schulz et al. 1993; Sieber 2002), ethanol (Dong et al. 1994; Gagné et al. 1987), and hydrogen peroxide (McInroy and Kloepper 1994; Misahgi and Donndelinger 1990; Sieber 2002). For health reasons, mercuric chloride should be used only very carefully and exclusively for plant tissues that cannot otherwise be sterilised (Gagné et al. 1987; Hollis 1951; O'Dell and Trappe 1992; Sriskandarajah et al. 1993). Note that sodium hypochlorite, an excellent oxidant, decomposes spontaneously in storage. Other, less frequently used, agents include propylene oxide vapour (Sardi et al. 1992) and formaldehyde (Schulz et al. 1993; Cao et al. 2002). Woody root tissues may also be dipped for 15 s in 95% ethanol and flame sterilised as employed for alfalfa roots by Gagné et al. (1987). In addition, the outer layer of woody root tissue can be aseptically peeled and the internal plant tissue can be excised using a sterile cork borer (Maifeld 1998; Reiter et al. 2002; Zinniel et al. 2002).

Combinations of agents can improve the effectiveness of surface sterilisation. For example, a combination of physical and chemical sterilisation has given good results with lignified roots (Table 17.1; Görke 1998). A common protocol involves a three-step procedure with ethanol, sodium hypochlorite, and then ethanol again (Schulz et al. 1993; Bills 1996; Sieber 2002). For example, for wheat roots, Coombs and Franco (2003b) applied 99% ethanol for 1 min, 3.1% sodium hypochlorite for 6 min and 99% ethanol for 30 sec. Sieber (2002) suggested omitting the initial soak in ethanol because it leads to contraction of the plant tissues, perhaps entrapping some epiphytic conidia, spores or mycelia. All the conidia of *Penicillium* sp. on roots of Norway spruce (artificially contaminated) were killed when surface sterilisation with hydrogen peroxide was used without the initial soak in ethanol (Sieber 2002). Examples of commonly used sterilising agents and conditions for their application are given in Table 17.1.

In general, a more stringent sterilisation procedure can be used with older and lignified plant tissue. Incubation times as well as concentrations of the sterilising agents directly affect the results. For example, an increase in the concentration of sodium hypochlorite from 3 to 6% active chlorine under otherwise constant conditions increases the number of root samples with complete sterilisation by 40%, but at the same time decreases the population densities of endophytic bacteria from 4.36×10^3 cfu/ml to

Table 17.1. Protocols for surface sterilisation of root tissue

Procedures and sterilising agents	Incubation time	Root type, age and/or diameter	Plant	Reference
Bacteria				
a) Wash under running tap water		5- to 7-week-old roots	*Cucumis sativus*, *Gossyium hirsutum*, *Pinus contorta* var. *latifolia*, *Zea mays*	Hallmann et al. 1998; McInroy and Kloepper 1995; Shishido et al. 1995
b) 1–3.1% Sodium hypochlorite[a]	1–2 min			
c) Two to four rinses in sterile water				
a) 99% Ethanol	1 min	1- to 3-month-old roots	*Triticum aestivum*	Coombs and Franco 2003a
b) 3.1% Sodium hypochlorite	6 min			
c) 99% Ethanol	0.5 min			
a) Wash with soapy water		4-to 28-month-old roots	*Medicago sativa*	Gagné et al. 1987
b) Wash thoroughly with tap water				
c) 0.2 M $HgCl_2$ in 50% ethanol	4 min			
d) Three rinses in sterile water	1 min			
a) Wash under running tap water	3 min	1- to 4-year-old roots	*Citrus jambhiri*	Gardner et al. 1982
b) Dip in 95% ethanol				
c) Flame				
d) Rinse in sterile water				
a) Wash under running tap water		Roots 1- to 5 mm in diameter	*Various herbaceous and woody plants*	Sardi et al. 1992
b) Exposure to propylene oxide vapor	1 h			
Physical sterilisation				
a) Shake vigorously in 0.9% NaCl solution containing 0.3 g acid-washed glass beads	20 min	13- to 14-week-old roots	*Solanum tuberosum*	Sessitsch et al. 2002
b) Five Rinses in sterile water				
Fungi				
36% Formaldehyde			*Musa acuminata*	Cao et al. 2002

Table 17.1. (continued)

Procedures and sterilising agents	Incubation time	Root type, age and/or diameter	Plant	Reference
a) Wash under running tap water b) 96% Ethanol c) 2–2.5% Sodium hypochlorite d) 96% Ethanol	 0.5–1.0 min 1–4 min 0.5 min	Nodule roots, fine hair roots, nonlignified roots, lignified roots	*Triticum aestivum*, *Erica carnea*	Oberholzer-Tschütscher 1982; Sieber et al. 1988; Crous et al. 1995[b]
a) Wash under running tap water b) 0.3–2% Sodium hypochlorite c) Rinse in sterile water	 1–3 min	Adventitious roots, fine hair roots	*Lolium perenne*, *Picia abies* various species of Ericaceae, various alpine plant species *Lycopersicon esculentum*	Skipp and Christensen 1989; Kattner and Schönhar 1990; Hambleton and Currah 1997; Stoyke and Currah 1991[b]. Hallmann and Sikora 1994
a) Wash under running tap water b) 70% Ethanol c) 1–3% Sodium hypochlorite d) 70% Ethanol	 0.5 min 1–3 min 0.5 min	Roots	*Hordeum vulgare*, *Phaseolus vulgarum*	Boyle et al. 2001
a) Wash under running tap water b) 75% Ethanol c) 3–5% Sodium hypochlorite[a] d) 75% Ethanol	1 min 3 min 0.5 min	Rhizomes; 6- to 9-month-old roots	*Pteridium aquilinum*; *Aphelandra tetragona*	Petrini et al. 1992; Werner et al. 1997[b]
0.01% Mercuric chloride	5–15 min	Fine roots	*Lupinus* spp., *Oxytropis campestris*	O'Dell and Trappe 1992[b]
5% Hydrogen peroxide	5–15 min	Fine roots	*Lupinus* spp., *Oxytropis campestris*	O'Dell and Trappe 1992[b]

Table 17.1. (continued)

Procedures and sterilising agents	Incubation time	Root type, age and/or diameter	Plant	Reference
a) Wash under running water		Adventitious roots	*Oryza sativa*	Fisher and Petrini 1992[b]
b) 75% Ethanol	1 min			
c) 20% Sodium hypochlorite	3 min			
d) 75% Ethanol	0.5 min			
a) Wash under running water		Non-ectomycorrhizal roots, 0–5.3 mm in diameter	*Abies alba*	Ahlich and Sieber 1996[b]
b) 99% Ethanol	1 min			
c) 35% Hydrogen peroxide	5 min			
d) 99% Ethanol	0.5 min			
Physical sterilisation				
a) Cut core into pieces		Boring cores from which bark has been removed	*Picea abies*	Maifeld 1998[b]
b) Move pieces quickly through flame of Bunsen burner				
Physical + chemical sterilisation				
a) Wash under running tap water		2- to 9-year-old roots undergoing secondary growth	*Fagus sylvatica, Picea abies, Pinus sylvestris, Betula pendula*	Görke 1998[b]
b) 70% Ethanol	1 min			
c) 10% Sodium hypochlorite	3 min			
d) 70% Ethanol	0.5 min			
e) Rinse twice in sterile water				
f) Discard bark				
g) Move pieces quickly through flame of Bunsen burner				

[a]Concentrations of sodium hypochlorite represent percent active chlorine
[b]Adapted from Sieber 2002

5.75 × 10 cfu/ml (A. Munif, personal communication). The effect of the sterilising agent penetrating the plant tissue on the bacterial cells was visually demonstrated by treating the root tissue with a tetrazolium-phosphate buffer solution (Patriquin and Döbereiner 1978). While metabolically active bacterial cells reduced tetrazolium and became stained, cells killed by the sterilisation remain unstained. Alternatively, apoplastic dyes such as 3-hydroxy-5,8,10-pyrenetrisulfonate (PTS) or 4.4′-bis (2-sulfostyryl) biphenyl can be used to monitor the relative progression of the applied agent into the plant tissue (Petersen et al. 1981).

17.2.3 Surfactants

Surfactants such as Tween 20 (Mahaffee and Kloepper 1997), Tween 80 (Sturz 1995) or Triton X-100 (Misaghi and Donndelinger 1990) added to the sterilising agent reduce the surface tension and allow the agent to reach into niches and grooves beyond the epidermal cells (Bills 1996; Hallmann et al. 1997a).

17.2.4 Rinsing

After each treatment, the plant tissue should be rinsed repeatedly in sterile water.

17.2.5 Sterility Check and Optimisation

Only if complete surface sterilisation of the root tissue is confirmed, can the isolated microorganisms be assumed to be endophytes. Validation of the surface sterilisation procedure can be done by (1) imprinting the surface-sterilised plant tissue onto nutrient media (Pleban et al. 1995; Shishido et al. 1995; Schulz et al. 1998), (2) culturing aliquots of water from the last rinsing onto nutrient media (McInroy and Kloepper 1994) or (3) dipping the roots into nutrient broth (Gagné et al. 1987). All three methods gave comparable results (Musson et al. 1995). If no microbial growth occurs on the medium, surface sterilisation is considered complete. A further check is to test the effect of the sterilising agent directly on fungal or bacterial cells. Roots are dipped into a fungal or bacterial suspension of known density, slightly dried, surface sterilised and subjected to a sterility check (Petrini 1984; Coombs and Franco 2003a).

17.3 Culture of Tissue and Plant Fluid of Sterilised Roots on Nutrient Medium

17.3.1 Segments

The surface-sterilised root is cut aseptically into segments, which are plated, i.e. pressed directly, onto an appropriate nutrient medium (see below). It is very important to cut the tissue into very small and thin segments, since colonisation may be very limited (Carroll 1995; Boyle et al. 2001). For fungal isolation, the nutrient medium should be supplemented with antibiotics to suppress bacterial growth. When isolating fungi, antifungal substances may also be added to retard the growth of fast-growing fungi; when isolating bacteria to inhibit growth of fungi (see below). Endophytic microorganisms that emerge from the root fragments are subsequently transferred to fresh media. This method is commonly used for endophytic fungi (Schulz et al. 1993; Hallmann and Sikora 1994; Bills 1996) and actinobacteria (Sardi et al. 1992; Coombs and Franco 2003a), but is rarely appropriate for endophytic bacteria (Araújo et al. 2001). This method selects for fast-growing microorganisms, therefore representing more a qualitative than a quantitative approach. When studying microbial diversity, slow growing microorganisms may be under-represented. In addition, concomitant growth of two or more microorganisms necessitates subsequent separation of the isolates and may also result in a lower number of colonies being recovered (Elvira-Recuenco and van Vuurde 2000).

17.3.2 Maceration of Root Tissue

Maceration is the preferred method for the isolation of endophytic bacteria from surface sterilised tissue, but may also be employed for isolating endophytic fungi (Sieber 2002), particularly slow growing ones. Theoretically, it captures all endophytic colonisers of the root tissue, i.e. colonisers of the root cortex as well as of the vascular tissue, systemic as well as local colonisers, and intercellular as well as intracellular colonisers. Maceration is useful for determining the broad spectrum of culturable bacterial and fungal endophytes; however, it excludes obligate biotrophs. To improve maceration and form a homogenous suspension, sterile water or sterile buffer solution is usually added to the surface-sterilised root tissue prior to maceration. Maceration of the surface-sterilised tissue can be performed with mortar and pestle or with mechanical devices such as a Klecco tissue pulveriser (Mahaffee and Kloepper 1997), a Polytron homogeniser (Zinniel et al. 2002) or a blender (Araújo et al. 2001), depending on sample size and the hardness of the plant material. Especially for woody plant tissues, processing by hand can be laborious and exhausting and mechanical devices are advisable. For the latter, optimal time and intensity of maceration need to be empirically determined. Following maceration of the surface-sterilised root tissue, the suspension is streaked onto nutrient medium. Sterility has to be guaranteed during all steps of this procedure. Since plant enzymes and toxins released during maceration can inactivate or kill endophytic microorganisms, temperature should not increase to lethal levels, necessitating cooling; the sample should be immediately diluted to decrease the concentrations of toxic compounds to ineffective levels. Alternatively, substances can be added to buffer the toxic compounds, e.g. polyvinylpyrrolidone (PVP) or EDTA. To minimise the time required for the entire procedure from surface sterilisation to maceration, Musson et al. (1995) described a microtitre plate method in which all steps are included in one plate. This method resulted in a higher detection limit and improved sterility, but was limited to small sample sizes.

17.3.3
Centrifugation of Root Tissue

Centrifugation is commonly used to collect the intercellular (apoplastic) fluid of plant tissue (De Wit and Spikman 1982; Boyle et al. 2001), but may also be used to extract endophytic microorganisms. This technique has been successfully applied for the isolation of endophytic bacteria from sugarcane stems (Dong et al. 1994), but supposedly is also suitable for the isolation of endophytic microorganisms from root tissue. Depending on sample size, sterile Eppendorf tubes or larger glass test tubes are used for centrifugation of the surface-sterilised tissues. Most of the apoplastic fluid will be removed at 3,000 *g* (Dong et al. 1994). The collected plant fluid is then streaked on nutrient medium for bacterial and fungal recovery. This approach avoids maceration of the root tissue, as demonstrated by cryo-scanning electron microscopy, which showed that the cells were still intact and thus no contamination with symplastic fluid had occurred (Dong et al. 1994). Although this technique seems to be time-consuming, requiring both surface sterilisation and centrifugation, several samples can be centrifuged at the same time and the overall time requirement may in effect be no more than for other methods.

Whereas surface sterilisation, followed by subsequent plating of macerated tissue segments or of centrifuged fluid, is a simple and valuable method that produces consistent results, it has its limitations: (1) it is laborious; (2) microorganisms can hide in niches that are not reached by the sterilising agent or "stick" to the host tissue, resulting in false identifications of endophytes; and (3) the sterilising agent may penetrate the root tissue and kill endophytes, so that microbial densities will be underestimated. These disadvantages can be avoided by using techniques to isolate microorganisms that do not involve surface sterilisation.

17.4 Vacuum and Pressure Extraction

To bypass the above mentioned problems, in particular involving surface sterilisation, alternative methods have been developed. These techniques use vacuum or pressure to collect plant fluid from the root tissue. Both approaches have in common that plant fluid of the conducting elements and adjacent intercellular spaces is collected, as discussed for centrifugation, thus reaching two niches often considered favourable for systemic colonisers (Hallmann et al. 1997a, 1997b). However, it does not isolate those endophytes that grow intracellularly. Bell et al. (1995) and Gardner et al. (1982) used an aseptic vacuum extraction technique to isolate bacteria from the xylem of grapevine and citrus roots. Cohen (1999) described an oversized vacuum filtration unit for large scale isolation of fungi from leaf tissue, which might also be adaptable for root tissue. The Scholander pressure bomb commonly used for measuring plant water status is also effective for the isolation of endophytic microorganisms (Fig. 17.1; Hallmann et al. 1997b). Von Tiedemann et al. (1983) described a technique for lignified root tissue. Sterile water is washed through the vascular system at low pressure to collect the endophytic microorganisms. To prevent damage of the plant

Fig. 17.1. Scholander pressure bomb. From left to right: Pressure bomb apparatus, inserting the root into the pressure cylinder, collection of plant sap from the root with a sterile Pasteur pipette

tissue, vacuum or pressure should be carefully increased to the maximum level tolerated by the respective tissue. In all cases, the plant extract is streaked onto nutrient media to allow recovery of the endophytic microorganisms. As with the surface sterilisation procedure, sterility checks need to be included to exclude the possibility that surface colonisers have been forced from the plant surface through the vascular system by the applied vacuum or pressure. This can be done by dipping the plant tissue immediately before extraction into a suspension with an indicator bacterium and checking for this bacterium after extraction. Additional sterility can be assured by surface sterilising the cut surface before applying pressure or vacuum. A sterile working environment can be easily achieved by treatment with ethanol.

Comparison of the pressure bomb method with maceration of surface sterilised roots resulted in slightly higher bacterial numbers for the latter technique (Hallmann et al. 1997b). However, the pressure bomb technique recovered a higher number of less commonly detected genera, resulting in higher indices for bacterial richness and diversity. Gram-positive species, such as *Bacillus* spp., were more frequently isolated with the maceration method than with the pressure method, but Gram-negative genera such as *Pseudomonas* and *Phyllobacterium* were recovered at similar levels using both methods. These results confirm the previously made assumptions that the pressure bomb method recovers predominantly colonisers of the vascular tissue, while the maceration method recovers bacteria of the vascular as well as the cortical root tissue. An advantage of vacuum and pressure extraction methods is the time saved by avoiding the time-consuming surface sterilisation procedure. However, limitations of the pressure bomb technique were encountered for young, fleshy root tissue of cucumber and tomato, which collapsed under the applied pressure. Those limitations did not apply to young roots of cotton, soybean and bean (Hallmann et al. 1997b).

17.5 Media

The recipes for most commonly used microbial growth media are listed for example on the web page of the German National Resource Centre for Biological Material: http://www2.dsmz.de/media/media.htm.

17.5.1 Media for Isolating Bacteria

The choice of the growth medium is crucial as it directly affects the number and type of endophytic microorganisms that can be isolated from the root tissue. For endophytic bacteria, commonly used media include tryptic soya agar (TSA), which supports growth of a broad range of bacteria (Gardner et al. 1982), R2A for bacteria requiring low levels of nutrients (oligotrophs; Reasoner and Geldreich 1985), nutrient broth-yeast extract medium for growth of less selective bacteria (Zinniel et al. 2002), King's B medium for growth of Pseudomonads (King et al. 1954; Misaghi and Donndelinger 1990) or SC – originally designed for coryneform bacteria, but also useful for fastidious endophytic bacteria (McInroy and Kloepper 1995). Comparing the different media, Elvira-Recuenco and van Vuurde (2000) obtained significantly higher bacterial densities on 5% TSA than on R2A and SC, while McInroy and Kloepper (1995) found higher bacterial densities on R2A and SC than on full strength TSA. However, full strength TSA seems to be less suitable, due to the fact that its high nutrient concentration allows fast growing bacteria to overgrow slower growing ones. Compared with SC medium, plate counts from R2A were more accurate due to less colony overgrowth and smaller colony size (McInroy and Kloepper 1995).

17.5.2 Media for Isolating Fungi

For isolating endophytic fungi, standard media include PDA (potato dextrose agar; Philipson and Blair 1957), malt extract–peptone–yeast extract (Schulz et al. 1995) and biomalt agar (Schulz et al. 1995), but minimal media such as SNA (synthetic nutrient agar; Boyle et al. 2001) may also be used. To isolate host-specific fungi, host tissue or extracts thereof may be included in a minimal medium (Arnold and Herre 2003). To increase the diversity of fungi isolated, it is wise to use a number of different media for each host plant, varying factors such as pH, temperature of cultivation, and also aeration.

17.5.3 Supplements

Antibiotics, fungicides or specific nutrients are commonly added to media to stimulate or suppress growth of certain microbial groups. In spite of the fact that fungal growth media are usually at slightly acidic pH values, antibacterial agents, e.g. oxytetracycline, streptomycin sulfate, penicillin, and/or novobiocin (Tsao 1970, Schulz et al. 1993), should always be included in the isolation media. Sublethal doses of fungicides restrict radial growth of fungal colonies, preventing overgrowth (Bills 1996). For example, 1–2 mg/l Cyclosporin A (active component of Sandimmune; Novartis, Basel, Switzerland) may be added to the growth medium to suppress fast-growing fungi (Dreyfuss and Chapela 1994). Even at low concentrations in agar (1–10 mg/l), it causes ascomycetous fungi to become unusually compact, and restricts growth of mucoraceous fungi (Bills 1996). Another technique is to "weed" out the ubiquitous fungi with a (hot) inoculating needle (Dreyfuss and Chapela 1994).

17.5.4 Selective Media

Selective culture techniques are used for microorganisms with specialised physiological capabilities. For example, nitrogen-free enrichment media is used for the isolation of endophytic diazotrophs (Hartmann et al. 2000), while media containing chitin, pectin or cellulose as sole nutrient source are used to select for fungi and bacteria with chitinolytic, pectinolytic or cellulytic activity, respectively (Chernin et al. 1995; Bills 1996; Berg et al. 2002).

17.6 Cultivation-Independent Methods

Cultivation-dependent methods such as plate counts underestimate microbial numbers as they do not record viable but non-culturable cells, e.g. obligate biotrophs, and microorganisms with unknown growth requirements. Non-culturable microorganisms can be detected using molecular methods. Their DNA sequences can be compared with those in databanks to identify the microorganisms. Alternatively, RNA may be used to identify the active genes. In most cases target sequences are amplified using the polymerase chain reaction (PCR). The 16S rRNA gene (rDNA) has become a frequently employed phylogenetic marker with which to describe bacterial diversity in natural environments (Vossbrinck et al. 1987). Similarly, 18S rDNA is frequently employed for determining fungal diversity (Kowalchuk et al. 1997; Smalla 2004). However, since it is not very variable, it may lead to amplification of metazoa (Zuccaro et al. 2003), oomycetes and diatoms (Nikolcheva et al. 2003). Thus, it is usually more reliable to amplify 28S rDNA or the internal transcribed spacer (ITS) regions (Schulz and Boyle 2005).

Fingerprinting techniques on the basis of PCR-amplified 16S rDNA (bacteria) or 18S and 28S rDNA (fungi) genes allow for analysis of the structure of the whole microbial community and their spatial and temporal variations in relation to environmental factors (Smalla 2004; Zuccaro et al. 2003; Nikolcheva et al. 2003; Nikolcheva and Bärlocher 2005). Such techniques include denaturing or temperature gradient gel electrophoresis (DGGE/TGGE; Kowalchuk et al. 1997; Heuer and Smalla 1997; Zuccaro 2003), terminal restriction fragment length polymorphism (T-RFLP; Liu et al. 1997; Nikolcheva et al. 2003), and PCR-single-strand-conformation-polymorphism (SSCP; Schwieger and Tebbe 1998). All these techniques have been successfully applied to the analysis of endophytic communities, e.g. T-RFLP (Reiter et al. 2002; Krechel et al. 2002), SSCP (A. Krechel et al. unpublished data) and DGGE (Garbeva et al. 2001) to characterise endophytes of potato, DGGE for those from marrum grass (Kowalchuk et al. 1997), citrus plants (Araújo et al. 2002), and fungi associated with decaying leaves (Nikolcheva and Bärlocher 2005) and algae (Zuccaro et al. 2003). Using eubacterial primers, it has been estimated that a population must represent about 1% of the total community to be detectable in a fingerprint (Smalla 2004). Group-specific primers are available for *Burkholderia* (Salles et al. 2001), Proteobacteria (Sessitsch et al. 2002), actinomycetes (Heuer et al. 1997) and many other taxa, allowing a sensitive analysis of the microbial community. Prominent bands can be excised, cloned and used for sequence determination in order to obtain further information as to the identity (Zuccaro et al. 2003) and phylogeny of dominant ribotypes, as shown for endophytic bacteria of citrus rootstocks by Araújo et al. (2002).

Another cultivation-independent approach is cloning of 16S rDNA, 18S rDNA or 28S rDNA fragments amplified from community or environmental DNA, with subsequent sequencing. This approach was applied by Sessitsch et al. (2002) to the analysis of diversity of potato-associated endophytes. However, DNA analysis does not generally allow conclusions regarding the metabolic activity of members of the microbial community or their gene expression to be drawn. This information may be obtained only through analysis of RNA (Griffith et al. 2000). Different methodological approaches have therefore been developed to link information on metabolic activity to certain ribotypes, such as the incorporation of bromide oxyuridine (BrdU) or ^{13}C into the nucleic acids of growing cells (Borneman 1999). Furthermore, fluorescence in situ hybridization (FISH) as well as marker and reporter genes are valuable tools with which to study the metabolic activity of endophytes (Martinez-Inigo et al. 2003; see Chap. 18 by Bloemberg et al.).

However, problematic with molecular results is that: (1) the databanks are far from complete, and (2) the entries are not always accurate; in the case of fungi, 20% of entries are estimated to be false (Bridge et al. 2003). Thus, in order to determine the taxon of a particular microorganism it is important not to rely only on one method (e.g. Zuccaro et al. 2003; Nikolcheva and Bärlocher 2005), but to evaluate multiple characteristics, i.e. morphology and physiology as well as molecular results.

17.7 Quantification of Colonisation

None of the methods for quantifying the degree of colonisation of endophytic microorganisms within their hosts is optimal. Colonisation can be quantified using visual methods, e.g. by direct counts of infections (Stone 1987); however, this is difficult for bacteria and yeasts. With fungi, biomass can be correlated with the concentration of fungal specific ergosterol (Newell et al. 1988; Weete and Ghandi 1996). This method may give variable results because ergosterol concentration varies with the age of the mycelium (Olsson et al. 2003). Phospholipid fatty acids have been used for quantification, but their concentration varies between fungal (Olsson et al. 2003) and bacterial (Olsson and Persson 1999) genera. Fungal biomass can also be measured using monoclonal antibodies. Here, the difficulty lies in developing an antibody specific enough to accurately quantify single endophytic taxa. Real-time PCR (Vaitilingom et al. 1998; Schena et al. 2004) is presumably the most accurate method for quantifying microbial colonisation within the host and has been successfully employed, e.g. by Winton et al. (2002), to quantify the density of fungal colonisation by *Phaeocrytopus gaeumannii* within the needles of Douglas-fir and by Oliveira and Vallim (2002) to quantify colonisation of the bacterial pathogen *Xylella fastidiosa* in the leaves of citrus trees.

17.8 Conclusions

The decision as to which isolation procedure is best for a given host depends on its physical niche, on factors that are plant-dependent, e.g. species, age and tissue, but also on the available resources and the total number of samples to be processed (Hallmann et al. 1997a). A polyphasic approach as suggested in Fig. 17.2 is recommended for the analysis of endophytic communities. Combinations of different techniques, e.g. surface sterilisation, vacuum and pressure extraction, cultivation independent (molecular) methods and cultivation dependent (isolation and morphological) identification, increase the likelihood of completely analysing microbial diversity, and consequently also enhancing our understanding of the interactions of microorganisms with the roots of their plant hosts.

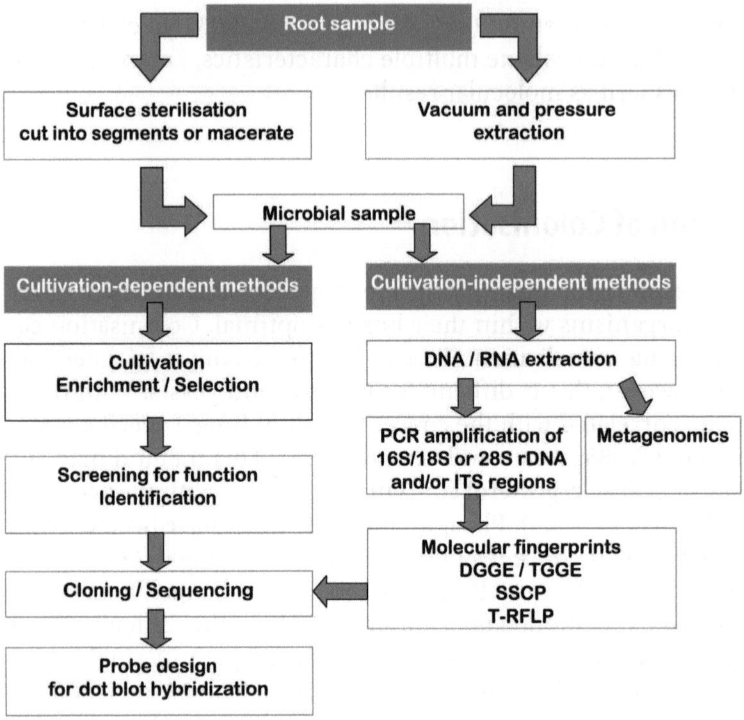

Fig. 17.2. Analysis of endophytic communities using a polyphasic approach

References

Ahlich K, Sieber T (1996) The profusion of dark septate endophytic fungi in non-ectomycorrhizal fine roots of forest trees and shrubs. New Phytol 132:259–270

Araújo WL, Maccheroni W, Aguilar-Vildoso CI, Barroso PAV, Saridakis HO, Azevedo JL (2001) Variability and interactions between endophytic bacteria and fungi isolated from leaf tissue of citrus rootstocks. Can J Microbiol 47:229–236

Araújo WL, Marcon J, Maccheroni W, Van Elsas JD, Van Vuurde JW, Azevedo JL (2002) Diversity of endophytic bacterial populations and their interaction with *Xylella fastidiosa* in citrus plants. Appl Environ Microbiol 68:4906–4914

Arnold AE, Herre EA (2003) Canopy cover and leaf age affect colonisation by tropical fungal endophytes: ecological pattern and process in *Theobroma cacao* (Malvaceae). Mycologia 95:388–398

Bell CR, Dickie GA, Harvey WLG, Chan JWYF (1995) Endophytic bacteria in grapevine. Can J Microbiol 41:46–53

Berg G, Roskot N, Steidle A, Eberl L, Zock A, Smalla K (2002) Plant-dependent genotypic and phenotypic diversity of antagonistic rhizobacteria isolated from different *Verticillium* host plants. Appl Environ Microbiol 68:3328–3338

Bills GF (1996) Isolation and analysis of endophytic fungal communities from woody plants. In: Redlin SC, Carris LM (eds) Endophytic fungi in grasses and woody plants: systematics, ecology and evolution. APS Press, St. Paul, MN, pp 31–65

Borneman J (1999) Culture-independent identification of microorganisms that respond to specified stimuli. Appl Environ Microbiol 65:3398–3400
Boyle C, Götz M, Dammann-Tugend U, Schulz B (2001) Endophyte-host interactions III. Local vs. systemic colonisation. Symbiosis 31:259–281
Bridge PD, Roberts PJ, Spooner BM, Panchal G (2003) On the unreliability of published DNA sequences. New Phytol 160:43–48
Cao LX, You JL, Zhou SN (2002) Endophytic fungi from *Musa acuminata* leaves and roots in South China. World J Microbiol Biotechnol 18:169–171
Carroll GC (1995) Forest endophytes: pattern and process. Can J Bot 73(S1):1316–1324
Chernin L, Ismailov Z, Haran S, Chet I (1995) Chitinolytic *Enterobacter agglomerans* antagonistic to fungal plant pathogens. Appl Environ Microbiol 61:85–92
Cohen SD (1999) Technique for large scale isolation of *Discula umbrinella* and other foliar endophytic fungi from *Quercus* species. Mycologia 91:917–922
Coombs JT, Franco CMM (2003a) Isolation and identification of actinobacteria from surface-sterilised wheat roots. Appl Environ Microbiol 69:5603–5608
Coombs JT, Franco CMM (2003b) Visualization of an endophytic *Streptomyces* species in wheat seed. Appl Environ Microbiol 69:4260–4262
Crous PW, Petrini O, Marais GF, Pretorius ZA, Rehder F (1995) Occurrence of fungal endophytes in cultivars of *Triticum aestivum* in South Africa. Mycoscience 36:739–752
DeWit PJGM, Spikman G (1982) Evidence for the occurrence of race and cultivar-specific elicitors of necrosis in intercellular fluids of compatible interactions of *Cladosporium fulvum* and tomato. Physiol Plant Pathol 21:1–11
Dong Z, Canny MJ, McCully MW, Roboredo MR, Cabadilla CF, Ortega E, Rodés R (1994) A nitrogen-fixing endophyte of sugarcane stems. Plant Physiol 105:1139–1147
Dreyfuss MM, Chapela IH (1994) Potential of fungi in the discovery of novel, low-molecular weight pharmaceuticals. In: Gullo VP (ed) The discovery of natural products with therapeutic potential, Butterworth-Heinemann, Stoneham, MA, pp 49–80
Elvira-Recuenco M, Vuurde van JWL (2000) Natural incidence of endophytic bacteria in pea cultivars under field conditions. Can J Microbiol 46:1036–1041
Fisher PJ, Petrini O (1992) Fungal saprobes and pathogens as endophytes of rice (*Oryza sativa* L.). New Phytol 120:137–143
Gagné S, Richard C, Rousseau H, Antoun H (1987) Xylem-residing bacteria in alfalfa roots. Can J Microbiol 33:996–1005
Garbeva P, van Overbeek LS, van Vuurde JWL, van Elsas JD (2001) Analysis of endophytic bacterial communities of potato by plating and denaturing gradient gel electrophoresis (DGGE) of 16S rDNA based PCR fragments. Microb Ecol 41:369–383
Gardner JM, Feldman AW, Zablotowicz RM (1982) Identity and behavior of xylem-residing bacteria in rough lemon roots of Florida citrus trees. Appl Environ Microbiol 43:1335–1342
Griffith RI, Whitely AS, O'Donell AG, Bailey MJ (2000) Rapid method for coextraction of DNA and RNA from natural environments for analysis of ribosomal DNA and rRNA-based microbial community composition. Appl Environ Microbiol 66:5488–5491
Görke C (1998) Mykozönosen von Wurzeln und Stamm von Jungbäumen unterschiedlicher Bestandsbegründungen. Bibl Mycol 173:1–462
Hallmann J, Sikora RA (1994) Occurrence of plant parasitic nematodes and non-pathogenic species of *Fusarium* in tomato plants in Kenya and their role as mutualistic synergists for biological control of root-knot nematodes. Int J Pest Manage 40:321–325
Hallmann J, Quadt-Hallmann A, Mahaffee WF, Kloepper JW (1997a) Bacterial endophytes in agricultural crops. Can J Microbiol 43:895–914
Hallmann J, Kloepper JW, Rodríguez-Kábana R (1997b) Application of the Scholander pressure bomb to studies on endophytic bacteria of plants. Can J Microbiol 43:411–416

Hallmann J, Quadt-Hallmann A, Rodríguez-Kábana R, Kloepper JW (1998) Interactions between *Meloidogyne incognita* and endophytic bacteria in cotton and cucumber. Soil Biol Biochem 30:925–937

Hambleton S, Currah RS (1997) Fungal endophytes from the roots of alpine and boreal Ericaceae. Can J Bot 75:1570

Hartmann A, Chatzinotas A, Assmus B, Kirchhof G (2000) Molecular microbial ecology studies on diazotrophic bacteria associated with non-legumes with special reference to endophytic diazotrophs. In: Subba Rao NS, Dommergues YR (eds) Microbial interactions in agriculture and forestry, vol II. Science, Enfield, pp 1–14

Heuer H, Smalla K (1997) Application of denaturing gradient gel electrophoresis (DGGE) and temperature gradient gel electrophoresis (TGGE) for studying soil microbial communities. In: van Elsas JD, Wellington EMH, Trevors JT (eds) Modern soil microbiology. Dekker, New York, pp 353–373

Heuer H, Krsek M, Baker P, Smalla K, Wellington EMH (1997) Analysis of actinomycete communities by specific amplification of genes encoding 16S rRNA and gel-electrophoretic separation in denaturing gradients. Appl Environ Microbiol 63:3233–3241

Holdenrieder O, Sieber T (1992) Fungal associations of serially washed healthy non-mycorrhizal roots of *Picea abies*. Mycol Res 96:151–156

Hollis JP (1951) Bacteria in healthy potato tissue. Phytopathology 41:350–367

Kattner D, Schönhar S (1990) Untersuchungen über das Vorkommen mikroskopischer Pilze in Feinwurzeln optisch gesunder Fichten (*Picea abies* Karst.) auf verschiedenen Standorten. Mitt Ver Forstl Standortskde Forstpflanzenzücht 35:39–43

King EO, Ward MK, Raney DE (1954) Two simple media for the demonstration of pyocyanin and fluorescein. J Lab Clin Med 44:301–307

Kowalchuk GA, Gerards S, Woldendorp JW (1997) Detection and characterisation of fungal infections of *Ammophila arenaria* (marram grass) roots by denaturing gradient gel electrophoresis of specifically amplified 18S rDNA. Appl Environ Microbiol 63:3858–3865

Krechel A, Faupel A, Hallmann J, Ulrich A, Berg G (2002) Potato-associated bacteria and their antagonistic potential towards plant-pathogenic fungi and the plant-parasitic nematode *Meloidogyne incognita* (Kofoid & White) Chitwood. Can J Microbiol 48:772–786

Liu WT, Marsh TL, Cheng H, Forney LJ (1997) Characterisation of microbial diversity by determining terminal restriction fragment length polymorphisms of genes encoding 16S rRNA. Appl Environ Microbiol 63:4516–4522

Mahaffee WF, Kloepper JW (1997) Temporal changes in the bacterial communities of soil, rhizosphere, and endorhiza associated with field-grown cucumber (*Cucumis sativus* L.). Microb Ecol 34:210–223

Maifeld D (1998) Endophytische Pilze der Fichte (*Picea abies* (L.) Karst.) – Neue Aspekte zur biologischen Kontrolle von *Heterobasidion annosum* (FR.) Bref. PhD dissertation, Eberhard-Karls-Universität Tübingen, Germany

Martinez-Inigo MJ, Lboe MC, Garbi C, Martin M (2003) Applicability of fluorescence in situ hybridisation to monitor target bacteria in soil samples. In: Del Re AAM, Capri E, Padovani L, Trevisan M (eds) Pesticides in air, plant, soil and water system. Proceedings of the XII Symposium Pesticide Chemistry, Piacenza, Italy, 4–6 June 2003, pp 609–615

McInroy JA, Kloepper JW (1994) Studies on indigenous endophytic bacteria of sweet corn and cotton. In: O'Gara F, Dowling DN, Boesten B (eds) Molecular ecology of rhizosphere microorganisms. VCH, Weinheim, Germany, pp 19–28

McInroy JA, Kloepper JW (1995) Survey of indigenous bacterial endophytes from cotton and sweet corn. Plant Soil 173:337–342

Misaghi IJ, Donndelinger CR (1990) Endophytic bacteria in symptom-free cotton plants. Phytopathology 80:808–811

Musson G, McInroy JA, Kloepper JW (1995) Development of delivery systems for introducing endophytic bacteria into cotton. Biocontrol Sci Technol 5:407–416

Newell SY, Arsuffi TL, Fallon RD (1988) Fundamental procedures for determining ergosterol content of decaying plant material by liquid chromatography. Appl Environ Microbiol 54: 1876–1879

Nikolcheva LC, Bärlocher F (2005) Seasonal and substrate preferences of fungi colonising leaves in streams: traditional versus molecular evidence. Environ Microbiol 7:270–80

Nikolcheva LC, Cockshutt AM, Bärlocher F (2003) Determining diversity of freshwater fungi on decaying leaves: comparison of traditional and molecular approaches. Appl Environ Microbiol 69:2548–54

Oberholzer-Tschütscher B (1982) Untersuchungen über endophytische Pilze von *Erica carnea* L. PhD dissertation, Swiss Federal Institute of Technology, Zürich, Switzerland

O'Dell TE, Trappe JM (1992) Root colonisation of *Lupinus latifolius* Agardhl and *Pinus contorta* Dougl. by *Phialocephala fortinii* Wang & Wilcox. New Phytol 124:93–100

Oliveira AC, Vallim MA (2002) Quantification of *Xylella fastidiosa* from citrus trees by real-time polymerase chain reaction assay. Phytopathology 92:1048–1054

Olsson S, Persson P (1999) The composition of bacterial populations in soil fractions differing in their degree of adherence to barley roots. Appl Soil Ecol 12:205–215

Olsson PA, Larsson L, Bago B, Wallander H, van Aarle IM (2003) Ergosterol and fatty acids for biomass estimation of mycorrhizal fungi. New Phytol 159:1–10

Patriquin DG, Döbereiner J (1978) Light microscopy observations of tetrazolium-reducing bacteria in the endorhizosphere of maize and other grasses in Brazil. Can J Microbiol 24:734–742

Petersen CA, Emanuel ME, Humphreys GB (1981) Pathway of movement of apoplastic fluorescent dye tracers through the endodermis at the site of secondary root formation in corn (*Zea mays*) and broad bean (*Vicia faba*). Can J Bot 59:618–625

Petrini O (1984) Endophytic fungi in British Ericaceae. Trans Br Mycol Soc 83:510–512

Petrini O, Fisher PJ, Petrini LE (1992) Fungal endophytes of bracken (*Pteridium aquilinum*), with some reflections on their use in biological control. Sydowia 44:282–293

Pleban S, Ingel F, Chet I (1995) Control of *Rhizoctonia solani* and *Sclerotium rolfsii* in the greenhouse using endophytic *Bacillus* spp. Eur J Plant Pathol 101:665–672

Philipseon MN, Blair ID (1957) Bacteria in clover root tissue. Can J Microbiol 3:125–129

Quadt-Hallmann A, Hallmann J, Kloepper JW (1997) Bacterial endophytes in cotton: location and interaction with other plant-associated bacteria. Can J Microbiol 43:254–259

Reasoner DJ, Geldreich EE (1985) A new medium for the enumeration and subculture of bacteria from potable water. Appl Environ Microbiol 49:1–7

Reiter B, Pfeifer U, Schwab H, Sessitsch A (2002) Response of endophytic bacterial communities in potato plants to infection with *Erwinia carotovora* subsp. *atroseptica*. Appl Environ Microbiol 68:2261–2268

Salles JF, De Souza FA, van Elsas JD (2001) Molecular method to assess the diversity of *Burkholderia* species in environmental samples. Appl Environ Microbiol 68:1595–1603

Sardi P, Sarachhi M, Quaroni S, Petrolini B, Borgonovi GE, Merli S (1992) Isolation of endophytic *Streptomyces* strains from surface-sterilised roots. Appl Environ Microbiol 58:2691–2693

Schena L, Nigro F, Ippolito A, Gallitelli D (2004) Real-time quantitative PCR: a new technology to detect and study phytopathogenic and antagonistic fungi. Eur J Plant Pathol 110: 893–908

Schulz B, Boyle C (2005) The endophytic continuum. Mycol Res 109:661–687

Schulz B, Wanke U, Draeger S, Aust H-J (1993) Endophytes from herbaceous plants and shrubs: effectiveness of surface sterilisation methods. Mycol Res 97:1447–1450

Schulz B, Sucker J, Aust HJ, Krohn K, Ludewig K, Jones PG, Döring D (1995) Biologically active secondary metabolites of endophytic *Pezicula* species. Mycol Res 99:1007–1015

Schulz B, Guske S, Dammann U, Boyle C (1998) Endophyte-host interactions II. Defining symbiosis of the endophyte-host interaction. Symbiosis 25:213–227

Schulz B, Boyle C, Draeger S, Römmert A-K, Krohn K (2002) Endophytic fungi: a source of biologically active secondary metabolites. Mycol Res 106:996–1004

Schwieger F, Tebbe CC (1998) A new approach to utilize PCR-single-strand-conformation polymorphism for 16S rRNA gene-based microbial community analysis. Appl Environ Microbiol 64:4870–4876

Sessitsch A, Reiter B, Pfeifer U, Wilhelm E (2002) Cultivation-independent population analysis of bacterial endophytes in three potato varieties based on eubacterial and *Actinomycetes*-specific PCR of 16S rRNA genes. FEMS Microbiol Ecol 39:23–32

Shishido M, Loeb BM, Chanway CP (1995) External and internal root colonisation of lodgepole pine seedlings by two growth-promoting *Bacillus* strains originated from different root microsites. Can J Microbiol 41:707–713

Sieber TN (2002) Fungal root endophytes. In: Waisel Y, Eshel A, Kafkafi U (eds) Plant roots: the hidden half. Dekker, New York, pp 887–917

Sieber TN, Riesen TK, Müller E, Fried PM (1988) Endophytic fungi in four wheat cultivars (*Triticum aestivum* L.) differing in resistance against *Stagonospora nodorum* (Berk.) Berk. J Phytopathol 122:289–306

Skipp RA, Christensen MJ (1989) Fungi invading roots of perennial rye grass (*Lolium perenne* L.) in pasture. N J Agric Res 32:423–431

Smalla K (2004) Culture-independent microbiology. In: Bull AT (ed) Microbial diversity and bioprospecting. ASM, Washington DC, pp 88–99

Sriskandarajah S, Kennedy IR, Yu DG, Tchan YT (1993) Effects of plant growth regulators on acetylene-reducing associations between *Azospirillum brasilense* and wheat. Plant Soil 153:165–178

Stone JK (1987) Initiation and development of latent infections by *Rhabdocline parkeri* on Douglas-fir. Can J Bot 65:2614–2621

Stoyke G, Currah RS (1991) Endophytic fungi from the mycorrhizae of alpine ericoid plants. Can J Bot 69:347–352

Strobel GA (2003) Endophytes as sources of bioactive products. Microbes Infect 5:535–544

Sturz AV (1995) The role of endophytic bacteria during seed piece decay and potato tuberization. Plant Soil 175:257–263

Tiedemann A von, Wolf G, Wilbert S (1983) Eine einfache Methode zur gezielten Isolierung von gefäßbesiedelnden Mikroorganismen aus holzigen Pflanzen. Phytopathol Z 107:87–91

Tsao PH (1970) Selective media for isolation of pathogenic fungi. Annu Rev Phytopathol 8:157–186

Vaitilingom M, Gendre F, Brignon P (1998) Direct detection of viable bacteria, molds, and yeasts by reverse transcriptase PCR in contaminated milk samples after heat treatment. Appl Environ Microbiol 64:1157–1160

Vossbrinck CG, Maddox JV, Friedman S, Debrunner-Vossbrinck BA, Woese CG (1987) Ribosomal RNA sequence suggests microsporidia are extremely ancient eukaryotes. Nature 326:411–414

Werner C, Petrini O, Hesse M (1997) Degradation of the polyamine alkaloid aphelandrine by endophytic fungi isolated from *Aphelandra tetragona*. FEMS Microbiol Lett 155:147–153

Weete JD, Ghandi SR (1996) Biochemistry and molecular biology of fungal sterols. In: Brambl R, Marzluf G (eds) The Mycota III: Biochemistry and molecular biology. Springer, Berlin Heidelberg New York, pp 421–438

Winton LM, Stone JK, Watrud LS, Hansen EM (2002) Simultaneous one-tube quantification of host and pathogen DNA with real-time polymerase chain reaction. Phytopathology 92:112–116

Zinniel DK, Lambrecht P, Harris NB, Feng Z, Kuczmarski D, Higley P, Ishimaru CA, Arunakumari A, Barletta RG, Vidaver AK (2002) Isolation and characterisation of endophytic colonising bacteria from agronomic crops and prairie plants. Appl Environ Microbiol 68:2198–2208

Zuccaro A, Schulz B, Mitchell J (2003) Molecular detection of ascomycetes associated with *Fucus serratus*. Mycol Res 107:1451–1466

18 Microbial Interactions with Plants: a Hidden World?

Guido V. Bloemberg, Margarita M. Camacho Carvajal

18.1 Introduction

Endophytic microorganisms, e.g. bacteria and fungi, are not only present within the plant, but may also colonise epiphytically before and during infection. Some of these organisms have important functions for the plant, such as nitrogen fixation (e.g. *Rhizobiaceae*, *Herbaspirillum*, *Azoarcus*), protection against pathogens (e.g. *Pseudomonas*), and provision of essential nutrients from the soil (e.g. mycorrhizae), but they can also be parasitic (e.g. *Agrobacterium*). Stages in the development of an interaction between endophyte and plant include attachment to the plant surface, infection and invasion, colonisation within the plant's tissues, proliferation (including sporulation), expression of various phenotypic traits, and spreading to other plant individuals. Several approaches can be taken to unravel the complex interactions between the plant and its endophytes. Genetics and measurement of enzymatic activities are powerful tools with which to discover the genes and traits involved in these interactions. However, these processes can only be understood in detail if the microorganisms and the expression of relevant genes can be visualised on and in the plant, where the endophytes encounter different tissues and conditions. Visualisation of the interactions between plant and microorganism on the cellular level provides a more detailed understanding of the basic principles of how endophytes function. In this chapter, methods of microscopy and techniques useful for the analysis of the spatio-temporal behaviour of endophytes on and in the plant will be evaluated, with emphasis on root-colonising microorganisms.

Guido V. Bloemberg, Margarita M. Camacho Carvajal: Leiden University, Institute of Biology, Wassenaarseweg 64, 2333AL Leiden, The Netherlands, E-mail: bloemberg@rulbim.leidenuniv.nl

Margarita M. Camacho Carvajal: Laboratory for Physiological Chemistry and Centre for Biomedical Genetics, Utrecht 3584CG, The Netherlands (Present Address)

Soil Biology, Volume 9
Microbial Root Endophytes
B. Schulz, C. Boyle, T. N. Sieber (Eds.)

18.2 Microscopic Techniques for Studying Plant-Microbe Interactions

Various techniques of microscopy are available for visualising microorganisms in the plant environment. We will briefly discuss some of the available techniques and give examples of how these techniques have been valuable in studying the interactions of microorganisms with the plant and its microflora.

18.2.1 Light Microscopy and Enzymatic Reporters

Light microscopes belong to the standard equipment of every microbiology laboratory and are relatively inexpensive. Several reporter genes with enzymatic functions such as β-glucuronidase (GUS) encoded by *gusA* and β-galactosidase encoded by *lacZ* are used to visualise bacterial cells and gene expression (Sambrook and Russel 2001). The advantages of these reporters are their comparatively low costs and high sensitivity. A clear disadvantage of using *gusA* or *lacZ* as reporters is that plants have to be fixed, and staining reagents have to penetrate into the plant to reach the bacteria, which is time consuming, can make detection less efficient, and is artefact-prone. Another disadvantage can be background staining of the plant tissue, which should be checked for before using these reporters.

We have used *lacZ* as a marker to facilitate visualisation of colonisation by single bacterial cells and bacterial microcolonies of the rhizosphere of both tomato and *Arabidopsis thaliana*, and to monitor gene expression (Fig. 18.1). No background β-galactosidase activity from tomato or *Arabidopsis* cells was observed. In particular we focused on the differences between the rhizosphere colonisation pattern of the biocontrol strain *Pseudomonas fluorescens* WCS365 on the model plant *Arabidopsis* and on tomato. WCS365 cells preferentially colonised the interjunctions between root cortex cells and the sites where lateral roots arise (Fig. 18.1A, B). At such sites, cells are disrupted by the emerging lateral root and nutrients are released. Cracks in the outer root cortex or plant surface are also very suitable sites for entering the plant and infecting the internal plant environment. The colonisation patterns of *P. fluorescens* WCS365 on tomato and *Arabidopsis* are more similar at early (first 3 days after inoculation) than at later stages, although the number of colony forming units (cfu) per centimetre of root tip is comparable. It seems that the bacteria around the root of a 7-day-old tomato plant are embedded in a "mesh" of root hairs and possibly plant and/or bacterial extracellular material, to which root

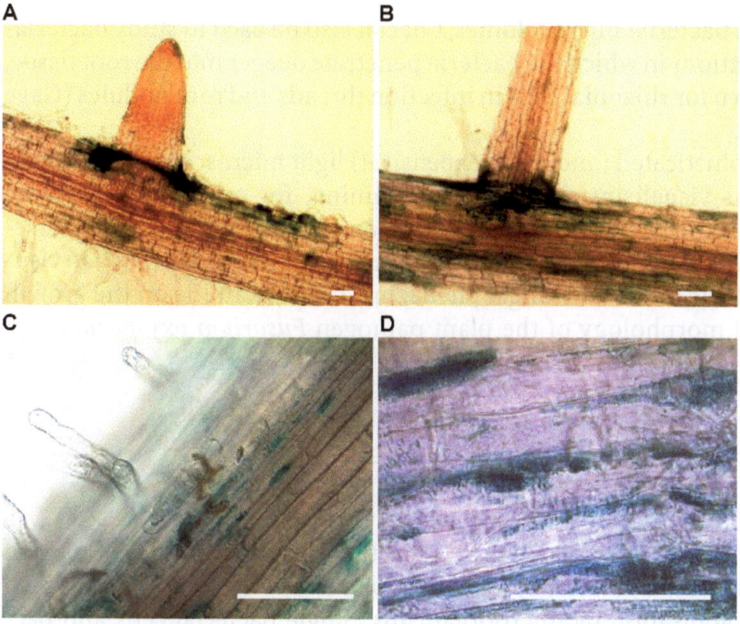

Fig. 18.1. A–D Light microscopy analysis of tomato and *Arabidopsis thaliana* root colonisation by *Pseudomonas fluorescens* strain WCS365 constitutively expressing the *lacZ* reporter gene encoding β-galactosidase. Sterile seedlings were inoculated with WCS365 2 days after germination and placed in a gnotobiotic growth system (tomato) or on solid plant nutrient solution (PNS) agar plates (*A. thaliana*) for 7 days. After staining for β-galactosidase activity, roots were examined using light microscopy. **A, B** Colonisation of the sites of lateral root emergence from tomato roots. **C, D** Microcolonies on the root surface of *A. thaliana* at the interjunctions of root cells. *Bars* 100 μm

border cells (BRD) or their excreted proteins (Brigham et al. 1995) may be major contributors. In contrast, in *Arabidopsis* the bacteria are located on the root surface and in between the cell walls of adjacent epidermal cells (Fig. 18.1C, D). The difference in colonisation patterns between tomato and *Arabidopsis* is most likely due to the difference in the rhizosphere environment created by the two plants. Many plants release BRD from the root tip into the rhizosphere (Hawes 1990). By definition, BRD are cells that become dispersed into suspension in response to gentle agitation in water (Hawes and Lin 1990). The number and properties of BRD vary among different kinds of plants and are conserved among families (Hawes 1990) Most plants have BRD but *A. thaliana* does not (Hawes 1990). BRD are one of the main components of root exudate (Hawes 1990; Hawes and Lin 1990), which is the major nutrient source for rhizosphere micro-organisms.

lacZ is not only a useful tool in determining the colonisation pattern of rhizobacteria in the rhizosphere of different plants and to observe single cells within bacterial microcolonies, but can also be used to study bacteria-root associations in which the bacteria penetrate deeper into the root tissue, as was shown for rhizobial cells in infection threads and root nodules (Gage et al. 1996).

More sophisticated (and more expensive!) light microscopes can be very valuable for visualising cells without staining, for example endophytic colonisation and bacteria-fungus interactions. We successfully used phase-contrast microscopy and differential interference microscopy (DIC) to visualise the effects of the antifungal phenazine-1-carboxamide on the growth and hyphal morphology of the plant pathogen *Fusarium oxysporum* f.sp. *radicis lycopersici* (Bolwerk et al. 2003).

18.2.2 Scanning Electron Microscopy

Scanning electron microscopy (SEM) has the advantage of high resolution and is, therefore, a powerful tool with which to follow the process of seed and root colonisation by microorganisms. Single bacterial cells and bacterial microcolonies can be visualised. Using SEM, bacteria were shown to preferentially colonise grooves on the seed coat and the root surface, where they formed densely packed microcolonies. Microcolonies are ideal environments for quorum sensing, which is used by bacteria as a regulatory process controlling, for example, the production of antifungal metabolites in the rhizosphere (Fig. 18.2, Chin-A-Woeng et al. 1997). Interestingly, Fig. 18.2D suggests that the microcolonies are covered by a mucous layer, possibly consisting of exopolysaccharides, which could form an additional diffusion barrier.

SEM is also very suitable for showing the morphological differences within endogenous communities of microorganisms, including obligately biotrophic bacteria and fungi that cannot be cultured on standard growth media. A nice example of such a study was given by Fett and Cooke (2003), who visualised a population of bacteria with different morphologies on mung bean sprouts.

SEM is ideally suited to visualising the total microflora since it does not require the tagging of microorganisms with reporters. However, it has the disadvantage that (1) the living material has to be fixed, and (2) bacterial species of similar shape and size cannot be differentiated. In addition the costs of purchasing an electron microscope and its maintenance are relatively high.

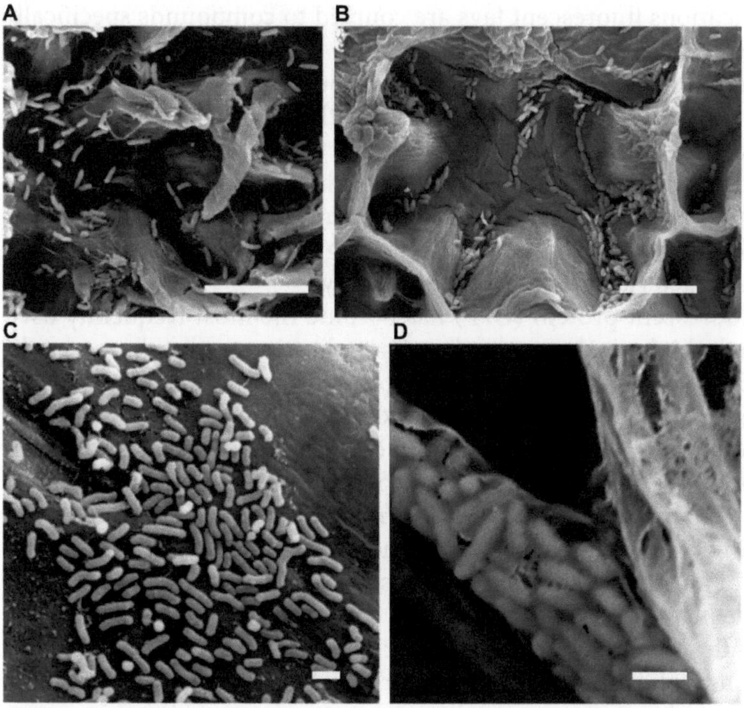

Fig. 18.2. A–D Scanning electron microscopy (SEM) analyses of tomato seed and root colonisation by *Pseudomonas putida* WCS358. **A** Seed coat colonisation 1 day after inoculation. **B** Seed coat colonisation monitored 3 days after inoculation. **C** Attachment on and early colonisation of the root surface. **D** Microcolony formed at an interjunction of root cells. *Bars* **A**, **B** 10 μm; **C**, **D** 1 μm. Images provided by T.F.C. Chin-A-Woeng, Leiden University, Leiden, The Netherlands

18.2.3 Epifluorescence Microscopy and the Application of Auto-Fluorescent Proteins

Fluorescence microscopy is based on the presence of fluorescent compounds, including proteins, which, after excitation with light of a certain wavelength, will emit light of a longer wavelength due to energy loss during the process of absorption and excitation. Confocal laser scanning microscopy (CLSM) is a highly sophisticated form of fluorescence microscope. The use of CLSM for visualising fluorescent molecules results in higher resolution and lower autofluorescence background compared to traditional fluorescence microscopy In addition, the resolution and sharpness of the digital images produced by CLSM can be improved by the use of deconvolution software that corrects for small defects in the optical lenses. In many applications fluorescent tags are coupled to compounds specifically binding to certain molecules, such as DNA or RNA, in the living cell. Such compounds are able to diffuse or to be transported into the cell. Some will diffuse or penetrate differentially through the cell membranes of Gram positive and Gram negative bacteria. Various fluorescent staining kits are commercially available. For example, the ViaGram Red kit (Invitrogen Molecular Probes, Breda, The Netherlands) contains three fluorescent compounds that offer the possibility to distinguish between Gram positive and negative bacteria and to test their viability.

Green fluorescent protein (GFP) has become the most frequently used reporter in the biological sciences since its application as a marker was published by Chalfie et al. (1994). GFP was isolated from the jellyfish *Aequorea victoria* and is extensively applied to mark cells, to visualise proteins within cells and to monitor gene expression. In contrast to the use of fluorescent stains, antibodies or probes targeted to 16S ribosomal RNA genes, the use of autofluorescent proteins as markers does not require any preparatory steps such as fixing with formaldehyde or ethanol, which might affect the biological material and the biological processes studied. In many cases, fixation results in death of the living cells. GFP as a reporter has many advantages, which include (1) stability (due to its barrel protein structure), (2) species-independent application (pro- and eukaryotes), (3) non-invasive analysis without the need for exogenous substrates or energy, and (4) in vivo monitoring while preserving the integrity of the living cells. Colour- and optimised variants of GFP, such as enhanced GFP (EGFP), enhanced cyan fluorescent protein (ECFP) and enhanced yellow fluorescent protein (EYFP), with shifted excitation and emission maxima, and increased brightness and stability have been developed and used for multiple colour imaging (Yang et al. 1998; Tsien 1998; Matus 1999; Ellenberg 1999). Useful information on the properties and commercial availability of autofluorescent proteins can be found on the following website: www.clontech.com/clontech/.

The development of GFP variants with altered excitation and emission wavelengths and the red fluorescent protein (RFP or DsRed), which was isolated from the coral *Discosoma* (Matz et al. 1999), allows for differential staining of bacterial strains for simultaneous visualisation using CLSM in one system. We made use of broad host-range cloning vectors for Gram negative bacteria, which are stably maintained without antibiotic pressure (Heeb et al. 2000) to tag *Pseudomonas* and *Rhizobium* with autofluorescent proteins encoded by e*gfp* (green), e*cfp* (cyan), e*yfp* (yellow), e*bfp* (blue) and *rfp* (red)(Stuurman et al. 2000; Bloemberg et al. 2000). Subsequently, we were able to visualise simultaneously and clearly differentiate up to three different bacterial populations at the single cell level in the rhizosphere (Bloemberg et al. 2000). GFP was shown to be an excellent marker for monitoring root colonisation at the single cell level by *Pseudomonas* spp. (Bloemberg et al. 1997, 2000), the interactions of *Rhizobium* with leguminous plants (Stuurman et al. 2000), the infection process of *Fusarium oxysporum* f.sp. *radicis lycopersici* (Lagopodi et al. 2002) and the interactions between biocontrol strains of *Pseudomonas* sp. and *F oxysporum* f.sp. *radicis lycopersici*. (Bolwerk et al. 2003). Besides whole plasmid systems, valuable transposons have been constructed for integration of *gfp* into bacterial chromosomes under constitutive expression (Burlage et al. 1995; Tombolini et al. 1997; Unge et al. 1997).

18.3 Visualisation of Bacterium-Plant Interactions

Studying microbial communities and their interactions with plants has been highly facilitated by using combinations of GFP, its colour variants cyan and yellow fluorescent protein and DsRed as markers. Tomato root colonisation studies showed that *Pseudomonas* cells adhere to the root surface preferentially at the junctions between cells, which are presumed sites of root exudation, where they proliferate and divide, resulting in the formation of microcolonies (Bloemberg et al. 1997, 2000; Fig. 18.3A). SEM and fluorescence microscopy studies revealed that microcolonies are covered with a mucoid-like layer of unknown origin (Chin-A-Woeng et al. 1997; Bloemberg et al. 1997). It can be hypothesised that this layer contains bacterial exopolysaccharides, the production of which is increased in bacterial biofilms (O'Toole et al. 2000; Sutherland 2001) and which could form a barrier for signal molecules such as acyl homoserine lactones. Interestingly, we observed that some plant cells were colonised intracellularly (Fig. 18.3B). Bacterial cells were also observed close to the openings of the stomata (Fig. 18.3C), which could form another site of endophytic colonisation, although we have not observed endophytic colonisation by the *Pseudomonas* strains we have used.

Microcolonies are most frequently initiated by one cell, but cells from external sources can attach to the colony as revealed by a study performed with a mixture of *P. fluorescens* WCS365 cells expressing *ecfp, egfp* and *rfp* (Bloemberg et al 2000). It is also expected that cells will detach from the microcolony to colonise other parts of the growing root. This shows that bacterial microcolonies are not static, but dynamic entities that can be invaded by external bacterial cells after the formation of the initial microcolony. It is of great fundamental interest to find out which genes, molecules and traits are involved in the attachment and departure of cells. Studies of an inoculant containing a mixture of *P. chlororaphis* PCL1391 and *P. fluorescens* WCS365 differentially labelled with auto-fluorescent proteins has shown that (1) predominantly mixed colonies are present on and in

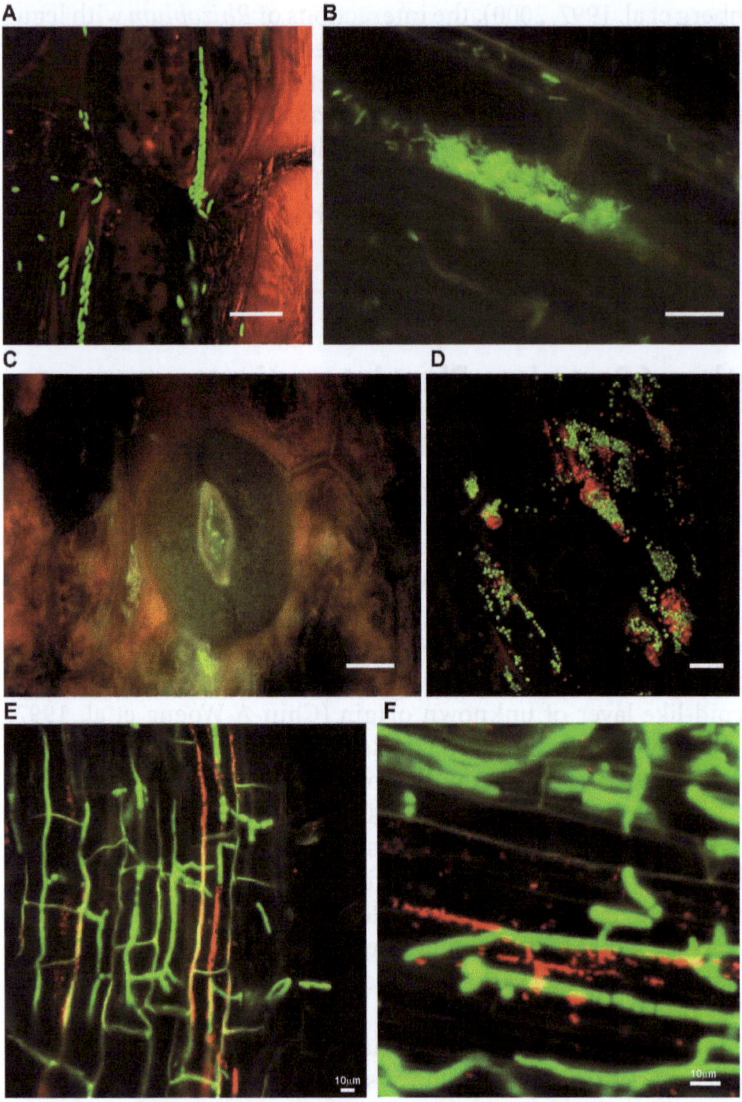

Fig. 18.3. A–F Confocal laser scanning microscopy (CLSM) analysis of tomato surfaces colonised by *Pseudomonas* spp. and *Fusarium oxysporum* f. sp. *radicis lycopersici* labelled with autofluorescent proteins. Tomato seedlings were grown upon inoculation in a gnotobiotic sand system. Plants were taken out after 7 days of growth and examined for the presence of green fluorescent protein (*gfp*)- or red fluorescent protein (*rfp*)-expressing organisms. **A** Colonisation of the root surface by *P. fluorescens* expressing enhanced GFP (*egfp*). **B** Colonisation of the lumen of a root cell. **C** Colonisation of a stoma. **D** Simultaneous imaging of root surface colonisation by a mixture of *P. putida* PCL1444 expressing *egfp* and *P. putida* PCL1446 expressing *rfp*. **E, F** Interactions between *P. chlororaphis* PCL1391 expressing *rfp* and the phytopathogenic fungus *Fusarium oxysporum* f. sp. *radicis lycopersici* expressing *gfp* during colonisation. *Bars* 10 μm. **D** courtesy of E. Lagendijk; **E, F** courtesy of T. Lagopodi and A. Bolwerk

older parts of the root, and (2) in a mixture of PCL1391 and WCS365, PCL1391 predominantly epiphytically colonised the root hairs (Dekkers et al. 2000), indicating differences in attachment and colonising abilities between these two closely related *Pseudomonas* spp. Differential labelling of strains has also been used to study the colonisation pattern of polyaromatic hydrocarbon (PAH)-degrading bacteria, which had been isolated for bioremediation studies. Interestingly, it was shown that these bacterial strains can form communities on the root as a response to the presence of PAHs in the soil (Fig. 18.3D).

Results of SEM and CLSM studies show that rhizobacteria, such as *Pseudomonas* spp. used for biocontrol, colonise the seed and root surface at the same positions as phytopathogenic fungi (Chin-A-Woeng et al. 1997; Bloemberg et al. 1997; Tombolini et al. 1999; Lugtenberg et al. 2001; Lagopodi et al. 2002; Bolwerk et al. 2003; Fig. 18.3). Visualisation is required to explore and fully understand the interactions between organisms. For example, visualisation of the relationship among *P. fluorescens* CHA0, carrot roots, and mycorrhizal mycelium showed that mucus-producing mutant strains of CHA0 can better adhere to the root, indicating that acidic extracellular polysaccharides contribute to root colonisation (Biancotto et al. 2001).

Invasion of plant tissue has been extensively studied for rhizobial infection that results in the nitrogen-fixing symbiosis with leguminous plants. Tagging with GFP made it possible to follow early nodulation events, for example of *Sinorhizobium meliloti* on alfalfa (Gage et al. 1996), and to follow rhizobial cells in the infection thread during its growth into the root cortex. Visualisation of rhizobia in the infection thread made it possible to calculate the rhizobial growth rate (Gage et al. 1996). GFP tagging was also successfully applied to study the movement of *Rhizobium* bacterothe root nodules (Stuurman et al. 2000). To achieve optimal fluorescence of the bacterial cells, sectioning of the plant material was required to prevent loss of light intensity during transmission through the plant tissue. Sectioning of the plant tissue was also successful in studying the endophytic colonisation of (1) rice roots and shoots by *Herbaspirillum* sp. (Elbeltagy et al. 2001) and (2) *Vitis vinifera* by the bacterial pathogen *Xylella fastidiosa* (Newman et al. 2003). If preservation is preferred or necessary, plant tissue sections can be fixed with paraformaldehyde, which does not affect the folding and the fluorescence of GFP (Stuurman et al. 2000; Elbeltagy et al. 2001).

Due to different emission and excitation spectra, a combination of GFP and DsRed is very suitable for visualisation of two populations of cells and for providing the possibility of studying competition events. Infection threads can contain mixed *S. meliloti* populations that can give rise to mixed populations of bacteroids in the root nodules (Gage 2002). Since not every laboratory is equipped with state of the art confocal laser scanning microscopes, combinations of different reporter genes should be considered.

Such combinations have been shown to be powerful tools for studying root colonisation and gene expression in the rhizosphere. Useful constructs have been made to deliver *gfp* and *gusA* in mini-Tn5 transposons or in plasmids (Ramos et al. 2002) for chromosomal insertion (Xi et al. 1999). Applying these constructs to the study of *Azospirillum brasilense* on and in wheat roots showed that *A. brasilense* preferentially colonises intercellular spaces and points of lateral root emergence where it is expected that relatively large quantities of nutrients will be released from the root cells (Xi et al. 1999; Ramos et al. 2002). Others have combined immunofluorescence and an rRNA-targeting probe to monitor the presence of organisms and metabolic activity in the rhizosphere. For example, *P. fluorescens* DR54 cells were analysed in the sugar beet rhizosphere, showing that cells of *P. fluorescens* at the root tip were metabolically most active and that bacteria from the surrounding soil population entered the rhizosphere 2 days after seed inoculation (Lübeck et al. 2000). Another dual marker system was developed with *gfp* and the *luxAB* genes encoding bacterial luciferase, which as a biomarker is dependent on cellular energy status. This construct was used to show that metabolic activity of *P. fluorescens* SBW25 was detectable on all parts of wheat and that this strain colonises specific sites of the seed (Unge et al. 1999; Unge and Jansson 2001).

18.4 Most Recent Developments in Visualising Plant-Microorganism Interactions

The stability of GFP can be regarded as one of its advantages. However, this makes GFP unsuitable for studying transient gene expression. GFP derivatives carrying at their C-terminus amino acid tags for the recognition of specific proteases have reduced half-life times of GFP to 1–1.5 h (Andersen et al. 1999). The use of such unstable GFP variants has made it possible to analyse transient gene expression in the rhizosphere, for example, the monitoring of ribosomal activity in *P putida* cells (Ramos et al. 2000). These variants have recently been applied to the study of several aspects of interactions between plants and microorganisms. For example, a system was constructed for the detection of acyl homoserine lactones (AHL), showing that quorum sensing and cross talk occur in microcolonies in the rhizosphere (Andersen et al. 2001; Steidle et al. 2001). Examples of GFP-based expression systems for studying the interaction of the bacterium with the plant are given by (1) Leaveau and Lindow (2001), who showed that foliar growth of *Erwinia herbicola* on bean is driven by the utilisation of sugars, e.g. fructose and/or sucrose; and (2) Aldon et al. (2000), who showed that the strong induction of *hrp* genes in the presence of the host plant cell depends specifically on physicalcontact between the bacterium and its target cell. The development and application of such reporter systems contribute significantly to increasing our fundamental knowledge of bacterial physiology and behaviour on the plant (Leveau and Lindow 2002).

The addition of a constitutively expressed *rfp* on the reporter construct or delivered in trans on a second plasmid should make it possible to follow the presence of bacterial cells and their gene expression. DsRed is specifically suited to this since it has been isolated from a different organism and its homology with GFP is extremely low, which excludes the possibility of recombination when *gfp* and *rfp* are present in one cell. DsRed has a longer folding (maturation) time than GFP, which might reduce or delay its detection after it is produced. However, an enhanced RFP that matures rapidly has recently been developed (Sörensen et al. 2003) and will improve the usefulness of *rfp*.

In order to identify genes expressed in the plant environment, systems based on differential fluorescence induction and optical trapping microscopy have been developed (Allaway et al. 2001). Several genes have been isolated that have led to new insights into bacterial life in the rhizosphere, such as the identification of a putative ABC transporter of putrescine (Allaway et al. 2001). Interestingly, uptake of putrescine from the rhizosphere environment was identified as an important trait for competitive root colonisation (Kuiper et al. 2001). New systems for promoter trapping based on the combination of an antibiotic resistance marker and GFP further extend the possibilities for researchers to identify genes specifically expressed in the plant environment (Izallalen et al. 2002).

18.5 Visualisation of Plant-Fungus Interactions

Fungi frequently colonise the internal and external plant environment as pathogens, mutualists or organisms without apparent effect on the plant. Fungal structures, including hyphae, fruiting bodies and spores can be visualised by light- and electron-microscopy with the advantages and disadvantages of these techniques as discussed above. The use of reporters such as *lacZ* or *gfp* requires genetic transformation, which is much more complicated for fungi than for bacteria. However, there is a growing number of fungal species for which transformation methods, including conventional protoplast transformation, ballistic bombardment and the more recently developed highly successful *Agrobacterium*-based transformation method for filamentous fungi (de Groot et al. 1998), are available Two examples will be discussed to illustrate the use of GFP in visualisation of fungi interacting with plants and microorganisms in the plant environment.

In order to visualise tomato root colonisation and infection processes in vivo we marked *F oxysporum* f. sp. *radicis-lycopersici*, which causes tomato foot and root rot, with GFP. A protocol to produce *Fusarium* protoplasts with the use of a mixture of hydrolytic enzymes was optimised and successfully applied for cotransformation of the protoplasts with two plasmids, one of which harboured *sgfp*, which is an optimised *gfp* variant (Lagopodi et al. 2002). This resulted in an efficient tagging and the constitutive *sgfp* expression was stable for at least nine subcultures. Homogeneity of the fluorescent signal was clearly visible in the hyphae as well as in chlamydospores and conidia. Since, after transformation, the *sgfp*-harbouring plasmid is integrated randomly into the chromosome, hyphal morphology, growth rate and pathogenicity were tested and found to be unaffected in the transformants tested CLSM was used to analyse colonisation, infection and disease processes of the isolate in detail on/in tomato roots, including the following interesting aspects: (1) an overview of the complete colonisation pattern of the tomato rhizosphere; (2) the very first steps of contact with the root, which took place in the root hair zone by mingling and attachment of hyphae to the root hairs, suggesting a chemotactic response towards the root hairs; (3) the preferential colonisation of the grooves along the junctions of the epidermal cells, which is similar to the colonisation patterns of rhizobacteria; (4) the absence of specific infection sites, such as sites of emergence of secondary roots, root tips or wounded tissue and the absence of specific infection structures, e.g. appressoria (Lagopodi et al. 2002). This study illustrates the powerful use of GFP as a marker as a non-invasive, convenient, fast and effective approach for studying plant-fungus interactions.

Fungi interact with other microorganisms in the plant environment. The use of different autofluorescent proteins is an excellent tool with which to distinguish microorganisms from each other and to visualise their interactions. Visualisation of interactions between phytopathogenic fungi and biocontrol agents will help to understand these interactions and facilitate the development of efficient biocontrol applications. We have studied the interaction between *F. oxysporum* f. sp. *radicis-lycopersici* tagged with GFP and the biocontrol strain *P. chlororaphis* PCL1391 tagged with DsRed (Bolwerk et al. 2003). CLSM studies of the single strains showed the following similarities: (1) the sites of colonisation of the tomato root surface by *Fusarium* are strikingly similar to those of the *Pseudomonas* biocontrol strains PCL1391 and WCS365; (2) both preferentially colonise sites between the junctions of two root cells. Studying the interactions between *Pseudomonas* biocontrol strains and *Fusarium* in the rhizosphere by differential labelling showed that (1) penetration of the fungal hyphae was not observed where bacterial microcolonies were present, and (2) *Pseudomonas* bacteria attached to and colonised *Fusarium* hyphae (Fig. 18.3F, G). The latter could be a novel biocontrol trait and preliminary studies showed that *Pseudomonas* strains show a chemotactic response towards *Fusarium* (De Weert et al. 2004). (3) Many stress responses in the fungal hyphae were observed in the presence of the biocontrol agent including increased vacuole formation, loss of growth directionality, curly growth, "swollen bodies" and increased hyphal branching. Similar responses could be induced by applying the purified antifungal metabolite phenazine-1-carboxamide produced by *P. chlororaphis* PCL1391.

Studying the molecular basis of the interactions between fungi and bacteria is an emerging field with great relevance for plant microbiology and plant pathology. Future studies will benefit greatly from tools developed for visualisation of these organisms.

18.6 Future Perspectives

The development of highly sophisticated microscopy tools and optimisation of autofluorescent proteins as reporters are making substantial contributions to a better fundamental understanding of how microorganisms interact with plants and the endemic microflora. More reasonably priced tools for microscopy will facilitate and stimulate such studies in the future. Visualisation is, however, dependent on whether the microorganims can be cultured and transformed with genetic constructs that harbour reporter gene(s). Some important endophytes cannot be cultured outside the plant, which is severely hampering their study. Discovery and understanding of plant growth requirements, including chemical signals and physical properties for providing suitable conditions for culturing, as well as genetic accessibility of these microorganisms, represents a major challenge.

Acknowledgements. We thank all the members of the Microbiology section and Gerda Lamers of the Institute of Biology Leiden for their contributions in valuable discussions for optimising and developing microscopic techniques, interpretation of results and technical assistance.

References

Aldon D, Brito B, Boucher C, Genin S (2000) A bacterial sensor of plant cell contact controls the transcriptional induction of *Ralstonia solanacearum* pathogenicity genes. EMBO J 19:2304–2314

Allaway D, Schofield NA, Leonard ME, Gilardoni L, Finan TM, Poole PS (2001) Use of differential fluorescence induction and optical trapping to isolate environmentally induced genes. Environ Microbiol 3:397–406

Andersen JB, Sternberg C, Poulsen LK, Bjorn SP, Givskov M, Molin S (1999) New unstable variants of green fluorescent protein for studies of transient gene expression in bacteria. Appl Environ Microbiol 64:2240–2246

Andersen JB, Heydorn A, Hentzer M, Eberl L, Geisenberg O, Christensen BB, Molin S, Givskov M (2001) *Gfp*-Based *N*-acyl homoserine-lactone sensor systems for detection of bacterial communities. Appl Environ Microbiol 67:575–585

Biancotto V, Andreotti S, Balestrini R, Bonfante P, Perotto S (2001) Extracellular polysaccharides are involved in the attachment of *Azospirillum brasilense* and *Rhizobium leguminosarum* to arbuscular mycorrhizal structures. Eur J Histochem 45:39–49

Bloemberg GV, O'Toole G, Lugtenberg BJJ, Kolter R (1997) Green fluorescent protein as a marker for *Pseudomonas* spp. Appl Environ Microbiol 63:4543–4551

Bloemberg GV, Wijfjes AHM, Lamers GEM, Stuurman N, Lugtenberg BJJ (2000) Simultaneous imaging of *Pseudomonas fluorescens* WCS365 populations expressing three different autofluorescent proteins in the rhizosphere; new perspectives for studying microbial communities. Mol Plant-Microbe Interact 13:1170–1176

Bolwerk A, Lagopodi AL, Wijfjes AHM, Lamers GEM, Chin-A-Woeng TFC, Lugtenberg BJJ, Bloemberg GV (2003) Interactions in the tomato rhizosphere of two Pseudomonas biocontrol strains with the phytopathogenic fungus *Fusarium oxysporum* f. sp. *radicis-lycopersici*. Mol Plant-Microbe Interact 16:983–993

Brigham LA, Woo HH, Nicoll SM, Hawes MC (1995) Differential expression of proteins and mRNAs from border cells and root tips of pea. Plant Physiol 109:457–463

Burlage RS, Yang ZK, Mehlhorn T (1995) A transposon for green fluorescent protein transcriptional fusions: applications for bacterial experiments. Gene 173:53–58

Chalfie M, Tu Y, Euskirchen G, Ward WW, Prasher DC (1994) Green fluorescent protein as a marker for gene expression. Science 263:802–805

Chin-A-Woeng TFC, de Priester W, van der Bij AJ, Lugtenberg BJJ (1997) Description of the colonisation of a gnotobiotic tomato rhizosphere by *Pseudomonas fluorescens* biocontrol strain WCS365, using scanning electron microscopy. Mol Plant-Microbe Interact 10:79–86

De Groot MJ, Bundock P, Hooykaas PJ, Beijersbergen AG (1998) *Agrobacterium tumefaciens*-mediated transformation of filamentous fungi. Nat Biotechnol 16:839–842

Dekkers LC, Mulders IH, Phoelich CC, Chin-A-Woeng TFC, Wijfjes AHM, Lugtenberg BJJ (2000) The *sss* colonisation gene of the tomato-*Fusarium oxysporum* f.sp. *radicis lycopersici* biocontrol strain *Pseudomonas fluorescens* WCS365 can improve root colonisation of other wild-type *Pseudomonas* spp. bacteria. Mol Plant-Microbe Interact 13:1177–1183

De Weert S, Kuiper I, Lagendijk EL, Lamers GEM, Lugtenberg BJJ (2004) Role of chemotaxis toward fusaric acid in colonisation of hyphae of *Fusarium oxysporum* f.sp. *radicis-lycopersici* by *Pseudomonas fluorescens* WCS365. Mol Plant-Microbe Interact 16:1185–1191

Elbeltagy A, Nishioka K, Sato T, Suzuki H, Ye B, Hamada T, Isawa T, Mitsui H, Minamisawa K (2001) Endophytic colonisation and in planta nitrogen fixation by a *Herbaspirillum* sp. isolated from wild rice species. Appl Environ Microbiol 67:5285–5293

Ellenberg J, Lippincott SJ, Presley JF (1999) Dual-colour imaging with GFP variants. Trends Cell Biol 9:52–56

Fett WF, Cooke PH (2003) Scanning electron microscopy of native biofilms on mung bean sprouts. Can J Microbiol 49:45–50

Gage DJ (2002) Analysis of infection thread development using Gfp- and DsRed-expressing *Sinorhizobium meliloti*. J Bacteriol 184:7042–7046

Gage DJ, Bobo T, Long SR (1996) Use of green fluorescent protein to visualise the early events of symbiosis between *Rhizobium meliloti* and alfalfa (*Medicago sativa*). J Bacteriol 178:7159–7166

Hawes MC (1990) Living plant cells released from the root cap: a regulator of microbial populations in the rhizosphere? Plant Soil 129:19–27

Hawes MC, Lin HC (1990) Correlation of pectolytic enzyme activity with the programmed release of cells from the root cap of *Pisum sativum*. Plant Physiol 94:1855–1859

Heeb S, Itoh Y, Nishijyo T, Schnider U, Keel C, Wade J, Walsh U, O'Gara F, Haas D (2000) Small, stable shuttle vectors based on the minimal pVS1 replicon for use in gram-negative, plant-associated bacteria. Mol Plant-Microbe Interact 13:232–237

Izallalen M, Levesque RC, Perret X, Broughton WJ, Antoun H (2002) Broad-host-range mobilizable suicide vectors for promoter trapping in gram-negative bacteria. Biotechniques 33:1038–1043

Kuiper I, Bloemberg GV, Noreen S, Thomas-Oates JE, Lugtenberg BJJ (2001) Efficient uptake of putrescine in the rhizosphere is essential for competitive root colonisation by *Pseudomonas fluorescens* strain WCS365. Mol Plant-Microbe Interact 14:1096–1104

Lagopodi AL, Ram AFJ, Lamers GEM, Punt PJ, van den Hondel CAMJ, Lugtenberg BJJ, Bloemberg GV (2002) Confocal laser scanning microscopical analysis of tomato root colonisation and infection by *Fusarium oxysporum* f. sp. *radicis-lycopersici* using the green fluorescent protein as a marker. Mol Plant Microb Interact 15:172–179

Leveau JH, Lindow SE (2001) Appetite of an epiphyte: quantitative monitoring of bacterial sugar consumption in the phyllosphere. Proc Natl Acad Sci USA 98:3446–3453

Leveau JH, Lindow SE (2002) Bioreporters in microbial ecology. Curr Opin Microbiol 5:259–265

Lübeck PS, Hansen M, Sørensen J (2000) Simultaneous detection of the establishment of seed-inoculated *Pseudomonas fluorescens* strain Dr54 and native soil bacteria on sugar beet root surfaces using fluorescense antibody and in situ hybridization techniques. FEMS Microbiol Ecol 33:11–19

Lugtenberg BJJ, Dekkers L, Bloemberg GV (2001) Molecular determinants of rhizosphere colonisation by *Pseudomonas*. Annu Rev Phytopathol 39:461–490

Matus A (1999) GFP in motion CD-ROM. Introduction: GFP illuminates everything. Trends Cell Biol 9:43

Matz MMV, Fradkov AF, Labas YA, Savitsky AP, Zaraisky AG, Markelov ML, Lukyanov SA (1999) Fluorescent proteins from non-bioluminescent *Anthozoa* species. Nat Biotechnol 17:969–973

Newman KL, Almeida RP, Purcell AH, Lindow SE (2003) Use of a green fluorescent strain for analysis of *Xylella fastidiosa* colonisation of *Vitis vinifera*. Appl Environ Microbiol 69:7319–7327

O'Toole G, Kaplan HB, Kolter R (2000) Biofilm formation as microbial development. Annu Rev Microbiol 54:49–79

Ramos C, Molbak L, Molin S (2000) Bacterial activity in the rhizosphere analyzed at the single-cell level by monitoring ribosome contents and synthesis rates. Appl Environ Microbiol 66:801–809

Ramos HJ, Roncato-Maccari LD, Souza FM, Soares-Ramos JR, Hungria M, Pedrosa FO (2002) Monitoring *Azospirillum*-wheat interactions using the *gfp* and *gusA* genes constitutively expressed from a new broad-host range vector. J Biotechnol 97:243–52

Sambrook J, Russell DW (2001) Molecular cloning: a laboratory manual. Cold Spring Harbor Laboratory Press, Cold Spring Harbor, NY

Sörensen M, Lippuner C, Kaiser T, Misslitz A, Aebischer T, Bumann D (2003) Rapidly maturing red fluorescent protein variants with strongly enhanced brightness in bacteria. FEBS Lett 552:110–114

Steidle A, Sigl K, Schuhegger R, Ihring A, Schmid M, Gantner S, Stoffels M, Riedel K, Givskov M, Hartman A, Langebartels C, Eberl L (2001) Visualisation of *N*-acylhomoserine lactone-mediated cell-cell communication between bacteria colonising the tomato rhizosphere. Appl Environ Microbiol 67:5761–5770

Stuurman N, Pacios Bras C, Schlaman CHRM, Wijfjes AHM, Bloemberg GV, Spaink HP (2000) The use of GFP color variants expressed on stable broad-host range vectors to visualise rhizobia interacting with plants. Mol Plant-Microbe Interact 13:1063–1069

Sutherland IW (2001) The biofilm matrix-an immobilized but dynamic microbial environment. Trends Microbiol 9:222–227

Tombolini R, Unge A, Davey ME, de Bruijn FJ, Jansson JK (1997) Flow cytometric and microscopic analysis of GFP-tagged *Pseudomonas fluorescens* bacteria. FEMS Microbiol Ecol 22:17–28

Tombolini R, van der Gaag DJ, Gerhardson B, Jansson JK (1999) Colonisation pattern of the biocontrol strain *Pseudomonas chlororaphis* MA342 on barley seeds visualised by using green fluorescent protein. Appl Environ Microbiol 65:3674–3680

Tsien RY (1998) The green fluorescent protein. Annu Rev Biochem 67:509–544

Unge A, Jansson JK (2001) Monitoring population size, activity, and distribution of *gfp-luxAB*-tagged *Pseudomonas fluorescens* SBW25 during colonisation of wheat. Microb Ecol 41:290–300

Unge A, Tombolini R, Möller A, Jansson JK (1997) Optimization of GFP as a marker for detection of bacteria in environmental samples. In: Hastings JW, Kricka LJ, Stanley PE (eds) Proceedings of the 9th international symposium on bioluminescence and chemiluminescence: bioluminescence and chemiluminescence: molecular reporting with photons. Wiley, Chichester, pp 391–394

Unge A, Tombolini R, Molbak L, Jansson JK (1999) Simultaneous monitoring of cell number and metabolic activity of specific bacterial populations with a dual *gfp-luxAB* marker system. Appl Environ Microbiol 65:813–821

Xi C, Lambrecht M, Vanderleyden J, Michiels J (1999) Bi-functional *gfp-* and *gusA*-containing mini-Tn5 transposon derivatives for combined gene expression and bacterial localization studies. J Microbiol Methods 35:85–92

Yang TT, Sina P, Green G, Kitts PA, Chen YT, Lybarger L, Chervenak R, Patterson GH, Piston DW, Kain SR (1998) Improved fluorescence and dual color detection with enhanced blue and green variants of the green fluorescent protein. J Biol Chem 273:8212–8216

19 Application of Molecular Fingerprinting Techniques to Explore the Diversity of Bacterial Endophytic Communities

Leo S. van Overbeek, Jim van Vuurde, Jan D. van Elsas

19.1 Introduction

Many bacteria associated with plants are able to penetrate and live as endophytes in roots and other tissues. Like most bacteria in natural environments, endophytic bacteria may be non-cultivable, and detectable only by microscopical and molecular techniques (Garbeva et al. 2001; Araújo et al. 2002; Sessitsch et al. 2002; Reiter et al. 2003b). Consequently, usage of molecular techniques reveals higher species diversity than classical isolation methods. Molecular fingerprinting of bacterial endophytic populations can expand our knowledge of populations that were previously inaccessible. So far, little information is available on the possibilities for employing molecular methods to detect bacterial populations inside plants. There is, however, overwhelming information from related ecosystems such as the soil and rhizosphere so that adaptation of methods to studies on bacterial endophyte communities should be feasible.

In this chapter we will briefly summarise the factors affecting bacterial endophyte populations in plants and provide an extended account of the available molecular detection and fingerprinting techniques, including their integration with other methods, to study such bacterial endophyte communities. Exploitation of molecular fingerprinting techniques is especially relevant for studying those endophytic populations that show clear effects on plant growth and development. Special emphasis will be placed on agricultural production systems.

Leo S. van Overbeek: Plant Research International B.V, Droevendaalsesteeg 1, 6708 PB, Wageningen, The Netherlands, E-mail: leo.vanoverbeek@wur.nl

Jim van Vuurde: Plant Research International B.V, Droevendaalsesteeg 1, 6708 PB, Wageningen, The Netherlands

Jan D. van Elsas: Groningen University, Department of Microbial Ecology, Biological Center, Kerklaan 30, 9751 NN, Haren, The Netherlands

Soil Biology, Volume 9
Microbial Root Endophytes
B. Schulz, C. Boyle, T. N. Sieber (Eds.)

19.2
Colonisation by Bacterial Endophytes

Bacterial populations residing on the surface or inside plants are commonly referred to as plant-associated bacteria. Most of these populations can colonise both internal and external tissue of plants and therefore a strict separation between 'genuine' endophytes and plant-associated bacteria cannot always be made. Endophytes thus comprise populations strictly bound to, and occasionally occupying, internal plant organs.

The initial step in the route to the internal plant system for soil and aerial bacteria is colonisation of the phytosphere (vis rhizosphere and phyllosphere). Endophytes occur in the rhizosphere (Sturz 1995; Hallman et al. 1997; Sturz and Nowak 2000) or the phyllosphere (Pillay and Nowak 1997). Colonisation of internal plant tissue requires adaptation of the bacterial cell to a highly specific and restrictive environment. Therefore, active colonisation, systemic spread and reproduction inside plants must be considered as important features of 'genuine' endophytic associations.

Certain plant organs may be environments in which only highly adapted bacterial species are able to become established. The plant xylem was demonstrated to be a selective environment for endophytic bacteria such as *Acetobacter diazotrophicus* (Dong et al. 1994), *Bacillus pumilus* (Benhamou et al. 1996, 1998), *Gluconacetobacter diazotrophicus* (Cocking 2003), *Herbaspirillum seropedicae* (James et al. 2002), *Klebsiella pneumoniae* (Dong et al. 2003) and *Serratia marcescens* (Gyaneshwar et al. 2001). Sieve tubes (phloem) are a habitat for a restricted group of pathogens, so called phytoplasms (Bové and Garnier 2003).

Endophytes can exert diverse effects on the performance of their host at different stages of growth and under different environmental conditions (Hallmann et al. 1997; Sturz and Nowak 2000). The effects described for endophytes inoculated into different plant species may be beneficial or detrimental [see Chaps. 3 (Kloepper and Ryu) and 4 (Berg and Hallmann)], although, most often, clear effects cannot be observed. The effect of bacterial populations associated with plants has been well described for pathogens and mutualistic symbionts; however, the role of species showing no obvious effects on plant health during the association may, in general, be underestimated.

19.3
Shifts of Bacterial Endophyte Communities

Being growing entities, plants have a dynamic interaction with their microbial inhabitants. Table 19.1 summarises studies on biotic and abiotic factors

Table 19.1. Factors influencing endophyte colonisation and community structure

Factors affecting endophytic community composition in plants	Example	Reference
Other micro-organisms	Colonisation and nodulation by *Rhizobium leguminosarum* in red clover was promoted by endophytic isolates of *Bacillus insolitus, Bacillus brevis* and *Agrobacterium rhizogenes*	Sturz and Christie 1996
Genotype	Higher diversity in root-endophyte composition was observed in recent versus ancient cultivars of wheat	Germida and Siciliano 2001
	Differences in endophyte composition and density were observed in different cotton cultivars	Adams and Kloepper 2002
Propagating material	*Enterobacter cloacae* applied to sterilised corn seeds resulted in endophytic colonisation of emerging plants	Hinton and Bacon 1995
	Optimal colonisation was obtained by inoculation of in vitro explants with endophytic strain *Pseudomonas* sp. PsJN resulting in increased resistance against *Verticillium dahliae* in tomato	Sharma and Nowak 1998
	Gradient in disease-suppressing endophytes observed from peel to the centre of potato tubers	Sturz et al. 1999
Agricultural practices	Crop rotation and tillage management of agricultural fields resulted in differences in number of disease suppressive endophytes in potato	Peters et al. 2003
	Endophytic communities in corn were influenced by herbicide, feritiliser and compost treatment of soil	Seghers et al. 2004
Localisation within plant	Non-nodular nitrogen fixing strain *Gluconacetobacter diazotrophicus* was isolated from apoplastic fluid of sugarcane	Dong et al. 1994
	Endophytic community structure in potato differed between the epidermis and internal stem and between lower and upper stem parts	Garbeva et al. 2001
Temperature	Endophytic colonisation of tomato by strain *Pseudomonas* sp. PsJN was highest in roots and shoots at 10°C and lowest at 30°C	Pillay and Nowak 1997

influencing endophytic populations. These studies make it clear that factors influencing plant growth also influence endophytic populations. The most important factors governing endophytic community structures are (1) the environment, (2) characteristics of the host plant species (genotype, growth stage, tissue type), and (3) endophyte populations already present in the plant.

Inoculation of in vitro plants with endophyte biocontrol strains aimed to suppress diseases in potato has been proposed by Nowak (1998). The rational for inoculation of in vitro transplants with endophytes was that already established populations will have a competitive advantage over organisms invading from the rhizosphere. The composition of populations in plants can shift depending on the intricate interplay of different factors. Eventually, climax populations of endophytes will become established.

19.4 Molecular methods to Study Bacterial Endophytes

Methods to detect beneficial and detrimental endophytic populations are important in studying the effects of agricultural management on the structure of endophyte communities and for determining the influence of various endophyte communities on plant health and crop yield. Genes responsible for beneficial or detrimental interactions may be the targets for detection. The structural nitrogen fixation (*nif*H) gene present in many nitrogen-fixing species has been detected in bacterial communities inside and near plants (Reiter et al. 2003a; Tan et al. 2003). Polymerase chain reaction (PCR)-based detection of genes involved in pathogen suppression has been developed for pyrrolnitrin and pyoluteorin (De Souza and Raaijmakers 2003) and ketosynthase (Metsä-Ketelä et al. 2002; Moffitt and Neilan 2003) genes. The flagellar subunit protein gene, *fli*C, which codes for an important protein responsible for host colonisation by the pathogen *Ralstonia solanacearum* (Tans-Kersten et al. 2001), was used to establish a sensitive PCR-based method to detect this pathogen in soil (Schönfeld et al. 2003). Although most of these detection systems were developed with the intention of studying bacterial communities in other habitats, they can easily be applied to study bacterial communities in plants.

The most commonly applied targets for molecular detection in environmental samples are genes that can be used for taxonomical differentiation, such as the small and large subunit genes of the ribosomal operon (16S and 23S, respectively), ribosomal RNA (rRNA) genes, and the RNA polymerase β subunit gene *rpo*B (Dahllöf et al. 2000). A specific 16S rDNA-based PCR system aimed at detecting *R. solanacearum* was, for instance, developed by Boudazin et al. (1999), whereas a 23S rDNA gene-directed probe was developed by Wullings et al. (1998), and applied for detection of the same pathogen using fluorescent in situ hybridisation (FISH) in tomato (Van Overbeek et al. 2002). A real-time PCR system based on the *rpo*B gene has been applied to detect *Bacillus anthracis* in different environmental samples (Qi et al. 2001).

Molecular techniques aimed at detecting specific genes are often restricted in their application to single targets. Degenerate primers suitable for PCR detection can overcome this restriction by targeting multiple genes as, for example, applied to the genes encoding ketosynthase (Metsä-Ketelä et al. 2002; Moffitt and Neilan 2003) and *nif*H (Steward et al. 2004). Multiple sequence detection is crucial for molecular community fingerprinting of environmental samples, e.g. using PCR coupled with denaturing gradient gel electrophoresis (PCR-DGGE). Molecular detection techniques are suitable tools for detecting both cultivable and non-cultivable endophytic populations in plants.

19.5 Molecular Fingerprinting of Endophyte Communities

19.5.1 Basic Concept of Molecular Fingerprinting

Shifts in endophytic populations and the effect of introduction of selected endophytes on bacterial communities have recently been studied using molecular fingerprinting techniques (Garbeva et al. 2001; Sessitsch et al. 2002; Araújo et al. 2002; Reiter et al. 2003b; Conn and Franco 2004; Seghers et al. 2004). Different methods have been applied, such as PCR-DGGE (Garbeva et al. 2001; Araújo et al. 2002; Reiter et al. 2003b) and PCR followed by terminal restriction fragment length polymorphism (PCR-T-RFLP; Sessitsch et al. 2002). Other molecular fingerprinting techniques, such as PCR followed by temperature gradient gel electrophoresis (PCR-TGGE; e.g., Felske et al. 1998b) and PCR-single strand conformational polymorphism (PCR-SSCP; Schwieger and Tebbe 1998; Schmalenberger and Tebbe 2003), have not yet been applied to microbial populations in plants. However, these methods have been successfully applied in studies of other environments, e.g. bulk and rhizosphere soils.

The principle of all these community fingerprinting techniques, i.e. competitive PCR amplification of a pool of 16S rRNA gene fragments, is the same, whereas final analyses of the fragments from environmental samples differs. The first step in all methods is extraction and purification of total community nucleic acids. This step is critical, as high molecular weight DNA pure enough for PCR amplification is required. Secondly, total microbial community DNA is PCR-amplified using primers spanning fragments of the 16S rDNA gene. Most commonly, primers that target regions in the 16S rDNA, which is conserved for all eubacterial species, are used. However, depending on the focus of the study, primers targeting specific groups of microorganisms, e.g. eubacteria, fungi or archeabacteria, should be applied. Separation of PCR products is performed in gels with denaturing gradients (DGGE) or temperature gradients (TGGE), or otherwise, e.g. SSCP and T-RFLP. A further requirement for PCR-DGGE and PCR-TGGE is the presence of a GC clamp: a stretch of around 40 nucleotides of mostly G and C residues. The GC clamp is required for immobilising denatured DNA fragments within the gel matrix. T-RFLP is recommended for studying habitats with low species richness (Engebretson and Moyer 2003), which is often the case for internal plant environments (Garbeva et al. 2001; Sessitsch et al. 2002). T-RFLP has appeared to be more sensitive for detection of weak bands than PCR-DGGE (Sessitsch et al. 2002; Conn and Franco 2004).

It is not our intention to describe the details of all methods in full. To obtain more details about the techniques described in this chapter, we refer readers interested in this topic to the *Molecular microbial ecology manual* (Akkermans et al. 2001). We will focus on critical steps in these fingerprinting techniques when studying bacterial endophyte populations as well as the potential offered by these methods. Further, the intricate tasks and challenges posed by the need to characterise non-cultivable endophytes will be considered.

19.5.2 Sample Preparation

To discriminate bacteria inside plants ('genuine' endophytes) from those attached or living adjacent to plants, a critical evaluation of methods used for sample preparation is required. Contamination of samples by bacterial cells from the outside of plants should be avoided. However, surface sterilisation of plant parts prior to nucleic acid extraction may not be sufficient to clean the material of surface nucleic acids. Even minor contamination may lead to false positive bands in fingerprints due to the sensitivity of the PCR amplification steps. Moreover, endospores from spore-forming bacterial species may survive these treatments – these are notorious contaminants in endophyte studies (Bent and Chanway 2002). Commonly, surface-sterilised stem parts are incubated on agar medium to check for the absence of colony growth from the outside (Araújo et al. 2002; Reiter et al. 2003b; see Chap. 17 by Hallman et al.). Colonies formed from the plant surface will develop adjacent to the plant part and not within it. This method is effective in determining the absence of cultivable cells and spores originating from the surface that may have survived surface sterilisation. However, non-cultivable contaminating bacteria will not be detected using this method.

For molecular detection, aseptic handling during sample preparation is extremely important. Surface sterilisation of plant parts followed by aseptic removal of the outer layer (epidermis) is an effective approach to remove any traces of external microbial life and thus avoid contamination of the sample from the outside (Garbeva et al. 2001; Sessitsch et al. 2002). However, Garbeva et al. (2001) and Sessitsch et al. (2002) concluded that this approach is feasible only with robust plant parts such as stems and (potato) tubers but not with fragile structures such as leaves and roots. A clear distinction between bacteria residing inside and outside these fragile structures thus cannot easily be made using molecular fingerprinting. This may be too restrictive when studying 'genuine' endophytes. A possible remedy to selectively isolate DNA and/or RNA from endophytes in finer structured organs may be a pretreatment with DNase and/or RNase prior to nucleic acid extraction.

19.5.3 Nucleic Acid Extraction

In principle, both DNA and RNA can be extracted from plant samples to evaluate endophytic populations. Comparison of fingerprints generated from DNA and RNA samples may reveal the activity of particular endophyte populations due to proposed higher ribosomal numbers in metabolically active cells. Felske et al. (1998b) and Duarte et al. (1998) used this approach in soil samples. For plant samples, this comparative approach has so far been applied only once, by Reiter et al. (2003b). The quality and purity of nucleic acid extracts from plants are the technical constraints for application of total plant microbial community DNA and RNA in molecular fingerprinting analyses.

Contamination by polymerase-inhibiting compounds from plants, such as (poly) phenolics, cannot always be avoided. These vary with the respective plant species. Community DNA extracts should be diluted in Tris-EDTA (pH 8) buffer prior to PCR amplification to avoid inhibition of polymerase activity during PCR.

Although DNA extraction procedures applied in different laboratories vary, they do not differ fundamentally. Most common procedures include pulverisation in liquid nitrogen followed by bead beating (Sessitsch et al. 2002; Reiter et al. 2003b) or bead beating of fresh and sliced plant samples (Garbeva et al. 2001; Araújo et al. 2002). Suspended cells are lysed in SDS solution and occasionally CTAB (cetyl trimethyl ammonium bromide) treatment is included to remove plant-derived exopolysaccharides (Garbeva et al. 2001; Reiter et al. 2003b). DNA is recovered using standard procedures; i.e. extraction with phenol and chloroform, precipitation in isopropanol and final wash steps in 70% ethanol. Crude extracts may be further purified using Wizard DNA clean up (Promega, Leiden, The Netherlands) prior to PCR amplification (Garbeva et al. 2001).

The procedures described above have proven suitable for obtaining nucleic acids for subsequent molecular fingerprinting. Commercially available extraction kits, such as the Mo Bio UltraClean soil DNA isolation kit (Mo Bio Laboratories, BIOzym TC, Landgraaf, The Netherlands), appeared to be less efficient in the recovery of high quality DNA than both methods described by Garbeva et al. (2001), and was also our experience in our laboratories. Although the Mo Bio soil DNA isolation system proved its validity for recovery of DNA from different soils, it appeared to be limited to the recovery of plant DNA. Garbeva et al. (2001) also applied an alternative protocol, in which DNA was extracted from bacterial cells dislodged by incubating sliced plant parts in buffer. Cells were collected by centrifugation of the incubation buffer and DNA was extracted from the cell pellet. The two protocols – DNA extraction from macerated plants and dislodged cells – were compared by PCR-DGGE analysis (Garbeva et al. 2001). The latter method yielded a clearer pattern with more distinctive bands in the DGGE gel. A modification of the standard protocol in which endophytic cells were collected by centrifugation resulted in higher quality DNA, presumably because there was less contamination with plant-derived phenolics.

Both methods described by Garbeva et al. (2001) were successfully applied to DNA extraction from different plants (tomato, leek, chrysanthemum, lettuce) followed by endophytic fingerprinting with PCR-DGGE in our laboratories. It is advisable to optimise the nucleic acid extraction protocols for each newly studied plant species or plant part, although we found both methods applicable to stems and roots of different plants without further adaptation. Only optimised sample preparation and DNA extraction procedures can guarantee successful molecular fingerprinting analysis.

19.5.4 PCR and Molecular Community Fingerprinting

The target sequences present in the total plant-extracted nucleic acid samples serve as templates for PCR. To assess microbial community structures in environmental samples, primers that target conserved regions of 16S rDNA genes are most commonly used. Molecular community fingerprints will, in principle, reveal all bands from 16S rDNA genes present in total plant nucleic acid extracts. Each band should represent one taxon. In practice, bands representing more than one taxon have been reported (Schmalenberger and Tebbe 2003). Also, single isolates represented by more than one band in fingerprints have been observed, as shown for *Bacillus* sp. strain Sal1 (Garbeva et al. 2001). The number of individual bands in fingerprints cannot thus simply be translated to the number of species present in the plant.

Estimation of the relative population size in molecular fingerprints is possible by measuring band intensities. However, linear amplification of individual target sequences in complex DNA extracts will probably not occur. Therefore, individual bands cannot be used in a straightforward manner for direct quantification. For additional information about cell numbers of individual populations, other methods are required. Therefore, molecular fingerprint analysis is most valuable to demonstrate microbial community shifts and to compare microbial community structures in different samples.

Plant cell organelles, such as chloroplasts and mitochondria, also possess 16S rDNA genes, and primers targeted to bacteria will also amplify these genes. Therefore, extra bands may be expected upon fingerprinting of the amplified products. Identification of individual PCR fragments in random clone libraries may fail as a result of the high abundance of cell organelles in total plant DNA extracts. A strategy to exclude 16S rDNA amplicons of eukaryotic origin is based on pre-amplification with a primer (799F, *Escherichia coli* numbering) that does not anneal to chloroplast DNA (Chelius and Triplett 2001). Clone libraries made in our laboratories by PCR amplification with primers 799F and 1401R with potato community DNA extract as template revealed that chloroplast amplicon numbers were lower in comparison with clone libraries made with eubacterial primers 968F and 1401R. PCR-DGGE analysis from the same amplicons revealed lower band intensity at the position where chloroplast bands were expected when primers 799F and 1401R were applied. Application of primers that target specific microbial groups also exclude amplicons of chloroplast and mitochondrial origin (Sessitsch et al. 2002; Reiter et al. 2003b). Group-specific primers may therefore be best suited for endophyte community structure analyses.

19.5.5 Group-Specific Molecular Community Fingerprinting

Candidate bacterial endophytes have been characterised for improvement of plant resistance and increased nutrient acquisition, e.g. by nitrogen fixation, for several important crops. These cultivable endophytic beneficials, which will to a great extent control plant health, belong to a limited number of taxonomic groups. Therefore, molecular tools for detection of these beneficial populations in plants will certainly gain importance in the near future. Taxonomic groups representing beneficial endophytes are *Aceto-*

bacter and *Gluconacetobacter* spp. (Dong et al. 1994; Cocking 2003), *Actinomyces* spp. (Zinniel et al. 2002; Castillo et al. 2003; Coombs et al. 2004), *Bacillus* spp. (Benhamou et al. 1998; Bacon and Hinton 2002; Reva et al. 2002), *Burkholderia* spp. (Balandreau et al. 2001), *Bradyrhizobium* (Chaintreuil et al. 2000), *Enterobacter* spp. (Hinton and Bacon 1995), *Pseudomonas* spp. (Duijff et al. 1997; Rediers et al. 2003), *Rhizobium* and *Sinorhizobium* spp. (Reiter et al. 2003a) and *Serratia* spp. [Press et al. 1997; Benhamou et al. 2000; Tan et al. 2001; Kamensky et al. 2003; see Chaps. 2 (Hallmann and Berg) and 3 (Kloepper and Ryu)]. However, some of these taxonomic groups also contain members with non-beneficial or even deleterious properties, as was the case for certain *Serratia marcescens* isolates (Gyaneshwar et al. 2001; Bruton et al. 2003). Endophytic taxa with potentially negative impacts on human health may belong to the group of *Enterobacteriaceae*. Association and endophytic colonisation of plants with human bacterial pathogens such as *Salmonella* spp. has recently been reported (Guo et al. 2001, 2002; Cooley et al. 2003; Dong et al. 2003). The lactic acid bacteria, on the contrary, represent a group of bacteria that are presumed to have positive effects on human health. A group-specific primer system for lactic acid bacteria was developed by Heilig et al. (2002). Nevertheless, endophytic colonisation by lactic acid bacteria has not yet been reported, although it is known that representatives of this group live in close association with plants (Ennahar et al. 2003).

Group-specific PCR systems that may be applied for molecular fingerprint analysis of taxonomically important groups of bacterial endophytes are presented in Table 19.2. The most important taxons amplified by primers described in literature are *Pseudomonas* spp., *Bacillus* spp, *Burkholderia* spp. and *Actinomyces* spp. Fingerprints from group-specific PCR will give an impression of all species with close taxonomic relationships to important known endophytes. For pathogen suppression, taxonomically related species may be the best competitors for water, nutrients and available space. For instance, the plant pathogen *R. solanacearum* was suppressed by a closely related species, *R. picketti*, on the rhizoplane of tomato (Shiomi et al. 1999). Group-specific primers covering all β-proteobacteria, including *Ralstonia* sp., appeared to be suitable for studying *R. solanacearum* and its taxonomically nearest relatives (Gomes et al. 2001). Primers covering α-proteobacteria (Gomes et al. 2001) are important because the group of α-proteobacteria harbours pathogenic species, such as *Agrobacterium tumefaciens*, as well as beneficial nitrogen-fixing species such as *Rhizobium* and *Bradyrhizobium* spp.

Table 19.2. Group-specific primer systems for detection of some endophytic bacterial taxa (groups)

Bacterial group	Nested primer system	Reference
Pseudomonas spp.	First step: *Pseudomonas* spp. specific PsR: 5′-GGTCTGAGAGGATGATCAGT-3′ and pSf: 5′-TTAGCTCCACCTCGCGGC-3′ Second step: *Pseudomonas* spp. specific f968: 5′-AACGCGAAGAACCTTAC- 3′ with GC clamp and PsR	Garbeva et al. 2004
	First step: eubacterial 8f: 5′-AGAGTTTGATCCTGGCTCCAG-3′ and 926r: 5′-CCGTCAATTCCTTT(AG)AGTTT-3′ Second step: *Pseudomonas* spp. specific 8f with CG clamp and PSMGx: 5′-CCTTCCTCCCAACTT-3′	Reiter et al. 2003b
Burkholderia spp.	First step: *Burkholderia* spp. specific Burk3: 5′-CTGCGAAAGCCGGAT-3′ and BurkR: 5′-TGCCATACTCTAGCYYGC-3′ Second step: *Burkholderia* spp. specific Burk3 with GC clamp and r1378: 5′-CGGTGTGTACAAGGCCCGGGAACG-3′	Salles et al. 2002
Actinomyces spp.	First step: *Actinomyces* spp. specific f243: 5′-GGATGAGCCCGCGGCCTA-3′ and r1378 Second step: eubacterial f984: 5′-AACGCGAAGAACCTTAC-3′ with GC clamp and r1378	Heuer et al. 1997
Bacillus spp.	First step: *Bacillus* spp. specific Bacf: 5′-GGGAAACCGGGGCTAATACCGGAT-3′ and r1378 Second step: eubacterial f968 with GC clamp and r1378	Garbeva et al. 2003
α-Proteobacteria	First step: α-proteobacteria specific f203 α: 5′-CCGCATACGCCCTACGGGGGAAAGATTTAT-3′ and r1494: 5′-CTACGG(T/C)TACCTTGTTACGAC-3′ Second step: eubacterial f984 with GC clamp and r1378	Gomes et al. 2001
β-Proteobacteria	First step: β-proteobacteria specific f203 β: 5′-CGCACAAGCGGTGGATGA-3′ and r1494 Second step: eubacterial f984 with GC clamp and r1378	Gomes et al. 2001

19.5.6 Molecular Identification of Species and Genes

Bands in molecular fingerprints reveal neither the identity nor the function of individual species. For identification of bands of noncultivable species, individual bands are isolated and, if necessary, the DNA is cloned. Subsequent sequencing and phylogenetic analysis theoretically enables identification, provided that the sequences found correspond to known sequences in the databanks. From the sequence information, new probes can be constructed for follow-up studies using FISH (e.g. Felske et al. 1998a; Amann and Ludwig 2000). Noncultivable species may be identified, though their function and ecological roles will remain unresolved.

Metagenome analysis may represent a new and promising approach in endophyte research (Rondon et al. 2000). Using this approach the function of genes from the noncultivable fraction present in different habitats can be unravelled. Large fragments (>50 kb) are required, prepared from soil or other environmental DNA samples and cloned into bacterial artificial chromosome (BAC) vectors. Expression studies and sequence data of genes and operons of cloned fragments should reveal some of the genetic properties of noncultivable species in the ecosystem under study. Due to the high incidence of beneficial bacteria among endophytes and other plant-associated bacteria commonly observed in many different studies, we expect that the phytosphere will become an interesting object for metagenome analysis studies. In the near future, use of the plant-metagenomic approach may result in new compounds for the development of pharmaceutical and agrochemical products.

19.6 Integration of Detection Techniques

19.6.1 Polyphasic Approach

Molecular analysis of plant ecosystems has limitations with respect to the detection, function, activity and ecological behavior of individual populations. Complementary data are often required, e.g. the use of conventional techniques of cultivation and microscopy (Van Elsas et al. 1998).

In an unpublished study we used a combination of in situ microbial activity staining, cultivation and PCR-DGGE to isolate and identify cultivable and non-cultivable bacteria associated with storage and vascular tissue of potato tubers. Sliced potato tubers were incubated in water agar containing triphenyl tetrazolium chloride, a colourless dye that changes into red formazan in the presence of microbial activity. Vascular bundles in tubers appeared to be hot spots for microbial activity and small samples extracted from the bundles were used for isolation and PCR-DGGE analysis. Two dominant isolates from the vascular bundle were identified by partial 16S rDNA gene sequence analysis as *Pantoea agglomerans* and *Bacillus pumilis* Comparison of the bands of these isolates with those of the PCR-DGGE fingerprints from total vascular community DNA revealed that three dominant bands did not match any of the isolates. A combination of plating and PCR-DGGE made it clear that dominant noncultivable species must be present in the vascular bundles of potato tubers. Additionally, the in situ metabolic activity staining facilitated sampling of potentially interesting areas by demonstrating hot spots of microbial activity in tuber tissue.

The application of multiple detection methods will aid in increasing our knowledge of plant ecosystems. Molecular fingerprinting methods must be considered as supplementary tools to the already existing panel of detection techniques. The advantage of molecular fingerprinting techniques over other techniques is the possibility to detect entire bacterial communities instead of only cultivable populations. For some ecosystems, such as soil and water, molecular fingerprinting is nowadays considered a routine method. A challenge for future endophyte research will be the integration of molecular fingerprinting methods for cultivable and non-cultivable bacteria and traditional culture-based endophyte research as complementary routines.

19.7 Conclusions

Molecular fingerprinting methods are required as tools to supplement conventional methods in endophyte research to study cultivable and noncultivable populations in plants. The intricacies of plants make these techniques difficult to perform. The presence of chloroplast and mitochondrial DNA in total plant DNA extracts, the abundance of plant DNA versus bacterial DNA, and the presence of (poly) phenolics may hamper subsequent analyses. Group (taxon)-specific primer systems can circumvent co-amplification of 16S rDNA sequences from plant cell organelles. The plant metagenome concept offers great perspectives for isolating new substances from plant-associated microorganisms, which may be important for the pharmaceutical and crop protection industries.

References

Adams PD, Kloepper JW (2002) Effect of host genotype on indigenous bacterial endophytes of cotton (*Gossypium hirsutum* L.). Plant Soil 240:181–189

Akkermans ADL, van Elsas JD, de Bruijn JJE (2001) Molecular microbial ecology manual. Kluwer, Dordrecht

Amann R, Ludwig W (2000) Ribosomal RNA-targeted nucleic acid probes for studies in microbial ecology. FEMS Microbiol Rev 24:555–565

Araújo WL, Marcon J, Maccheroni W, Van Elsas JD, Van Vuurde JWL, Azevedo JL (2002) Diversity of endophytic bacterial populations and their interaction with *Xylella fastidiosa* in citrus plants. Appl Environ Microbiol 68:4909–4914

Bacon CW, Hinton DM (2002) Endophytic and biological control potential of *Bacillus mojavensis* and related species. Biol Control 23:274–284

Balandreau J, Viallard V, Cournoyer B, Coenye T, Laevens S, Vandamme P (2001) *Burkholderia cepacia* genomovar III is a common plant-associated bacterium. Appl Environ Microbiol 67:982–985

Benhamou N, Kloepper JW, Quadt-Hallman A, Tuzun S (1996) Induction of defense-related ultrastructural modifications in pea root tissues inoculated with endophytic bacteria. Plant Physiol 112:919–929

Benhamou N, Kloepper JW, Tuzun S (1998) Induction of resistance against *Fusarium* wilt of tomato by combination of chitosan with an endophytic bacterial strain: ultrastructure and cytochemistry of the host response. Planta 204:153–168

Benhamou N, Gagné S, Le Quéré D, Dehbi L (2000) Bacterial-mediated induced resistance in cucumber: beneficial effect of the endophytic bacterium *Serratia plymuthica* on the protection against infection by *Pythium ultimum*. Phytopathology 90:45–56

Bent E, Chanway CP (2002) Potential for misidentification of a spore-forming *Paenibacillus polymyxa* isolate as an endophyte by using culture-based methods. Appl Environ Microbiol 68:4650–4652

Boudazin G, Le Roux AC, Josi K, Labarre P, Jouan B (1999) Design of division specific primers of *Ralstonia solanacearum* and application to the identification of European isolates. Eur J Plant Pathol 105:373–380

Bové JM, Garnier M (2003) Phloem-and xylem-restricted plant pathogenic bacteria. Plant Sci 164:421–438

Bruton BD, Mitchell F, Fletcher J, Pair SD, Wayadande A, Melcher U, Brady J, Bextine B, Popham TW (2003) *Serratia marcescens*, a phloem-colonising, squash bug-transmitted bacterium: causal agent of cucurbit yellow vine disease. Plant Dis 87:937–944

Castillo U, Harper JK, Strobel GA, Sears J, Alesi K, Ford E, Lin J, Hunter M, Maranta M, Ge H, Yaver D, Jensen JB, Porter H, Robison R, Millar D, Hess WM, Condron M, Teplow D (2003) Kakadumycins, novel antibiotics from *Streptomyces* sp NRRL 30566, an endophyte of *Grevillea pteridifolia*. FEMS Microbiol Lett 224:183–190

Chaintreuil C, Giraud E, Prin Y, Lorquin J, Bâ A, Gillis M, Lajudie P, Dreyfus B (2000) Photosynthetic Bradyrhizobia are natural endophytes of the African wild rice *Oryza breviligulata*. Appl Environ Microbiol 66:5437–5447

Chelius MK, Triplett EW (2001) The diversity of archaea and bacteria in association with the roots of *Zea mays* L. Microb Ecol 41:252–263

Cocking EC (2003) Endophytic colonisation of plant roots by nitrogen-fixing bacteria. Plant Soil 252:169–175

Conn VM, Franco CMM (2004) Analysis of the endophytic actinobacterial population in the roots of wheat (*Triticum aestivum* L.) by terminal restriction fragment length polymorphism and sequencing of 16S rRNA clones. Appl Environ Microbiol 70:1787–1794

Cooley MB, Miller WG, Mandrell RE (2003) Colonisation of *Arabidopsis thaliana* with *Salmonella enterica* and enterohemorrhagic *Escherichia coli* O157: H7 and competition by *Enterobacter asburiae*. Appl Environ Microbiol 69:4915–4926

Coombs JT, Michelsen PP, Franco CMM (2004) Evaluation of endophytic actinobacteria as antagonists of *Gaeumannomyces graminis* var. *tritici* in wheat. Biol Control 29:359–366

Dahllöf I, Baillie H, Kjelleberg S (2000) *rpo*B-based microbial community analysis avoids limitations inherent in 16S rRNA gene intraspecies heterogeneity. Appl Environ Microbiol 66:3376–3380

De Souza JT, Raaijmakers JM (2003) Polymorphisms within the *prn*D and *plt*C genes from pyrrolnitrin and pyoluteorin-producing *Pseudomonas* and *Burkholderia* spp. FEMS Microbiol Ecol 43:21–34

Dong YM, Iniguez AL, Ahmer BMM, Triplett EW (2003) Kinetics and strain specificity of rhizosphere and endophytic colonisation by enteric bacteria on seedlings of *Medicago sativa* and *Medicago truncatula*. Appl Environ Microbiol 69:1783–1790

Dong ZM, Canny MJ, McCully ME, Roboredo MR, Cabadilla CF, Ortega E, Rodés R (1994) A nitrogen-fixing endophyte of sugarcane stems – a new role for the apoplast. Plant Physiol 105:1139–1147

Duarte GF, Rosado AS, Seldin L, Keijzer-Wolters AC, van Elsas JD (1998) Extraction of ribosomal RNA and genomic DNA from soil for studying the diversity of the indigenous bacterial community. J Microbiol Methods 32:21–29

Duijff BJ, Gianinazzi-Pearson V, Lemanceau P (1997) Involvement of the outer membrane lipopolysaccharides in the endophytic colonisation of tomato roots by biocontrol *Pseudomonas fluorescens* strain WCS417r. New Phytol 135:325–334

Engebretson JJ, Moyer CL (2003) Fidelity of select restriction endonucleases in determining microbial diversity by terminal-restriction fragment length polymorphism. Appl Environ Microbiol 69:4823–4829

Ennahar S, Cai YM, Fujita Y (2003) Phylogenetic diversity of lactic acid bacteria associated with paddy rice silage as determined by 16S ribosomal DNA analysis. Appl Environ Microbiol 69:444–451

Felske A, Akkermans ADL, De Vos WM (1998a) In situ detection of an uncultured predominant Bacillus in Dutch grassland soils. Appl Environ Microbiol 64:4588–4590

Felske A, Wolterink A, Van Lis R, Akkermans ADL (1998b) Phylogeny of the main bacterial 16S rRNA sequences in Drentse A grassland soils (The Netherlands). Appl Environ Microbiol 64:871–879

Garbeva P, van Overbeek LS, van Vuurde JWL, van Elsas JD (2001) Analysis of endophytic bacterial communities of potato by plating and denaturing gradient gel electrophoresis (DGGE) of 16S rDNA based PCR fragments. Microb Ecol 41:369–383

Garbeva P, van Veen JA, van Elsas JD (2003) Predominant *Bacillus* spp. in agricultural soil under different management regimes detected via PCR-DGGE. Microb Ecol 45:302–316

Garbeva P, van Veen JA, van Elsas JD (2004) Assessment of the diversity, and antagonism towards *Rhizoctonia solani* AG3, of *Pseudomonas* species in soil from different agricultural regimes. FEMS Microbiol Ecol 47:51–64

Germida JJ, Siciliano SD (2001) Taxonomic diversity of bacteria associated with the roots of modern, recent and ancient wheat cultivars. Biol Fert Soils 33:410–415

Gomes NCM, Heuer H, Schönfeld J, Costa R, Mendonca-Hagler L, Smalla K (2001) Bacterial diversity of the rhizosphere of maize (*Zea mays*) grown in tropical soil studied by temperature gradient gel electrophoresis. Plant Soil 232:167–180

Guo X, Chen JR, Brackett RE, Beuchat LR (2001) Survival of *Salmonellae* on and in tomato plants from the time of inoculation at flowering and early stages of fruit development through fruit ripening. Appl Environ Microbiol 67:4760–4764

Guo XA, van Iersel MW, Chen JR, Brackett RE, Beuchat LR (2002) Evidence of association of Salmonellae with tomato plants grown hydroponically in inoculated nutrient solution. Appl Environ Microbiol 68:3639–3643

Gyaneshwar P, James EK, Mathan N, Reddy PM, Reinhold-Hurek B, Ladha JK (2001) Endophytic colonisation of rice by a diazotrophic strain of *Serratia marcescens*. J Bacteriol 183:2634–2645

Hallmann J, Quadt-Hallmann A, Mahaffee WF, Kloepper JW (1997) Bacterial endophytes in agricultural crops. Can J Microbiol 43:895–914

Heilig HGHJ, Zoetendal EG, Vaughan EE, Marteau P, Akkermans ADL, de Vos WM (2002) Molecular diversity of *Lactobacillus* spp. and other lactic acid bacteria in the human intestine as determined by specific amplification of 16S ribosomal DNA. Appl Environ Microbiol 68:114–123

Heuer H, Krsek M, Baker P, Smalla K, Wellington EMH (1997) Analysis of actinomycete communities by specific amplification of genes encoding 16S rRNA and gel-electrophoretic separation in denaturing gradients. Appl Environ Microbiol 63:3233–3241

Hinton DM, Bacon CW (1995) *Enterobacter cloacae* is an endophytic symbiont of Corn. Mycopathologia 129:117–125

James EK, Gyaneshwar P, Mathan N, Barraquio WL, Reddy PM, Iannetta PPM, Olivares FL, Ladha JK (2002) Infection and colonisation of rice seedlings by the plant growth-promoting bacterium *Herbaspirillum seropedicae* Z67. Mol Plant-Microbe Interact 15:894–906

Kamensky M, Ovadis M, Chet I, Chernin L (2003) Soil-borne strain IC14 of *Serratia plymuthica* with multiple mechanisms of antifungal activity provides biocontrol of *Botrytis cinerea* and *Sclerotinia sclerotiorum* diseases. Soil Biol Biochem 35:323–331

Metsä-Ketelä M, Halo L, Munukka E, Hakala J, Mäntsälä P, Ylihonko K (2002) Molecular evolution of aromatic polyketides and comparative sequence analysis of polyketide ketosynthase and 16S ribosomal DNA genes from various *Streptomyces* species. Appl Environ Microbiol 68:4472–4479

Moffitt MC, Neilan BA (2003) Evolutionary affiliations within the superfamily of ketosynthases reflect complex pathway associations. J Mol Evol 56:446–457

Nowak J (1998) Benefits of in vitro "biotisation" of plant tissue cultures with microbial inoculants. In Vitro Cell Dev Biol-Plant 34:122–130

Peters RD, Sturz AV, Carter MR, Sanderson JB (2003) Developing disease-suppressive soils through crop rotation and tillage management practices. Soil Tillage Res 72:181–192

Pillay VK, Nowak J (1997) Inoculum density, temperature, and genotype effects on in vitro growth promotion and epiphytic and endophytic colonisation of tomato (*Lycopersicon esculentum* L) seedlings inoculated with a pseudomonad bacterium. Can J Microbiol 43:354–361

Press CM, Wilson M, Tuzun S, Kloepper JW (1997) Salicylic acid produced by *Serratia marcescens* 90-166 is not the primary determinant of induced systemic resistance in cucumber or tobacco. Mol Plant-Microbe Interact 10:761–768

Qi YA, Patra G, Liang X, Williams LE, Rose S, Redkar RJ, DelVecchio VG (2001) Utilisation of the *rpo*B gene as a specific chromosomal marker for real-time PCR detection of *Bacillus anthracis*. Appl Environ Microbiol 67:3720–3727

Rediers H, Bonnecarrère V, Rainey PB, Hamonts K, Vanderleyden J, De Mot R (2003) Development and application of a *dap*B-based in vivo expression technology system to study colonisation of rice by the endophytic nitrogen-fixing bacterium *Pseudomonas stutzeri* A15. Appl Environ Microbiol 69:6864–6874

Reiter B, Bürgmann H, Burg K, Sessitsch A (2003a) Endophytic *nif*H gene diversity in African sweet potato. Can J Microbiol 49:549–555

Reiter B, Wermbter N, Gyamfi S, Schwab H, Sessitsch A (2003b) Endophytic *Pseudomonas* spp. populations of pathogen-infected potato plants analysed by 16S rDNA- and 16S rRNA-based denaturating gradient gel electrophoresis. Plant Soil 257:397–405

Reva ON, Smirnov VV, Pettersson B, Priest FG (2002) *Bacillus endophyticus* sp *nov.*, isolated from the inner tissues of cotton plants (*Gossypium* sp.). Int J Syst Evol Microbiol 52:101–107

Rondon MR, August PR, Bettermann AD, Brady SF, Grossman TH, Liles MR, Loiacono KA, Lynch BA, MacNeil IA, Minor C, Tiong CL, Gilman M, Osburne MS, Clardy J, Handelsman J, Goodman RM (2000) Cloning the soil metagenome: a strategy for accessing the genetic and functional diversity of uncultured microorganisms. Appl Environ Microbiol 66:2541–2547

Salles JF, De Souza FA, van Elsas JD (2002) Molecular method to assess the diversity of *Burkholderia* species in environmental samples. Appl Environ Microbiol 68:1595–1603

Schmalenberger A, Tebbe CC (2003) Bacterial diversity in maize rhizospheres: conclusions on the use of genetic profiles based on PCR-amplified partial small subunit rRNA genes in ecological studies. Mol Ecol 12:251–261

Schönfeld J, Heuer H, van Elsas JD, Smalla K (2003) Specific and sensitive detection of *Ralstonia solanacearum* in soil on the basis of PCR amplification of *fli*C fragments. Appl Environ Microbiol 69:7248–7256

Schwieger F, Tebbe CC (1998) A new approach to utilise PCR-single-strand-conformation polymorphism for 16s rRNA gene-based microbial community analysis. Appl Environ Microbiol 64:4870–4876

Seghers D, Wittebolle L, Top EM, Verstraete W, Siciliano SD (2004) Impact of agricultural practices on the *Zea mays* L. endophytic community. Appl Environ Microbiol 70:1475–1482

Sessitsch A, Reiter B, Pfeifer U, Wilhelm E (2002) Cultivation-independent population analysis of bacterial endophytes in three potato varieties based on eubacterial and *Actinomycetes*-specific PCR of 16S rRNA genes. FEMS Microbiol Ecol 39:23–32

Sharma VK, Nowak J (1998) Enhancement of Verticillium wilt resistance in tomato transplants by in vitro co-culture of seedlings with a plant growth promoting rhizobacterium (*Pseudomonas* sp. Strain PsJN). Can J Microbiol 44:806–806

Shiomi Y, Nishiyama M, Onisuka T, Marumoto T (1999) Comparison of bacterial community structures in the rhizoplane of tomato plants grown in soils suppressive and conducive towards bacterial wilt. Appl Environ Microbiol 65:3996–4001

Steward GF, Jenkins BD, Ward BB, Zehr JP (2004) Development and testing of a DNA macroarray to assess nitrogenase (*nif*H) gene diversity. Appl Environ Microbiol 70:1455–1465

Sturz AV (1995) The role of endophytic bacteria during seed piece decay and potato tuberisation Plant Soil 179:303–303

Sturz AV, Christie BR (1996) Endophytic bacteria of red clover as agents of allelopathic clover-maize syndromes. Soil Biol Biochem 28:583–588

Sturz AV, Nowak J (2000) Endophytic communities of rhizobacteria and the strategies required to create yield enhancing associations with crops. Appl Soil Ecol 15:183–190

Sturz AV, Christie BR, Matheson BG, Arsenault WJ, Buchanan NA (1999) Endophytic bacterial communities in the periderm of potato tubers and their potential to improve resistance to soil-borne plant pathogens. Plant Pathol 48:360–369

Tan ZY, Hurek T, Gyaneshwar P, Ladha JK, Reinhold-Hurek B (2001) Novel endophytes of rice form a taxonomically distinct subgroup of *Serratia marcescens*. Syst Appl Microbiol 24:245–251

Tan XY, Hurek T, Reinhold-Hurek B (2003) Effect of N-fertilisation, plant genotype and environmental conditions on *nif*H gene pools in roots of rice. Environ Microbiol 5:1009–1015

Tans-Kersten J, Huang HY, Allen C (2001) *Ralstonia solanacearum* needs motility for invasive virulence on tomato. J Bacteriol 183:3597–3605

Van Elsas JD, Duarte GF, Rosado AS, Smalla K (1998) Microbiological and molecular biological methods for monitoring microbial inoculants and their effects in the soil environment. J Microbiol Methods 32:133–154

Van Overbeek LS, Cassidy M, Kozdroj J, Trevors JT, van Elsas JD (2002) A polyphasic approach for studying the interaction between *Ralstonia solanacearum* and potential control agents in the tomato phytosphere. J Microbiol Methods 48:69–86

Wullings BA, van Beuningen AR, Janse JD, Akkermans ADL (1998) Detection of *Ralstonia solanacearum*, which causes brown rot of potato, by fluorescent in situ hybridisation with 23S rRNA-targeted probes. Appl Environ Microbiol 64:4546–4554

Zinniel DK, Lambrecht P, Harris NB, Feng Z, Kuczmarski D, Higley P, Ishimaru CA, Arunakumari A, Barletta RG, Vidaver AK (2002) Isolation and characterisation of endophytic colonising bacteria from agronomic crops and prairie plants. Appl Environ Microbiol 68:2198–2208

Subject Index

MIX
Papier aus verantwortungsvollen Quellen
Paper from responsible sources
FSC® C105338

If you have any concerns about our products,
you can contact us on
ProductSafety@springernature.com

In case Publisher is established outside the EU,
the EU authorized representative is:
Springer Nature Customer Service Center GmbH
Europaplatz 3, 69115 Heidelberg, Germany

Printed by Libri Plureos GmbH
in Hamburg, Germany